Computational Epidemiology

Ellen Kuhl

Computational Epidemiology

Data-Driven Modeling of COVID-19

 Springer

Ellen Kuhl
Mechanical Engineering
Stanford University
Stanford, CA, USA

ISBN 978-3-030-82892-9   ISBN 978-3-030-82890-5   (eBook)
https://doi.org/10.1007/978-3-030-82890-5

This Springer imprint is published by the registered company Springer Nature Switzerland AG
The registered company address is: Gewerbestrasse 11, 6330 Cham, Switzerland

To my mom, with whom I had the most scientific discussions about COVID-19.

Foreword

Nothing will ever be quite the same.

The Great COVID-19 Pandemic that started in December 2019 will be remembered as the main event of the first part of the 21st Century. It affected all parts of the world, it laid bare social inequalities, revealed the excesses of modern life, shook all strata of society, and changed profoundly how individuals, groups, and nations interact with each other.

Early in the crisis, it was realized that theoretical epidemiology and modeling are essential tools to confront the problem. As Daniel Bernoulli stated in the first-ever epidemiological studies of 1760: "*I simply wish that, in a matter which so closely concerns the wellbeing of the human race, no decision shall be made without all the knowledge which a little analysis and calculation can provide*". By January 2020, the first 'little analysis and calculation' provided by epidemiological studies revealed the seriousness of the situation and scientists around the world rang the alarm bell. As the crisis escalated, the attention of the media, governments, and the public turned towards modelers with burning questions: how bad is it? how will it evolve? how many cases should we anticipate? will the hospitals be overwhelmed? how many deaths? what can we do?

Like many people in academia, Ellen Kuhl found herself naturally interested in the problem. As an extreme athlete, a marathoner, a triathlete and an iron-(wo)man, Ellen is fearless. Early on, she realized that the methods she had been developing to understand neurodegenerative diseases could be readily adapted to the evolving crisis. As a scientist, Ellen combines a unique ability for modeling, great technical skills, and a wonderful intuition for good problems. She quickly built elegant data-driven models for how the disease spreads around different parts of the world through the airline network that soon became landmark studies. As an educator, she also realized that students in Stanford would gain from learning more about the science behind the disease. Rapidly, she put together a new course that was taught in the 2020 winter term, as the second wave was gathering strength. In her course that I attended and enjoyed tremendously, she taught both the basics of epidemiology and her own research as well as many topics related to the wider social and political impacts of the crisis. This book is the combined result of her course, state-of-the-art research, and general reflections.

The first part of the book is a fast-paced self-contained review of the fundamental ideas of epidemiology. It is full of wonderful stories, biographies, and anecdotes that brings life to the topic. It introduces and defines important concepts such as the reproduction number, herd immunity, and vaccine efficacy. On the mathematical side, it is centered around the development of the the famous SIR model that captures the evolution of three homogeneous populations of people susceptible (S), Infected (I_s) and Recovered (R) going back to the work of Kermack and McKendrick in 1927. The model will tell you that for a reproductive number R larger than one, after an initial exponential growth, there will be saturation when enough of the population has been infected before the disease eventually runs its course.

This classic material is updated for our computational world in Part II where Ellen shows how to implement efficiently the classic theory and how it can be generalized to problems that are directly relevant to the COVID crisis. In particular, in Chapter 7 and 8 two extra populations, the exposed (E) and asymptomatic (I_a), are introduced to capture the propagation of the disease through populations that do not display yet signs of infections, a crucial feature of this disease. As Ellen shows, this simple system of four or five differential equations is already enough to capture many features of the disease and can be used to understand early disease outbreaks.

The third part of the book brings the topic of epidemiology to current research level. One of the key features of a pandemic is the propagation of the disease from one region to another. In our hyper-connected world there are many ways diseases can travel. At the global level, the airline network is a natural way for the disease to spread especially through people who show no symptoms. Ellen shows how the SEIR model can then be extended to a network and efficiently implemented. Remarkably, this simple idea is sufficient to understand the early spread and different phases of the disease.

Yet, a key question remains. How do we use daily data about the number of cases to validate and infer parameters for the various models? Part IV of the book addresses directly this question and brings the topic to the forefront of current data research by presenting Bayesian inference methods and how it can be efficiently applied to the COVID crisis. This data-driven approach combined with the network models bring state-of-the-art techniques to the study of the disease dynamics and shows how secondary data, such as mobility data from cell phones, can be used as a barometer to predict the development of the disease.

Through the fog of war, governments have made many mistakes that have cost countless lives. As Anne Frank wrote: *"What is done cannot be undone, but at least one can keep it from happening again"*. How do we keep it from happening? We learn from our mistakes, sharpen our tools and models and confront the current and next crises with the full power of science. Ellen's wonderful book gives us the ideas, methods, tools, knowledge, and concepts to understand the current pandemic and be prepared for the next one.

Oxford, May 2021   *Alain Goriely*

Preface

The objective of this book is to understand the outbreak dynamics of the COVID-19 pandemic through the lens of computational modeling. Computational modeling can provide valuable insights into the dynamics and control of a global pandemic to guide public health decisions. So... why didn't it? Why did dozens of computational models make predictions that were orders of magnitude off? This book seeks to answer this question by integrating innovative concepts of mathematical epidemiology, computational modeling, physics-based simulation, and probabilistic programming. We illustrate how we can infer critical disease parameters–in real time–from reported case data to make informed predictions and guide political decision making. We critically discuss questions that COVID-19 models can and cannot answer and showcase controversial decisions around the early outbreak dynamics, outbreak control, and gradual return to normal. As scientists, it is our ethic responsibility to educate the public to ask the right questions and to communicate the limitations of our answers. Throughout this book, we will create data-driven models for COVID-19 to do so.

Who is this book for? If you are a student, educator, basic scientist, or medical researcher in the natural or social sciences, or someone passionate about big data and human health: This book is for you! Don't worry, this book is introductory and doesn't require a deep knowledge in epidemiology. A fascination for numbers and a general excitement for physics-based modeling, data science, and public health are a lot more important. And this is why this book is both, a textbook for undergraduates and graduate students, and a monograph for researchers and scientists.

As a *textbook*, this book can be used in the mathematical life sciences suitable for courses in applied mathematics, biomedical engineering, biostatistics, computer science, data science, epidemiology, health sciences, machine learning, mathematical biology, numerical methods, and probabilistic programming.

As a *monograph*, this book integrates the basic fundamentals of mathematical and computational epidemiology with modern concepts of data-driven modeling and probabilistic programming. It serves researchers in epidemiology and public health, with timely examples of computational modeling, and scientists in the data science and machine learning, with applications to COVID-19 and human health.

What does this book cover? This book consists of four main parts that gradually build from mathematical epidemiology via computational and network epidemiology to data-driven epidemiology.

Part I. Mathematical Epidemiology introduces the basic concepts of epidemiology in view of the COVID-19 pandemic with the objectives to understand the cause of an infectious disease, predict its outbreak dynamics, and design strategies to control it. We introduce the paradigm of compartment modeling and revisit analytical solutions for the classical SIS, SIR, and SEIR models using outbreak data of the COVID-19 pandemic. Typical problems include estimating the herd immunity threshold and efficacy of different COVID-19 vaccines and the growth rate, basic reproduction number, contact period of the COVID-19 outbreak at different locations.

Part II. Computational Epidemiology introduces numerical methods for ordinary differential equations and applies these methods to discretize, linearize, and solve the governing equations of SIS, SIR, SEIR, and SEIIR models to interpret the case data of COVID-19. Typical problems include simulating the first wave of the COVID-19 outbreak using reported case data and understanding the effects of early community spreading and outbreak control. We discuss why classical epidemiology models have failed to predict the outbreak dynamics of COVID-19 and introduce new concepts of data-driven dynamic contact rates and serology-informed asymptomatic transmission to address these shortcomings.

Part III. Network Epidemiology discusses numerical methods for partial differential equations and applies these methods to discretize, linearize, and solve the spreading of infectious diseases using discrete mobility networks and finite element models. Instead of solving the outbreak dynamics of COVID-19 locally for each region, state, or country, we now allow individuals to travel and populations to mix globally, informed by cell phone mobility data and air travel statistics. Typical problems include understanding the early outbreak patterns, the effect of travel restrictions, and the risk of reopening after lockdown.

Part IV. Data-driven Epidemiology covers the most timely methods and applications of this book. It focuses on probabilistic programming with the objectives to understand, predict, and control the outbreak dynamics of the COVID-19 pandemic. We integrate computational epidemiology and data-driven modeling to explore disease data in view of different compartment models using a probabilistic approach and quantify the uncertainties of our analysis. Typical problems include inferring the reproduction dynamics, visualizing the effects of asymptomatic transmission, and correlating case data and mobility.

This book is by no means complete. It does not cover agent-based modeling, age-dependent modeling, population mixing, purely statistical, stochastic, or probabilistic modeling, forecasting, and many other key aspects of computational epidemiology. Instead, it is a personal reflection on the role of data-driven modeling during the COVID-19 pandemic, motivated by the curiosity to understand it. Because Science.

Stanford, May 2021   *Ellen Kuhl*

Acknowledgements

Now, that's a wrap! I started writing this book on January 14, 2021, during the United States peak of the COVID-19 pandemic and finished on May 31, 2021, at the lowest incidence in 14 months. Dozens of people have contributed to this book, directly or indirectly, through endless discussions around the pandemic. I was fortunate to collaborate with an amazing international team, Alain Goriely from the UK, Francisco Sahli Costabal from Chile, Henry van den Bedem from the Netherlands, Kevin Linka from Germany, Mathias Peirlinck from Belgium, Paris Perdikaris from Greece, and Proton Rahman from Canada. The daily discussions of our local outbreak dynamics–while never meeting in person–were an essential part of my pandemic life made possible by Zoom. Thank you for sharing this unique experience!

I enjoyed regular COVID-19 meetings with my scientific friends around the world, Silvia Budday, Krishna Garikipati, Tom Hughes, Tinsley Oden, Paul Steinmann, Tarek Zohdi, and the IMAG/MSM Working Group for Multiscale Modeling and Viral Pandemics. But the true motivation for this book came from the students of the new course *Data-driven modeling of COVID-19* that we created at Stanford University in the Fall of 2020 to make online learning a bit more tangible. The enrollment ranged from undergraduates in biology, classics, computer science, engineering, ethics, human biology, mechanical engineering, and Spanish, to master and PhD students in aeronautics, astronautics, computer science, environment and resources, management science and engineering, and mechanical engineering. Together, we designed this course as the pandemic unfolded–in real time–and this book is a collection of class notes and feedback from students, guest speakers, and lecturers. We received support from the Stanford Bio-X Program and the School of Engineering COVID-19 Research and Assistance Fund. Massive thanks to all students in the class, to our amazing course assistants, Amelie Schäfer, Oguz Tikenogullari, Mathias Peirlinck, and Kevin Linka, and to my favorite guest speaker, Alain Goriely.

Last but not least, I thank the true heroes of the pandemic, Jasper and Syb, and all the kids who not only had to bear the uncertainties of a global pandemic, but also endure their parents working from home. Thank you, Henry, for getting through this together, with 354 miles of swimming, 3611 miles of biking, and 3279 miles of running. This has been a truly memorable time. I'm glad it's over–at least for now.

Contents

Part I
Mathematical epidemiology

Chapter 1
Introduction to mathematical epidemiology

Abstract Mathematical epidemiology is the science of understanding the cause of a disease, predicting its outbreak dynamics, and developing strategies to control it. Throughout the past century, mathematical epidemiology has become the cornerstone of public health. This success story is a result of the rapid advancement of mathematical models and the broad availability of massive amounts of data. Now, two fundamental things have changed that may forever affect our understanding of modern epidemiology: the outbreak of the COVID-19 pandemic, in the age of machine learning. Here, to embrace this opportunity, we introduce the concept of mathematical epidemiology in light of data-driven modeling and physics-based learning. We briefly review the traditional concepts of epidemiology, and discuss how our classical understanding of testing, reproduction, herd immunity, and immunization has changed in view of the COVID-19 pandemic. We compare the strengths and weaknesses of purely statistical and mechanistic models and illustrate how we can integrate the large volume of COVID-19 data into mechanistic compartment models to infer model parameters, learn correlations, and identify causation. The learning objectives of this chapter on mathematical epidemiology are to

- recognize the ever-present risk of infectious diseases, now and beyond COVID-19
- interpret epidemiological data from molecular, antigen, and antibody tests
- explain the basic reproduction number and its impact on public health decisions
- estimate herd immunity thresholds for given basic reproduction numbers
- demonstrate how vaccination can accelerate the path towards herd immunity
- calculate the efficacy and risk ratio of a vaccine
- understand the strengths and weaknesses of statistical and mechanistic models
- explain the importance of nonlinear feedback in compartment modeling
- find and prepare COVID-19 case data towards creating data-driven models
- discuss limitations of testing, reported data, and modeling

By the end of the chapter, you will be able to think in terms of data-driven modeling, know where to find and how to interpret your data, judge how to select your model, and understand why so many models have failed to explain and predict the outbreak dynamics of the COVID-19 pandemic.

E. Kuhl, *Computational Epidemiology*, https://doi.org/10.1007/978-3-030-82890-5_1

1.1 A brief history of infectious diseases

In the year 2020, everybody–at home on their couch–became an infectious disease expert. To calibrate all this newly gained knowledge, it seems a good idea to briefly reiterate the basic concepts and nomenclature of infectious diseases. Infectious diseases are spread by either bacterial or viral agents and are ever-present in society [8]. Every once in a while, this may result in outbreaks that have a significant impact on a local or global level. Depending on their spatial and temporal spread, we can classify outbreaks as endemic, epidemic, or pandemic [28]. *Endemic* outbreaks, for example chickenpox, are permanently present in a region or a population. *Epidemic* outbreaks, for example the seasonal flu, affects a lot of people in a short period of time, spread across several communities, and then disappear. *Pandemic* outbreaks, for example the Spanish flu, are epidemic outbreaks that affects a lot of people in a short period of time and spread across the entire world. Mathematical modeling of infectious diseases is important to understand their outbreak dynamics and inform political decision making to manage their spread [4].

Table 1.1 History of recent infectious disease outbreaks. Time period, type of disease, number of deaths, and location.

period	disease	deaths	location
1346 - 1350	Black Death	100,000,000	1/3 of Europe
1665 - 1666	Great Plague	100,000	1/4 of London
1918 - 1920	Spanish flu	50,000,000	worldwide
1980	measles	2,600,000	worldwide / year
2003	SARS	774	worldwide
2009 - 2010	H1N1	18,500	worldwide
2011	tuberculosis	1,400,000	worldwide / year
	HIV/AIDS	1,200,000	worldwide / year
	malaria	627,000	worldwide / year
2011	measles	160,000	worldwide (-94%)
2012 - 2020	MERS	866	worldwide
	seasonal flu	35,000	United States / year
2014 - 2016	ebola	11,000	Africa
2018	ebola	2,280	Congo
2019 -2021	COVID-19	3,300,000	worldwide

The COVID-19 pandemic. On March 11, 2020, Tedros Adhanom Ghebreyesus, the Director-General of the World Health Organization, declared the COVID-19 outbreak a global pandemic [49]. On that day, it had affected 126,702 people worldwide. What nobody could have foreseen is that within the following year, by March 11, 2021, this number had increased by three orders of magnitude, to 118.57 million [12]. Naturally, we often think of the COVID-19 pandemic as the deadliest and most devastating infectious disease in modern history.

Endemic. The word endemic is derived from the greek words *en* meaning in and *demos* meaning people. An infectious disease is endemic when it is constantly maintained at a baseline level in a geographic region without external inputs. For example, chicken pox is endemic in the United Kingdom. A person-to-person transmitted disease is endemic if each infected person passes the disease to one other person on average. The infection neither dies out nor increases exponentially, it is in an endemic steady state. Infectious disease experts ask:

- How many people are infectious at any give time? – What is $I(t)$?
- How fast do new infections arise? – What is dI/dt? – Stability analysis
- What are the effects of quarantine or vaccination? – What is R_0?
- Can we eradicate the disease? – What is $H = 1 - 1/R_0$? – Limit analysis

Epidemic. The word epidemic is derived from the greek words *epi* meaning upon and *demos* meaning people. An infectious disease is epidemic when it spreads rapidly across a large number of people in a short time period. For example, the seasonal flu is epidemic. Epidemics often come in waves with several recurring outbreaks, which can sometimes be seasonal. There is usually no increase in susceptibles and the disease dies out as the number of infectives decreases because a large enough fraction of the population has become immune. Infectious disease experts ask:

- How severe will the epidemic be? – What are I_{max}?
- When will it reach its peak? – What is $t(I_{max})$?
- How long will it last? – What are S_∞ and R_∞? What is $t(S_\infty)$?
- What are the effects of vaccination? – What are R_0 and $H = 1 - 1/R_0$ and I?

Pandemic. The word pandemic is derived from the greek words *pan* meaning all and *demos* meaning people. An infectious disease is pandemic when it spreads rapidly across a large region or worldwide in a short time period. A pandemic is a global outbreak. For example, smallpox, tuberculosis, the black death, the Spanish flu were pandemics, and COVID-19 is now. Infectious disease experts ask:

- How severe is the pandemic, when will it peak? – What are I_{max} and $t(I_{max})$?
- What are effective measures to manage the outbreak? – What is $R(t)$?
- What are the effects of quarantine or lockdown? – What is $\beta(t)$?
- What are the effects of travel restrictions? – What are κ and L_{ij}?
- How do we prioritize vaccination? – What are R_0 and $H = 1 - 1/R_0$ and I?

Table 1.1 summarizes recent infectious disease outbreaks ranging from the Black Death in the 14th century with an estimated 100 million deaths across Europe to the current COVID-19 pandemic with 3.3 million deaths worldwide to date. Notably, in 1980, the annual death toll of the measles with 2.6 million was of comparable size. Today, after massive vaccination campaigns, this number has dropped significantly, by 94% to 160,000 [5].

FRIDAY, MAY 30, 1919

SCIENCE

The Lessons of the Pandemic: MAJOR GEORGE
A. SOPER 501

The most astonishing thing about the pandemic was the complete mystery which surrounded it. Nobody seemed to know what the disease was, where it came from or how to stop it. Anxious minds are inquiring to-day whether another wave of it will come again.

Fig. 1.1 The Lessons of the Pandemic. In this famous Science publication, George A. Soper reflects on the Spanish flu that resulted in 50 million deaths worldwide in 1918 and 1919.

Figure 1.1 shows the cover page of the Science publication *The Lessons from the Pandemic* from May 1919, a reflection on the scientific understanding of the Spanish flu [29]. While some of it is specific to the nature of the 1918 influenza and the time during which it occurred, much of it still applies to the COVID-19 pandemic today: The Spanish flu lasted from 1918 to 1920 and, similar to COVID-19, occurred in multiple waves. Both, the Spanish flu and COVID-19, are contagious respiratory illnesses that spread from person to person, mainly by droplets, through cough, sneeze, or talk. Naturally, increased hygiene, mask wearing, and physical distancing in addition to strict isolation and quarantine are successful strategies to manage both conditions [10]. While COVID-19 is caused by a coronavirus, SARS-CoV-2, the 1918 influenza was caused by the H1N1 influenza A virus, a virus of avian origin. Within 25 months, it infected 500 million people, one third of the world's population, and resulted in more than 50 million deaths [5]. The major differences between the Spanish flu and COVID-19 are their high risk populations and the mechanisms of death: In contrast to the Spanish flu, which affected mainly healthy adults between 25 and 40 years of age, COVID-19 affects mainly individuals of 65 years and older with comorbidities. Victims of the 1918 influenza mainly died from secondary bacterial pneumonia, whereas victims of COVID-19 die from an overactive immune response that results in organ failure [32]. Nevertheless, comparing the COVID-19 outbreak to previous pandemics can provide insight and guidance to manage the current COVID-19 pandemic. An important element of this comparison are mathematical and computational tools that have been designed to quantify and explain the spreading mechanisms of infectious diseases. This is the objective of mathematical and computational epidemiology [3].

1.2 Introduction to epidemiology

Epidemiology is the study of distributions, patterns, and determinants of health-related events in human populations [3]. It is a cornerstone of public health , and shapes policy decisions by identifying risk factors for outbreaks and targets for prevention [28].

Epidemiology literally means the study of what is upon the people. It is derived from the greek words *epi*, meaning upon *demos*, meaning people, and *logos* meaning study, suggesting that it applies only to human populations. By this definition, epidemiology is the scientific, systematic, data-driven study of the distribution, i.e., where, who, and when, and determinants, i.e., causes and risk factors, of health-related patterns and events in specified populations, i.e., the world, a country, state, county, city, school, neighborhood. Epidemiology also includes the study of outbreak dynamics, outbreak control, and informing political decision making.

Agent, host, and environment. A key premise in epidemiology is that health-related events are not evenly distributed in a population; rather, they affect some individuals more than others [3]. An important goal of epidemiology is to identify the causes that put these individuals at a higher risk [2]. A simple but popular model to analyze and explain disease causation is the *epidemiologic triangle*. The epidemiologic triangle summarizes the interplay of the three components that contribute to the spread of an infectious disease: an external *agent*, a susceptible *host*, and an *environment* in which agent and host interact [9]. Descriptive epidemiologists characterize this interaction as the seed, the soil, and the climate [42]. Effective public health measures assess all three components and their interactions to control or prevent the spreading of a disease. Figure 1.2 illustrates the epidemiologic triangle of COVID-19 with agent, host, and environment. For COVID-19, the agent is the SARS-CoV-2 virus, the hosts are people, and the environment are droplets. Interventions between any two of these three components can help reduce the spread of the disease [20]. For example, reduced exposure, vaccination, and antiviral treatment can modulate agent-host interactions and reduce the number of new infections.

From descriptive to mathematical epidemiology. Mathematical models of infectious diseases date back to Daniel Bernoulli's model for smallpox in 1760 [5], and they have been developed and improved extensively since the 1920s [14]. In the middle of the 19th century, the English physician John Snow conducted a famous series of experiments of the cholera outbreak in London to discover the cause of the disease and to prevent its recurrence [46]. Because his research illustrates the classic sequence–from descriptive epidemiology and hypothesis generation, to mathematical epidemiology and hypothesis testing–John Snow is considered the father of modern epidemiology. Figure 1.4 summarizes the experiments of John Snow that ended the cholera outbreak in London by removing the handle of a public water

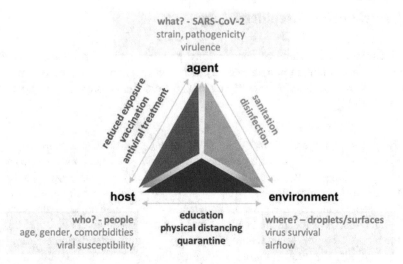

what? - SARS-CoV-2
strain, pathogenicity
virulence

agent

reduced exposure
vaccination
antiviral treatment

sanitation
disinfection

host   **environment**

who? - people
age, gender, comorbidities
viral susceptibility

education
physical distancing
quarantine

where? – droplets/surfaces
virus survival
airflow

Fig. 1.2 Epidemiologic triangle of COVID-19 with agent, host, and environment. The epidemiologic triangle illustrates the interplay of the three components that contribute to the spread of a disease: an external agent, a susceptible host, and an environment in which agent and host interact. For COVID-19, the agent is the SARS-CoV-2 virus, the hosts are people, and the environment are droplets. Interventions between any of these three components can reduce the spread of the disease.

Fig. 1.3 Daniel Bernoulli is considered the most famous mathematician of the Bernoulli family, although he actually studied medicine. He was born on February 8, 1700 in Groningen, the Netherlands, and is best known for his applications of mathematics to mechanics. To prove the efficacy of vaccination against smallpox, he proposed the first compartment model in 1760 [5]. It only had two compartments, but demonstrated the potential of modeling to understand the mechanisms of transmission, predict future spread, and control the outbreak through vaccination. Daniel Bernoulli died on March 27, 1782 in Basel, Switzerland.

pump. Today the field of epidemiology is considered a quantitative discipline that uses rigorous mathematical tools of probability and statistics to develop and test hypotheses to understand and explain health-related events [3]. In simple terms, today's epidemiologists count, scale, and compare. They *count* the number of cases; *scale* this count by a characteristic population to define fractions; and *compare* the evolution of these fractions over time [9]. Mathematical epidemiology has advanced significantly throughout the past decades [11], and models have become more sophisticated and complex [39]. This implies that in epidemiology today, is often no longer possible to solve epidemiological models analytically and in closed form [3].

From mathematical to computational epidemiology. Computational epidemiology is a multidisciplinary field that integrates mathematics, computer science, and public health to better understand central questions in epidemiology such as the

Fig. 1.4 The English physician John Snow is considered one of the founders of modern epidemiology, in part because of his research during the cholera outbreak in London in 1954 [46]. John Snow lived from March 15, 1813 to June 16, 1858. He questioned the widely accepted paradigm that cholera would spread by polluted bad air. He postulated that water would be the cause of the cholera outbreak in Soho, London in 1854. By talking to local residents, he identified the source of the outbreak as the public water pump on Broad Street. Although his chemical and microscopic observations were inconclusive, his spreading patterns of the cholera outbreak were convincing enough to persuade the local council to disable a street pump on Broad Street by removing its handle. There is a common belief that this action marked the ending of the outbreak.

spread of diseases or the effectiveness of public health interventions [22]. Real-time epidemiology is a rapidly developing area within computational epidemiology that seeks to support policy makers–in real time–as an outbreak is unfolding. An important application of real-time epidemiology is disease surveillance, the data-driven collection, analysis, and interpretation of large volumes of disease data from a variety of sources. This information can help to evaluate the effectiveness of control and preventative health measures [12]. Central to computational epidemiology is the accurate knowledge of people who are affected by an infectious disease at any given point in time. During the COVID-19 pandemic, for the first time in history, this information has been collected conscientiously, shared publicly, updated daily, and made freely available in real time [13].

Sensitivity and specificity. A diagnostic tests, for example, the nasal swab test, can be inaccurate in two ways: a **false positive** result erroneously labels a healthy person as infected, resulting in unnecessary quarantine and contact tracing; a **false negative** result oversees an infected person, resulting in the risk of infecting others.

sensitivity = true positives / all sick people

Sensitivity measures the proportion of positives, diseased people, that are correctly identified and not overlooked.

specificity = true negatives / all healthy people

Specificity measures the proportion of negatives, healthy people, that are correctly identified and not classified as diseased. A perfect test has 100% sensitivity and 100% specificity.

1.3 Testing, testing, testing

During a global pandemic, knowing how many people currently have the disease or have previously had it provides crucial information for healthcare providers and policy makers, both on the individual and population levels. Unfortunately, there is no method to provide this information with absolute accuracy. We usually use two measures to characterize the degree of accuracy of a test: *sensitivity*, the fraction of correctly identified positive, diseased individuals; and *specificity*, the fraction of correctly identified negative, healthy individuals. Understanding these limitations is important when modeling, simulating, and predicting the outbreak dynamics of a pandemic, especially in view of disease management and political decision making.

Table 1.2 Testing for COVID-19. Summary of the three most common tests for SARS-CoV-2. Diagnostic tests, including molecular and antigen tests, provide information about acute infection, whereas antibody or serology tests provide information about previous infection. Publicly available case data are based on diagnostic testing; seroprevalence reports are based on antibody testing.

| diagnostic testing | | antibody testing |
molecular testing	antigen testing	serology testing
provides information about acute COVID-19 infection	provides information about acute COVID-19 infection	does not provide information about acute infection
does not provide information about past infection	does not provide information about past infection	tests for antibody presence, does not guarantee immunity
can take hours to days	takes minutes to hours	takes minutes to hours
relatively accurate early in the infection	faster and cheaper than molecular tests	quick results
accuracy drops later in infection	less accurate than molecular tests	tests can vary in accuracy
point of care testing; collect swabs at home and mail in	point of care testing; collect swabs at home and mail in	blood test that can be done at doctor's office or at home

Testing for COVID-19. Since its outbreak in late 2019, the COVID-19 pandemic has generated an exponentially growing demand for testing, and diagnostic assays that enable mass screening have been developed at an unprecedented pace. To successfully use these tests to inform public health strategies, it is critical to understand their individual strengths and limitations. There are two different types of assays for SARS-CoV-2, the virus that causes COVID-19: diagnostic tests and antibody tests.

A *diagnostic test* can show if you have an active COVID-19 infection and need to take steps to quarantine or isolate yourself from others. Molecular tests and antigen tests fall under this category. A *molecular test* is a diagnostic test that detects genetic material from the virus using, for example, reverse transcription polymerase chain reaction or nucleic acid amplification. An *antigen test* is a diagnostic test that detects specific proteins made by the virus. Samples for diagnostic tests are typically collected with a nasal or throat swab or with saliva from spitting into a tube. Diagnostic testing is critical to provide early treatment, quarantine individuals sooner, and trace and isolate their contacts to reduce the spread of the virus.

Fig. 1.5 Testing for COVID-19. The objective of COVID-19
testing is to probe whether an individual is currently or
has previously been infected with SARS-CoV-2. A nasal or
throat swab test is a diagnostic test that provides information
about an acute infection. Diagnostic testing is critical to
provide early treatment, quarantine individuals, and trace
and isolate their contacts to reduce the spread of the virus. A
blood test is an antibody test that provides information about
a previous infection. Antibody tests are critical to estimate
the overall dimension of an outbreak. Throughout this book,
we use COVID-19 case data from public dashboards based
on confirmed positive diagnostic tests.

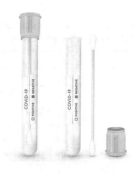

An *antibody test* can detect antibodies that are made by your immune system in
response to a previous infection with SARS-CoV-2. In contrast to genetic material
from the virus or specific proteins made by the virus, antibodies can take several
days or even weeks to develop, but they can remain present for several weeks or
months after recovery. A *serology test* is an antibody test that looks for antibodies in
blood samples. These venous blood samples are typically collected at the doctor's
office or in the clinic. Antibody tests should not be used to diagnose an active infection. Instead, antibody testing is important to understand the kinetics of the immune
response to infection, clarify whether an infection protects from future infection,
characterize how long immunity will last, and estimate the overall dimension of an
outbreak.

Table 1.2 contrasts the three most common types of tests for COVID-19. Throughout
this book, we use COVID-19 case data from public dashboards based on confirmed
positive diagnostic tests. Only in Chapters 8 and 13, where we explore the effects of
asymptomatic transmission, we use specialized models that combine case data from
diagnostic tests with seroprevalence data from antibody tests.

1.4 The basic reproduction number

The basic reproduction number is a powerful but simple concept to explain the
contagiousness and transmissibility of an infectious disease [6]. For decades, epidemiologists have successfully used the basic reproduction number to quantify how
many new infections a single infectious individual creates in an otherwise completely susceptible population [6]. During the COVID-19 pandemic, the public media, scientists, and political decision makers across the globe have adopted the basic
reproduction number as an illustrative metric to explain and justify the need for different outbreak control strategies [10]: An outbreak will continue for reproduction
numbers larger than one, $R_0 > 1$, and come to an end for reproduction numbers
smaller than one, $R_0 < 1$ [8]. However, especially in the midst of a global pandemic,
it is difficult–if not impossible–to measure R_0 directly [48].

Table 1.3 Basic reproduction numbers and herd immunity thresholds for common infectious diseases. Herd immunity is the indirect protection from an infectious disease that occurs when a large fraction of the population has become immune. The herd immunity threshold, $H = 1 - 1/R_0$, beyond which this protection occurs is a function of the basic reproduction number R_0.

disease	R_0	H	disease	R_0	H
measles	12 - 18	92 - 95%	mumps	4.0 - 7.0	75 - 86%
pertussis	12 - 17	92 - 94%	COVID-19	2.0 - 6.0	50 - 83%
rubella	6 - 7	83 - 86%	SARS	2.0 - 5.0	50 - 80%
smallpox	6 - 7	83 - 86%	ebola	1.5 - 2.5	33 - 60%
polio	5 - 7	80 - 86%	influenza	1.5 - 1.8	33 - 44%

Table 1.3 summarizes the basic reproduction numbers for common infectious diseases. It varies from $R_0 = 1.5 - 1.8$, for less contagious diseases like influenza, to $R_0 = 12 - 18$ for the measles and pertussis. Knowing the precise value of R_0 is important, but challenging, because of limited testing, inconsistent reporting, and incomplete data [6]. Throughout this book, instead of measuring the basic reproduction number directly, we estimate it using mathematical modeling and reported case data [17]. Mathematical models interpret the basic reproduction number R_0 as the ratio between the infectious period C, the period during which an infectious individual can infect others, and the contact period B, the average time it takes to come into contact with another individual [8],

$$R_0 = C/B. \tag{1.1}$$

The longer a person is infectious, and the more contacts the person has during this time, the larger the reproduction number [6]. While we cannot control the infectious period C, we can change our behavior to increase the contact period B [17]. This is precisely what community mitigation strategies and political interventions seek to attempt.

The reproduction number of COVID-19. Since the beginning of the coronavirus pandemic, no other number has been discussed more controversially than the reproduction number of COVID-19 [22]. The earliest COVID-19 study that followed the first 425 cases of the Wuhan outbreak via direct contact tracing reported a basic reproduction number of $R_0 = 2.2$ [15]. However, especially during the early stages of the outbreak, information was limited because of insufficient testing, changes in case definitions, and overwhelmed healthcare systems. While the concept of R_0 seems fairly simple, the reported basic reproduction numbers for COVID-19 vary hugely with country, culture, calculation, and time [17]. Most basic reproduction numbers of COVID-19 we see in the public media today are estimates of mathematical models. These estimates depend critically on the choice of the model, its initial conditions, and many other modeling assumptions [6]. To no surprise, the mathematically predicted basic reproduction numbers cover a wide range, from $R_0 = 2 - 4$ for exponential growth models to $R_0 = 4 - 7$ for more sophisticated compartment models [22].

Throughout this book, we identify or infer the reproduction number of COVID-19–and the uncertainty associated with it–using computational epidemiology [22], Bayesian analysis [10], and reported case data [12]: In the first example for the very early COVID-19 outbreak in China in Section 7.8, we identify a basic reproduction number of $R_0 = 12.58 \pm 3.17$ across 30 Chinese provinces. In the second example for the early outbreak in the United States in Section 10.2, we find a basic reproduction number of $R_0 = 5.30 \pm 0.95$ across all 54 states and territories. This value suggests that COVID-19 is less contagious than the measles and pertussis with $R_0 = 12 - 18$, as infectious as rubella, smallpox, polio, and mumps with $R_0 = 4 - 7$, slightly more infectious than SARS with $R_0 = 2 - 5$, and more infectious than an influenza with $R_0 = 1.5 - 1.8$. Our basic reproduction number for the United States is significantly lower than our basic reproduction number for China, which could be caused by an increased awareness of COVID-19 transmission a few weeks into the global pandemic. In our third example for the early outbreak in Europe in Section 12.2, the basic reproduction number takes similar values of $R_0 = 4.62 \pm 1.32$ across all 27 countries.

During these early stages of exponential growth, with new case numbers doubling within two or three days, the most urgent question amongst health care providers and political decision makers was: Can we reduce the reproduction number? For the broad public, this question became famously and illustratively rephrased as: Can we flatten the curve [15]? For the modeling community, the quest for a lower reproduction number all of a sudden meant that traditional epidemiology models were no longer suitable because of changes in the disease dynamics [12]. While traditional models with static parameters were well-suited to model the outbreak dynamics of unconstrained, freely evolving infectious diseases with fixed basic reproduction numbers in the early 20th century, they fail capture how behavioral changes and political interventions can modulate the reproduction number to manage the COVID-19 pandemic in the 21st century [17]. In fact, *static reproduction numbers* are probably the single most common cause of model failure in COVID-19 modeling [12].

Fortunately, several months into the pandemic, most countries have successfully managed to flatten the new-case-number curves and the reproduction numbers have dropped to values closer to or below one. To model these changes in disease dynamics and reproduction, in Section 7.7, we introduce a *dynamic reproduction number*, $R(t)$, that accounts for time-varying contact periods. In our fourth example for the European Union in Section 12.2, we show that the initial basic reproduction number of $R_0 = 4.22 \pm 1.69$ dropped to an effective reproduction number of $R(t) = 0.67 \pm 0.18$ by mid May 2020. Using machine learning, we correlate mobility and reproduction and identify the responsiveness between the drop in air traffic, driving, walking, and transit mobility and the drop in reproduction to $\Delta t = 17.24 \pm 2.00$ days. In the final examples in Sections 13.2 and 13.3 of nine locations across the world, we systematically infer the dynamic reproduction number $R(t)$ throughout a time window of 100 days while accounting for both symptomatic and asymptomatic transmission.

From the failure of traditional static epidemiology models [15], we have now learned that we need to introduce dynamic time-varying model parameters if we

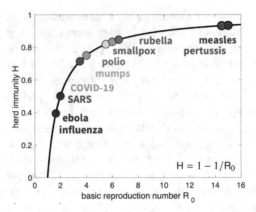

Fig. 1.6 Basic reproduction numbers and herd immunity thresholds for common infectious diseases. Herd immunity is the indirect protection from an infectious disease that occurs when a large fraction of the population has become immune, either through previous infection or through vaccination. The black line highlights the herd immunity threshold, $H = 1 - 1/R_0$, beyond which this protection occurs, as a function of the basic reproduction number R_0. The herd immunity threshold varies from 33-44% for influenza to 92-95% for the measles.

want to correctly model behavioral and political changes and reproduce the reported case numbers [17]. This naturally introduces a lot of freedom, a large number of unknowns, and a high level of uncertainty. However, in stark contrast to the epidemic outbreaks in the early 20th century, we now have thoroughly-reported case data and the appropriate tools [23] to address this challenge. The massive amount of COVID-19 case data, well documented and freely available, has induced a clear paradigm shift from traditional mathematical epidemiology towards data-driven, physics-based modeling of infectious disease [1]. This new technology naturally learns the most probable model parameters–in real time–from the continuously emerging case data, allows us to make projections into the future, and quantifies the uncertainty on the estimated parameters and predictions [26].

1.5 Concept of herd immunity

An important consequence of the basic reproduction number R_0 is the condition for herd immunity [9]. Herd immunity describes the indirect protection from an infectious disease that occurs when a large fraction of the population has become immune, either through previous infection or through vaccination [3]. The critical threshold at which the disease reaches this endemic steady state is called the *herd immunity* threshold,

$$H = 1 - 1/R_0 . \tag{1.2}$$

The larger the basic reproduction number R_0, the higher the herd immunity threshold H. Table 1.3 and Figure 1.6 summarize the basic reproduction numbers R_0 for several common infectious diseases along with the estimates of their herd immunity thresholds H. The herd immunity threshold varies from 33-44% for influenza to 92-95% for the measles. For the reported basic reproduction numbers of $R_0 = 2.0 - 6.0$ of COVID-19, the estimated herd immunity threshold would range from 50-83%. Recent studies that account for the emerging new and more infectious B.1.351 and B.1.1.7 variants of COVID-19 estimate these values to 75-95% [20].

1.6 Concept of immunization

To prevent or revert an epidemic outbreak, we need to ensure that, on average, every infectious individual infects less than one new individual. The concept of herd immunity describes the natural path towards ending an outbreak. However, if the basic reproduction number R_0 is large, the herd immunity threshold $H = 1 - 1/R_0$ is high, and waiting for herd immunity through infection alone can be quite devastating. Vaccination is a powerful strategy to accelerate the path towards herd immunity [1].

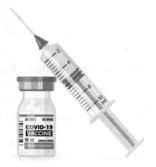

Fig. 1.7 Vaccination against COVID-19. The objective of COVID-19 vaccination is to provide immunity against severe acute respiratory syndrome coronavirus 2 or SARS-CoV-2, the virus that causes COVID-19 [30]. By May, 2021, thirteen vaccines were authorized for public use: two RNA vaccines (Pfizer–BioNTech and Moderna), five conventional inactivated vaccines (BBIBP-CorV, CoronaVac, Covaxin, WIBP-CorV and CoviVac), four viral vector vaccines (Sputnik V, Oxford–AstraZeneca, Convidecia, and Johnson & Johnson), and two protein subunit vaccines (EpiVacCorona and RBD-Dimer); and more than a billion doses of COVID-19 vaccines were administered worldwide.

Immunization threshold. Vaccination effectively reduces the susceptible population S. Successfully immunizing a fraction I of the population, reduces the susceptible population from S to $S_0^* = [1 - I] S_0$ and, with it, the reproduction number from R_0 to $R_0^* = [1 - I] R_0$. The critical immunization threshold I, below which the effective basic reproduction number is smaller than one, $R_0^* = [1 - I] R_0 < 1$, defines the fraction of the population that needs to be immunized to prevent or revert the outbreak of an epidemic,

$$I > 1 - 1/R_0. \tag{1.3}$$

From Table 1.3, we conclude that the required immunization fraction varies significantly, from $I > 92 - 95\%$ for the measles with a basic reproduction number of $R_0 = 12 - 18$ to $I > 33 - 44\%$ for the common influenza with a basic reproduction number of $R_0 = 1.5 - 1.8$. This explains, at least in part, why some infectious diseases are a lot more difficult to control through immunization than others.

Eradication through vaccination. Once enough individuals are immunized–either through infection or through vaccination–the outbreak stops. A disease that stops circulating in a specific region is considered *eliminated* in that region. Polio, for example, was eliminated in the United States by 1979 after widespread vaccination efforts. A disease that is eliminated worldwide is considered *eradicated*. Eradicating a disease through vaccination is a desirable but elusive goal [1]. Malaria has been a candidate for eradication, but although its incidence has been drastically reduced through vaccination, completely eradicating it remains challenging because infection does not result in life long immunity. Polio has been eliminated in most countries through massive vaccination efforts, but still remains present in some regions because its early symptoms often remain unnoticed and infected individuals continue to infect others. Measles have been the target of widespread vaccination, but although the disease is highly recognizable through its characteristic rash, a long latent period from exposure to the first onset of symptoms complicates outbreak control. To this day, smallpox is the only human infectious disease that has been successfully eradicated through vaccination. The eradication of smallpox is the result of focused surveillance, rapid identification, and ring vaccination [8]. In a massive vaccination campaign launched in 1967, anyone who could have possibly been exposed to smallpox was quickly identified and vaccinated to prevent its further spread. The last known case of smallpox occurred in Somalia in 1977. In 1980 World Health Organization declared smallpox eradicated. The eradication of smallpox remains one of the most notable and profound public health successes in history.

Efficacy and risk ratio. The *efficacy* e of a vaccine is the relative reduction in the disease attack rate between the unvaccinated placebo group n_{pla} and the vaccinated group n_{vac}. For a randomized trial with an equal allocation, meaning equally sized placebo and vaccinated groups, $n_{pla} = n_{vac}$, the efficacy is

$$e = \frac{n_{pla} - n_{vac}}{n_{pla}} \cdot 100\% = \left[1 - \frac{n_{vac}}{n_{pla}}\right] \cdot 100\% = 1 - r. \tag{1.4}$$

The *risk ratio* r is the ratio between the attack rate of the vaccinated group n_{vac} and the placebo group n_{pla},

$$r = \frac{n_{vac}}{n_{pla}} \cdot 100\% = 1 - e. \tag{1.5}$$

The efficacy is an important measure to characterize the success of a vaccine and define critical thresholds below which a vaccination trial should stop.

Table 1.4 Contingency table to quantify the significance of a vaccine. The table compares the total number of vaccinated and placebo individuals that developed and did not develop the disease.

	positive	negative	total
vaccine	a	b	a+b
placebo	c	d	c+d
total	a+c	b+d	n

Example: Efficacy of the first COVID-19 vaccine. On November 9, 2020, the Pfizer and BioNTech trial reported a number or COVID-19 cases of $n_{covid} = n_{pla} + n_{vac} = 94$ and an efficacy of $e > e_{min}$ with $e_{min} = 90\%$. The efficacy e of a vaccine is the relative reduction in the disease attack rate between the unvaccinated placebo group n_{pla} and the vaccinated group n_{vac},

$$e = \left[1 - \frac{n_{vac}}{n_{pla}}\right] \cdot 100\% = 1 - r \quad \text{with} \quad r = \frac{n_{vac}}{n_{pla}} \cdot 100\% = 1 - e.$$

where r is the risk ratio. The Pfizer and BioNTech trial was a randomized trial with an equal allocation [30]. Although it did not report detailed numbers, we can estimate the number of placebo and vaccinated cases n_{pla} and n_{vac} and the risk ratio r from the reported efficacy $e_{min} = 90\%$ with $n_{vac} = n_{covid} - n_{pla}$,

$$e = \left[1 - \frac{n_{covid} - n_{pla}}{n_{pla}}\right] \cdot 100\% = \left[2 - \frac{n_{covid}}{n_{pla}}\right] \cdot 100\% > e_{min}.$$

Solving for the number of placebo cases yields the general equation,

$$n_{pla} > \frac{n_{covid}}{2 - e_{min}},$$

and, for the Pfizer and BioNTech case, $n_{pla} > n_{covid}/1.1 = 85.45$. This implies that for a total of $n_{covid} = 94$ COVID-19 positive cases, at a number of placebo cases $n_{pla} = 86$ and vaccinated cases $n_{vac} = 8$, the efficacy is larger than 90%. Back-calculating the efficacy for these populations,

$$e = \left[1 - \frac{n_{vac}}{n_{pla}}\right] \cdot 100\% = \left[1 - \frac{8}{86}\right] \cdot 100\% = 90.7\% > 90\% = e_{min},$$

confirms the simulation. According to the protocol, Pfizer and BioNTech planned to take a look at the data at five stages with $n_{covid} = 32, 64, 92, 120, 164$ reported positive cases and only continue the trial if the efficacy was above $e > e_{crit}$ with $e_{crit} = 62.7\%$. From this information, we can calculate the critical numbers n_{pla} and n_{vac} below which the trial would have stopped at any of the five stages. Solving for the number of placebo cases yields the general equation,

$$n_{pla} > \frac{n_{covid}}{2 - e_{crit}},$$

and, for the Pfizer and BioNTech case, $n_{pla} > n_{covid}/1.373$. This implies that for $n_{covid} = 32, 64, 92, 120, 164$, the minimum number of placebo cases to continue the trial was $n_{pla} \geq 24, 47, 68, 88, 120$, for which the resulting efficacies of $e = 66.7\%, 63.8\%, 64.7\%, 63.6\%, 63.3\%$ would all have been slightly above the critical threshold of $e_{crit} = 62.7\%$.

Fisher's exact test and contingency tables. Fisher's exact test is a statistical significance test to analyze contingency tables that quantify the effects of a vaccine compared to a placebo. Table 1.4 illustrates a generic contingency table. From it, we can calculate the *case rate* across the entire trial as the ratio between the total number of positive cases and the total number of enrolled individuals,

$$v_{\text{disease}} = \frac{a+b}{a+b+c+d} \quad \text{with} \quad a+b+c+d = n_{\text{tot}}. \tag{1.6}$$

The case rates across the vaccinated and placebo groups are,

$$v_{\text{vac}} = \frac{a}{a+b} \quad \text{with} \quad a = n_{\text{vac}} \quad \text{and} \quad v_{\text{pla}} = \frac{c}{c+d} \quad \text{with} \quad c = n_{\text{pla}}. \tag{1.7}$$

Most vaccination trials are designed as a *randomized trial*, meaning they assign participants at random to a vaccinated or placebo group, with equal allocation, meaning they target an equal enrollment into both groups, $a+b \approx c+d \approx n_{\text{tot}}/2$. We can use the contingency table to calculate the *statistical significance p* of the deviation from a null hypothesis,

$$p = \frac{\binom{a+b}{a}\binom{c+d}{c}}{\binom{n}{a+c}} = \frac{\binom{a+b}{b}\binom{c+d}{d}}{\binom{n}{b+d}} = \frac{(a+b)!(c+d)!(a+c)!(b+d)!}{a!\,b!\,c!\,d!\,n!}, \tag{1.8}$$

in terms of binominal coefficients or factorial operators using a hypergeometric distribution. For a vaccination trial, the lower the *p*-value, the larger the effect on the vaccinated group compared to the placebo group.

Example: Statistical significance of the first COVID-19 vaccine. On November 9, 2020, the Pfizer and BioNTech trial reported $n_{\text{tot}} = 43{,}538$ enrolled participants. From this number, we can calculate the case rates across the entire trial, in the vaccinated group, and in the placebo group assuming a randomization at 1:1 between the vaccinated and placebo groups. The case rate of the entire trial is the ratio between the total number of COVID-19 positive cases and the total number of enrolled individuals,

$$v_{\text{covid}} = \frac{n_{\text{covid}}}{n_{\text{tot}}} = \frac{94}{43{,}538} = 0.190\%.$$

The case rates of the two groups are the ratios between the vaccinated and placebo COVID-19 positive cases and half of the enrolled cases,

$$v_{\text{vac}} = \frac{n_{\text{vac}}}{n_{\text{tot}}/2} = \frac{8}{43{,}538/2} = 0.037\% \qquad v_{\text{pla}} = \frac{n_{\text{pla}}}{n_{\text{tot}}/2} = \frac{86}{43{,}538/2} = 0.395\%.$$

Using Fisher's exact test, we can test the null hypothesis that vaccinated and placebo participants will equally likely contract COVID-19.

	positive	negative	total
vaccine	8	21,761	21,769
placebo	86	21,683	21,769
total	94	43,444	43,538

We can use the contingency table to estimate the efficiency of the Pfizer BioN-Tech vaccine. Fisher's exact test calculates the significance of the deviation from a null hypothesis,

$$p = \frac{21769!\,21769!\,94!\,43444!}{8!\,21761!\,86!\,21683!\,43538!},$$

and confirms, for the Pfizer and BioNTech case with $p < 0.00001$, that individuals in the vaccinated and placebo groups will not equally likely contract COVID-19.

1.7 Mathematical modeling in epidemiology

During the early onset of the COVID-19 pandemic, all eyes were on mathematical modeling with the general expectation that mathematical models could precisely predict the trajectory of the pandemic. Mathematical modeling rapidly became front and center to understanding the exponential increase of infections, the shortage of ventilators, and the limited capacity of hospital beds; too rapidly as we now know. Bold and catastrophic predictions not only initiated a massive press coverage, but also a broad anxiety in the general population [15]. However, within only a few weeks, the vastly different predictions and conflicting conclusions began to create the impression that all mathematical models are generally unreliable and inherently wrong [12]. While the failure of COVID-19 modeling–often by an order of magnitude and more–was devastating for policymakers and public health practitioners, initial mistakes are not new to the modeling community where an iterative cycle of prediction, failure, and redesign is common standard and best practice [26]. However, the successful use of mathematical models implies to set the expectations right [45]. Understanding what models can and cannot predict is critical to the Art of Modeling. Epidemiologists distinguish two kinds of models to understand the outbreak dynamics of an infectious disease: statistical models and mechanistic models [12]. Depending on the degree of complexity, the most popular mechanistic models are compartment models and agent-based models.

Statistical models, or more precisely, purely statistical models, use machine learning or regression to analyze massive amounts of data and project the number of infections into the future. The essential idea is to select a function $D(t)$, use statistical tools to fit its coefficients to reported case data $\hat{D}(t)$, and make projections into the future. The function can be quadratic, cubic, logistic, power-law, or exponential,

Table 1.5 Mathematical modeling of COVID-19. Summary of the three most common models.
Statistical or forecasting models fit nonlinear functions to case data over time, whereas mechanistic
models, including compartment and agent-based models, simulate outbreak and contact dynamics.
Throughout this book, we use compartment models to simulate the outbreak dynamics of infectious
diseases including COVID-19.

statistical models forecasting models	mechanistic models compartment models	agent-based models
use machine learning, statistics, regression, or method of least squares	use physics-based modeling based on nonlinear reaction-diffusion equations	use rule-based approaches to study the interaction of autonomous systems
model case numbers through a nonlinear function	model population through compartments	model every individual as an independent agent
formulate number of cases as a function of time	formulate rules by which individuals pass through the compartments	formulate simple rules by which individual agents interact
fit coefficients that are purely phenomenological	infer parameters that have a mechanistic interpretation	identify parameters that summarize human behavior
predicts case numbers from fitting a function	predict outbreak dynamics, characterize sensitivities, quantify uncertainties	predict outbreak dynamics as emergent collective behavior of individual agents
no feedback mechanisms	nonlinear feedback	discrete contact networks
predictions are inexpensive, but very unreliable	predictions are reliable only for a small time window	predictions are detailed, but computationally expensive

for example, $D(\vartheta, t) = \exp(c_0 + c_1 t + c_2 t^2 + c_3 t^3)$, where $D(\vartheta, t)$ are the modeled cumulative cases per day, $\vartheta = \{c_0, c_1, c_2, c_3\}$ are the model parameters, and t is the time. One of the simplest statistical tools to compare the model $D(\vartheta, t)$ to the data $\hat{D}(t)$ and identify values for its parameters ϑ is the method of least squares. It is important to understand that these parameters are purely phenomenological, they are derived purely by fitting a curve, and typically do not have a mechanistic interpretation. Early in an outbreak, when little is known about disease transmission, epidemiologists often use statistical models because they do not rely on any prior knowledge of the disease. An example for a statistical model of COVID-19 is the initial IHME model [23]. Early in the pandemic, the IHME model used case data from China and Italy to create similar curves, forecast case numbers in the United States, and inform the White House's response to the pandemic. Carefully constructed statistical frameworks can be used for short-time forecasting using machine learning or regression. This could potentially be useful to understand how to allocate resources or make rapid short-term recommendations. However, purely statistical models can neither capture the dynamics of disease transmission nor the effects of mitigation strategies. This explains, at least in part, why the early COVID-19 predictions based on purely statistical models were off by an order of magnitude or more. To address these serious limitations, several COVID-19 models have now been adjusted to combine both statistical modeling and mechanistic modeling.

Mechanistic models simulate the outbreak through interacting disease mechanisms by using local nonlinear population dynamics and global mixing of populations

[4]. The underlying idea is to identify fundamental mechanisms that drive disease dynamics, for example the duration of the infectious period or the number of contacts an infectious individual has during this time. Unlike purely statistical models, mechanistic models include important nonlinear feedback: The more people become infected, the faster the disease spreads. By their very nature, the parameters of mechanistic models are not just fitting parameters, they usually have a clear epidemiological interpretation. This makes mechanistic modeling a powerful strategy to explore different outbreak scenarios or study how an outbreak would change under various assumptions and political interventions [12]. Another advantage of mechanistic models is that we can adjust and improve them dynamically as more information becomes available. Throughout this book, we gradually improve a class of mechanistic models by adding new information. For example, we introduce time-varying dynamic contact rates that vary in different lockdown levels and add the effect of asymptomatic transmission [26]. Even if we do not precisely know the dimension of asymptomatic disease spread, we can use mechanistic models to study what-if scenarios: What would the disease landscape look like if two third of all infectious were asymptomatic? Mechanistic modeling naturally extends into sensitivity analysis and uncertainty quantification. As such, it not only provides valuable information about the robustness of the model, but also about the most effective parameters to modulate a disease outbreak [19]. Rather than studying a single one disease trajectory, we could explore a range of trajectories around the mean and characterize the best- and worst-case scenarios. The two most popular mechanistic models in epidemiology are compartment models and agent-based models, and both have been used to understand the outbreak dynamics of COVID-19. When choosing between compartment models and agent-based, it is important to understand the major strengths and weaknesses of each model.

Compartment models are the most common approach to model the epidemiology of an infectious disease [14]. Compartment models simulate the collective behavior of subgroups of the population through a number of compartments with labels, for example, SEIR for susceptible, exposed, infectious, and recovered. Individuals move between compartments and the order of the labels indicates the successive motion, for example, SEIS means susceptible, exposed, infectious, then susceptible again [4]. The underlying principle is to model the time evolution of these groups through a set of coupled ordinary differential equations, identify rate constants that characterize their interaction using reported case data, and vary these rate constants to probe different outbreak scenarios.

Figure 1.8 illustrates a compartment model that represents the characteristic timeline of COVID-19 through six compartments, the susceptible, exposed, infectious, recovered, hospitalized, and dead groups. This SEIRHD model is defined through a set of six ordinary differential equations that simulate how many individuals reside in each compartment throughout the duration of the outbreak. The model parameters of a compartment model define the transition rates between the individual compartments and, for more complex models, the fraction of individuals that transition into a particular path of the disease. For this example, the parameters $\vartheta = \{\beta, \alpha, \gamma, \nu_h, \nu_d\}$

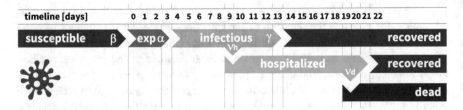

Fig. 1.8 Characteristic timeline of COVID-19. On day 0, a fraction of all *susceptible* individuals is exposed to the virus. After a latent period of $A = 1/\alpha = 3$ days, the *exposed* individuals become *infectious*. After an infectious period of $C = 1/\gamma = 10$ days, a fraction $(1 - \nu_h)$ transition to the *recovered* group, whereas a fraction ν_h develops severe symptoms and is hospitalized. Of the *hospitalized* individuals, a fraction $(1 - \nu_d)$ recovers, whereas a fraction ν_d becomes *dead*. We can simulate this behavior through an SEIRHD model with six compartments.

are the contact rate β, latent rate α, and infectious rate γ, and the hospitalized and dead fractions, ν_h and $\nu_h \cdot \nu_d$. From reported case numbers, hospitalizations, and deaths, we can identify or infer the set of model parameters ϑ, the rate constants and fractions, that best explain the model output using statistical tools. Importantly, in contrast to purely statistical models, compartment models are based on model parameters that have a clear physical interpretation. The most important parameters of any compartment model are the contact rate β and the infectious rate γ. Together, they define an important nonlinear feedback that is not present in purely statistical models: The more people become infected, the faster the spread of the disease. The basic reproduction number $R_0 = \beta/\gamma$, the ratio of the contact rate β and the infectious rate γ, characterizes the magnitude of this feedback. It is an easy-to-understand disease metric that explains how quickly susceptible individuals become infected and how fast a disease spreads across a population [6]. Some epidemiologists argue that, because of their mechanistic nature, compartment models are better suited for long-term predictions than purely statistical models. While this might be true for infectious diseases that develop freely, without any political intervention, the COVID-19 pandemic has taught us that long-term predictions of outbreak dynamics are challenging, even with the most sophisticated mechanistic models [15]. Understanding the potential and the limitations of compartment modeling is one of the main objectives of this book.

In Chapters 2 and 3, we introduce two simple classical compartment models before we introduce the most common compartment model for COVID-19 in Chapter 4. We show how we can use these models to estimate the reproduction number from reported COVID-19 case data. Knowing the precise reproduction number has important consequences for estimating the dimensions of herd immunity and immunization. Compartment models capture the fundamental dynamics of disease transmission and the effects of public health interventions; however, classical compartment models ignore the dynamics of the contact rate and its variation across a population. In Section 7.7 we introduce a compartment model that explicitly accounts for a time-varying dynamic contact rate, $\beta(t)$, and captures a varying contact behavior in different subgroups of the population. The dynamic nature of the contact rate naturally introduces a dynamic reproduction number, $R(t) = \beta(t)/\gamma$ and allows us to quantify the effectiveness of policy measures as we discuss in Section 12.2.

Agent-based models simulate individuals or agents interacting in various social settings and estimate the spread of a disease as these agents come into contact with one another. The underlying idea is to represent each agent individually, formulate relatively simple rules by which individual agents interact, and interpret the collective behavior across all agents as the emergent dynamics of an outbreak. A strength of agent-based models is that they simulate human behavior very granularly: They can assign different parameters or behavior patterns to each individual agent instead of simulating the collective behavior of entire populations. As such, agent-based models offer a lot more freedom than compartment models, but also require a lot more detail. For example, to formulate rules of interaction, agent-based models draw on social connectivity networks, from activity surveys, cell phone locations, public transportation, or airlines statistics. By their very nature, agent-based approaches are computationally expensive, especially for large populations. For small populations, agent-based modeling is a powerful strategy to predict how individual behavior, for example the violation of quarantine, leads to a collective behavior and modulates disease spread. For large populations, agent-based modeling can help rationalize collective model parameters, for example contact rates or reproduction numbers, that feed into more abstract, population level models. Above a certain population size, agent-based models simply become computationally unfeasible and most epidemiologists would turn to a more macroscopic approach that represents groups of individuals collectively as subgroups of the population. Since our objective is to design and discuss data-driven models for the COVID-19 pandemic, throughout the remainder of this book, we focus exclusively on compartment models.

When choosing between purely statistical, compartment, and agent-based models, it is important to know upfront which questions the model should address [45]. Table 1.5 compares the three models and summarizes their strengths and weaknesses. Throughout this book, we focus on mechanistic compartment modeling.

1.8 Data-driven modeling in epidemiology

One year after the World Health Organization had declared the COVID-19 outbreak a global pandemic, SARS-CoV-2 has resulted in more than 118 million reported cases across more than 180 countries and over 2.6 million deaths worldwide. Unlike any other disease in history, the COVID-19 pandemic has generated an unprecedented volume of data, well documented, continuously updated, and broadly available to the general public. There is a critical need for time- and cost-efficient strategies to analyze and interpret these data to systematically manage the pandemic on a global level. Yet, the precise role of physics-based modeling and machine learning in providing quantitative insight into the dynamics of COVID-19 remains a topic of ongoing debate.

Physics-based modeling is a successful strategy to integrate multiscale, multi-physics data and uncover mechanisms that explain the dynamics of specific outbreak characteristics. However, physics-based modeling alone often fails to efficiently combine large data sets from different sources and different levels of resolution [26]. Machine learning is as a powerful technique to integrate multimodality, multifidelity data, from cities, counties, states, and countries across the world, and reveal correlations between different disease phenomena. However, machine learning alone ignores the fundamental laws of physics and can result in ill-posed problems or non-physical solutions [1]. Throughout this book, we illustrate how data-driven modeling can integrate classical physics-based modeling and machine learning to infer critical disease parameters–in real time–from reported case data to make informed predictions and guide political decision making. As a valuable by product, this approach naturally lends itself in sensitivity analysis and uncertainty quantification. From the COVID-19 pandemic, we have learnt that even small inaccuracies in the model can trigger large changes in the number of cases. To understand the vulnerability of the model to these small changes, especially in view of the varying reporting practices of the COVID-19 case data, sensitivity analysis and quantifying uncertainty have become critical elements of robust predictive modeling.

Epidemiology data. In data-driven modeling, we need data to fit or infer our model parameters. Unlike earlier pandemics for which case data are often sparse, irregular, or incomplete, the COVID-19 pandemic is amazingly well documented [13]. On hundreds of public COVID-19 dashboards, we can find and download a vast variety of case numbers: daily new cases, active cases, recovered cases, seriously critical cases, cumulative cases, and deaths, at the level of cities, counties, states, countries, or the entire world [9, 12, 25]. Local dashboards often also share the number of hospitalizations and intensive care units, which were of great concern especially at the early onset of the pandemic. More recently, these dashboards have also included the number of tests and vaccines. Seroprevalence data with information about the history of the disease, both asymptomatic and symptomatic, are rare and often only available from scientific publications rather than governmental databases. These data are typically updated on a daily basis, and contain notable weekday-weekend alterations. When using the data to infer model parameters and learn about the outbreak behavior, we usually smoothen these alteration using seven-day moving averages. Finally, to compare data from different locations, we typically scale the reported case data by the population. A common metric for comparison is the seven-day-per-100,000 *incidence*, the number of new cases per 100,000 individuals across a seven-day window [38]. Policy makers across the globe use this incidence value to characterize the severity of the outbreak and justify the need for political interventions.

Figure 1.9 illustrates a typical data set for the COVID-19 outbreak that we use to infer our epidemiological model parameters. The orange lines summarize the outbreak dynamics worldwide throughout the year after the World Health Organization had declared COVID-19 a global pandemic, from March 11, 2020 to March 11, 2021. The light orange lines represent the reported daily new cases, which display notable weekday-weekend fluctuations associated with testing and reporting irreg-

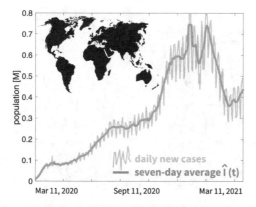

Fig. 1.9 Daily new cases of COVID-19 and their seven-day moving average. Daily new COVID-19 cases are reported on public dashboards worldwide. The light orange curve represents the raw data, the dark orange curve is the seven-day moving average, both reported throughout the year after the WHO had declared COVID-19 a global pandemic on March 11, 2020.

ularities. The dark orange lines are the associated seven-day *moving average* $\hat{I}(t)$, which smoothens the fluctuations and displays a much clearer trend in the disease dynamics. Within this one-year window, the absolute daily case numbers peaked on January 8, 2021 with 841,304 new cases. The moving seven-day average peaked on January 11, 2021 with $\Delta I_{max} = 745,404$ cases. If we assume an infectious period of $C = 1/\gamma = 7\,$days, this would result in a maximum infectious population of $I_{max} = 7 \cdot \Delta I_{max} = 7 \cdot 745,404 = 5,217,828$, meaning that mid January 2021, more than 5 million people were sick with COVID-19. For a total population of $N = 7.8\,$billion people, this corresponds to a peak seven-day-per-100,000 *incidence* of $I_{max} \cdot 100,000/N = 67$. There are many different ways to use the reported case data. The simplest way is to select a function, use statistical tools to fit its coefficients to the reported case data, and make projections into the future. However, not only the daily new cases, but also the seven-day average in Figure 1.9 display substantial fluctuation and it seems difficult to find a function that could explain the orange curves. For data-driven modeling, it is often easier to use the total cumulative case numbers, which always increase monotonically and tend to be more smooth in general.

Figure 1.10 shows the total cumulative cases of COVID-19 $\hat{D}(t)$ and a simple statistical model $D(t)$ to fit the data throughout the first year of the pandemic. During the very early stages of an outbreak, an exponential growth model, $D(t) = D_0 \exp(G\,t)$, with a growth rate G often provides a good approximation of the total number of cases $\hat{D}(t)$ and is easy to fit. Indeed, this, or similar exponential models, is what many early approaches used. While the initial phase of the outbreak is well represented by exponential growth models, they soon tend to overestimate the outbreak. This is why, during the early COVID-19 pandemic, there was a broad overestimate of the number of cases and of the number of ventilators and hospital beds needed to tread diseased individuals [15]. In this book, instead of using purely statistical models, we use mechanistic models like the compartment model in Figure

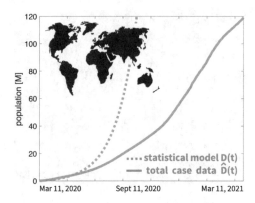

Fig. 1.10 Total cumulative cases of COVID-19 and a statistical model to fit the data. Cumulative COVID-19 cases, the sum of all daily new cases to that date, are reported on public dashboards worldwide. The solid orange curve represents the raw data, the dashed orange curve is an example of a statistical model, both reported throughout the year after the WHO had declared COVID-19 a global pandemic on March 11, 2020.

1.8. For this SEIRHD compartment model, the number of total cases is the sum of all infectious, hospitalized, recovered and dead individuals, $I(t) + R(t) + H(t) + D(t)$. Throughout this book, we use different types of compartment models, infer their model parameters using reported COVID-19 case data from public databases [9, 12, 25], seroprevalence data from scientific publications [26], and mobility data [2, 10, 10], and vary the parameters to probe different outbreak scenarios.

Problems

1.1 Testing, testing, testing. On June 6, 2020, Donald J. Trump, the President of the United States, famously said "Remember this, when you test more, you have more cases." What is wrong with this statement? What did he really mean to say? Discuss the implications of the testing frequency on reported case numbers, and, ultimately, on policy making.

1.2 Herd immunity. On September 15, 2020, Donald J. Trump, the President of the United States, claimed that COVID-19 would go away, without a vaccine. "You'll develop herd. Like a herd mentality." What did he really mean to say? Discuss the effects of vaccination on herd immunity.

1.3 Herd immunity. Assume the basic reproduction number of the swine flu, caused by the H1N1 virus, was on the order of $R_0 = 1.4 - 1.6$. Calculate its herd immunity threshold H and compare it against the herd immunity thresholds of other infectious diseases in Table 1.3 and Figure 1.6. Comment on whether this makes the swine flu a good candidate for eradication through vaccination.

1.4 COVID-19 variants B.1.351 and B.1.1.7. Assume the basic reproduction number of COVID-19 in December 2020 was $R_0 = 2.0$. Calculate the herd immunity threshold H. Now, assume that the variant B.1.351 with a 50% increased basic reproduction number has been introduced into the population [20]. How does the herd immunity threshold change? How would the variant B.1.1.7 with a 56% increased basic reproduction number change the herd immunity threshold? Comment on your results.

1. Avoid needless crowding—influenza is a crowd disease.

2. Smother your coughs and sneezes—others do not want the germs which you would throw away.

3. Your nose, not your mouth was made to breathe through—get the habit.

4. Remember the three C's—a clean mouth, clean skin, and clean clothes.

5. Try to keep cool when you walk and warm when you ride and sleep.

6. Open the windows—always at home at night; at the office when practicable.

7. Food will win the war if you give it a chance—help by choosing and chewing your food well.

8. Your fate may be in your own hands—wash your hands before eating.

9. Don't let the waste products of digestion accumulate—drink a glass or two of water on getting up.

10 Don't use a napkin, towel, spoon, fork, glass or cup which has been used by another person and not washed.

11. Avoid tight clothes, tight shoes, tight gloves—seek to make nature your ally not your prisoner.

12. When the air is pure breathe all of it you can—breathe deeply.

GEORGE A. SOPER
SANITARY CORPS,
U. S. A.
MAY 30, 1919

Fig. 1.11 The Lessons of the Pandemic. Twelve condensed rules to the the avoidance of unnecessary personal risks and to the promotion of better personal health by George A. Soper in reflection of the Spanish flu in 1919.

1.5 The Lessons of the Pandemic. In his Science publication *The Lessons of the Pandemic*, George A. Soper lists twelve public health measures to manage the Spanish flu in 1919. Read the twelve recommendations in Figure 1.11 and comment on which of them are still valid for the COVID-19 pandemic today.

1.6 Efficacy of the AstraZeneca COVID-19 vaccine. The randomized, equal allocation AstraZeneca trial enrolled 11636 participants in its interim efficacy analysis, 5807 in the vaccinated group and 5829 in the unvaccinated placebo group. Of the vaccinated group, 30 developed COVID-19, and of the placebo group 101. Calculate the efficacy and risk ratio of the AstraZeneca vaccine.

1.7 Case rates during the AstraZeneca COVID-19 trial. The AstraZeneca trial enrolled 11636 participants in its interim efficacy analysis, 5807 in the vaccinated group and 5829 in the unvaccinated placebo group. Of the vaccinated group, 30 developed COVID-19, and of the placebo group 101. Create a contingency table for the interim efficacy analysis. Calculate the overall case rate of the trial and the case rates in the vaccinated and placebo groups.

1.8 Efficacy of the Janssen COVID-19 vaccine. The randomized, equal allocation Janssen COVID-19 trial enrolled 39,321 participants, 19,630 received the vaccine and 19,691 received placebo. At least 14 days after vaccination, the trial recorded 116 COVID-19 cases in the vaccine group and 348 cases in the placebo group. Calculate the efficacy of the Janssen COVID-19 vaccine and compare it to the efficacy of the Pfizer BioNTech and AstraZeneca vaccines. How would you expect the efficacy to change after another 14 days?

1.9 Efficacy of the Janssen COVID-19 vaccine. The randomized, equal allocation Janssen COVID-19 trial enrolled 39,321 participants, 19,630 received the vaccine and 19,691 received placebo. At least 28 days after vaccination, the trial recorded 66 COVID-19 cases in the vaccine group and 193 cases in the placebo group. Calculate the efficacy of the Janssen COVID-19 vaccine after 28 days and compare it to the efficacy after 14 days. Comment on why you would or would not have expected this result.

1.10 Statistical models. Find and download the total cumulative COVID-19 case data $\hat{D}(t)$ from your own city, county, state, country, or the world. Plot the case data similar to Figure 1.10. Try to fit a linear function $D(t) = D_0 + c_1 t$ and a quadratic function $D(t) = D_0 + c_1 t + c_2 t^2$ to the very early stages of the outbreak. Which function is easier to fit? What does that tell you about the early outbreak?

1.11 Statistical models. Find and download the total cumulative COVID-19 case data $\hat{D}(t)$ from your own city, county, state, country, or the world. Plot the case data similar to Figure 1.10. Fit an exponential function $D(t) = D_0 \exp(G t)$ to the early stages of the outbreak. What is your growth rate G? When does the exponential function fail to describe the case data $\hat{D}(t)$?

1.12 Epidemiology data. Find and download the daily new COVID-19 case data from your own city, county, state, country, or the world. Plot the raw data similar to Figure 1.9. Explain the local fluctuations in the reported case data that occur on the order of days.

1.13 Epidemiology data. Find and download the daily new COVID-19 case data from your own city, county, state, country, or the world. Calculate and plot the seven-day moving average similar to Figure 1.9. How many waves can you identify? Explain the global fluctuations that occur on the order of weeks or months. Interpret the growth or decay in case numbers in view of specific events or political interventions.

1.14 Epidemiology data. Find the COVID-19 case data from your own state or country and your furthest away vacation destination. Compare the different outbreak dynamics. Do you think the reporting between both locations is consistent? Identify at least four potential sources of error in reporting daily COVID-19 case data.

1.15 Incidence. Find and download the daily new COVID-19 case data from your own city, county, state, country, or the world. Calculate the seven-day moving average. Identify the peak seven-day moving average ΔI_{max} within your simulation window. Assume an infectious period of $C = 1/\gamma = 7$ days. Calculate the maximum infectious population $I_{max} = 7 \cdot \Delta I_{max}$. Find the total population N of your location and calculate its peak seven-day-per-100,000 incidence, $I_{max} \cdot 100,000/N$.

1.16 Incidence and the effect of scale. Studying an outbreak at a more global scale tends to smoothen fluctuations and local peaks. Find and download the daily new COVID-19 case data from the next smaller or larger scale compared to the previous problem. If you have studied your state, now study your city, county, or country. Identify the peak seven-day moving average ΔI_{max} within your simulation window and calculate the maximum infectious population $I_{max} = 7 \cdot \Delta I_{max}$. Find the total population N of your location and calculate its peak seven-day-per-100,000 incidence, $I_{max} \cdot 100,000/N$. Compare you results against the seven-day-per-100,000 incidence at the smaller or larger scale. Interpret your results.

1.17 Epidemiology data and compartment models. The compartment model in Figure 1.8 introduces three possible disease paths from infection: direct recovery at a fraction $(1 - \nu_h)$, hospitalization and recovery at a fraction $\nu_h (1 - \nu_d)$, and hospitalization and death at a fraction $\nu_h \cdot \nu_d$. The fraction $\nu_h \cdot \nu_d$ is called the case fatality rate and is about 2% worldwide for COVID-19. Find extreme values for the case fatality rate. Discuss which factors influence the case fatality rate both globally and locally.

1.18 Epidemiology data and compartment models. Assume you want to learn parameters for your compartment model, for example the one in Figure 1.8, from reported case data, hospitalizations, and deaths. Early in the pandemic, when testing was slow and reporting was delayed, epidemiologists suggested to use deaths rather than daily new cases for model calibration. This initiated a controversial and still ongoing discussion how to count COVID-19 deaths. Discuss the difference between *death from* and *death with* COVID-19 in terms of absolute numbers, case fatality ratios, and model parameters.

References

1. Alber M, Buganza Tepole A, Cannon W, De S, Dura-Bernal S, Garikipati K, Karniadakis G, Lytton WW, Perdikaris P, Petzold L, Kuhl E (2019) Integrating machine learning and multiscale modeling: Perspectives, challenges, and opportunities in the biological, biomedical, and behavioral sciences. npj Digital Medicine 2:115.
2. Anderson RM, May RM (1982) Directly transmitted infectious diseases: control by vaccination. Science 215:1053-1060.
3. Anderson RM, May RM (1991) Infectious Diseases of Humans. Oxford University Press, Oxford.
4. Apple Mobility Trends. https://www.apple.com/covid19/mobility. accessed: June 1, 2021.
5. Bernoulli D (1760) Essay d'une nouvelle analyse de la mortalite causee par la petite verole et des avantages de l'inoculation pour la prevenir. Mémoires de Mathématiques et de Physique, Académie Royale des Sciences, Paris 1-45.
6. Brauer F, Castillo-Chavez C (2001) Mathematical Models in Population Biology and Epidemiology. Springer-Verlag New York.
7. Brauer F, van den Dreissche P, Wu J (2008) Mathematical Epidemiology. Springer-Verlag Berlin Heidelberg.

8. Brauer F (2017) Mathematical epidemiology: Past, present and future. Infectious Disease Modelling 2:113-127.

9. Brauer F, Castillo-Chavez C, Feng Z (2019) Mathematical Models in Epidemiology. Springer-Verlag New York.

10. Delamater PL, Street EJ, Leslie TF, Yang YT, Jacobsen KH (2019) Complexity of the basic reproduction number (R_0). Emerging Infectious Diseases 25:1-4.

11. Dieckmann O, Heesterbeek JAP (2000) Mathematical Epidemiology of Infectious Diseases: Model Building, Analysis and Interpretation. Wiley.

12. Dietz K (1993) The estimation of the basic reproduction number for infectious diseases. Statistical Methods in Medical Research 2:23-41.

13. Dong E, Gardner L (2020) An interactive web-based dashboard to track COVID-19 in real time. The Lancet Infectious Diseases 20:533-534.

14. European Centre for Disease Prevention and Control. Situation update worldwide. `https://www.ecdc.europa.eu/en/geographical-distribution-2019-ncov-cases` accessed: June 1, 2021.

15. Eurostat. Your key to European statistics. Air transport of passengers. `https://ec.europa.eu/eurostat` accessed: June 1, 2021.

16. Evans AS (1976) Viral Infections of Humans. Epidemiology and Control. Plenum Medical Book Company, New York and London.

17. Fauci AS, Lane HC, Redfield RR (2020) Covid-19–Navigating the uncharted. New England Journal of Medicine 382:1268-1269.

18. Fine PEM (1993) Herd immunity: history, theory, practice. Epidemiologic Reviews 15:265-302.

19. Gelman A, Carlin JB, Stern HS, Dunson DB, Vektari A, Rubin DB (2013) Bayesian Data Analysis. Chapman and Hall/CRC, 3rd edition.

20. Gorbalenya AE, Baker SC, Baric RS, de Groot RJ, Drosten C, Gulyaeva AA, Haagmans BL, Lauber C, Leontovich AM, Neuman BW, Penzar D, Perlman S, Poon LLM, Samborskiy D, Sidorov IA, Sola I, Ziebuhr J (2020) Severe acute respiratory syndrome-related coronavirus: the species and its viruses-a statement of the coronavirus study group. Nature Microbiology 5:536-544.

21. Hethcote HW (2000) The mathematics of infectious diseases. SIAM Review 42:599-653.

22. Holmdahl I, Buckee C. Wrong but useful–What Covid-19 epidemiolgic models can and cannot tell us. New England Journal of Medicine 383:303-305.

23. Institute for Health and Metrics Evaluation IHME. COVID-19 Projections. `https://covid19.healthdata.org`. assessed: July 27, 2020.

24. International Air Transport Association (2020) `https://www.iata.org`. accessed: July 9, 2020.

25. Ioannidis JPA, Cripps S, Tanner MA (2021) Forecasting for COVID-19 has failed. International Journal of Forecasting, in press.

26. Johns Hopkins University (2021) Coronavirus COVID-19 Global Cases by the Center for Systems Science and Engineering. `https://coronavirus.jhu.edu/map.html`, `https://github.com/CSSEGISandData/covid-19` assessed: June 1, 2021.

27. Kermack WO, McKendrick G (1927) Contributions to the mathematical theory of epidemics, Part I. Proceedings of the Royal Society London Series A 115:700-721.

28. Krämer A, Kretzschmar M, Krickeberg K (2010) Modern Infectious Disease Epidemiology. Springer-Verlag New York.

29. Kuhl E (2020) Data-driven modeling of COVID-19 – Lessons learned. Extreme Mechanics Letters 40:100921.

30. Kyriakidis NC, Lopez-Cortes A, Vasconez Gonzalez E, Barreto Grimaldos A, Ortiz Prado E. SARS-CoV-2 vaccines strategies: a comprehensive review of phase 3 candidates. npj Vaccines 6:28.

31. Li Q, Guan X, Wu P, Wang X, ... Feng Z (2020) Early transmission dynamics in Wuhan, China, of novel coronavirus-infected pneumonia. New England Journal of Medicine 382:1199-1207.

32. Liang ST, Liang LT, Rosen JM (2021) COVID-19: a comparison to the 1918 influenza and how we can defeat it. BMJ Postgraduate Medical Journal 97:273-274

33. Linka K, Peirlinck M, Kuhl E (2020) The reproduction number of COVID-19 and its correlation with public heath interventions. Computational Mechanics 66:1035-1050.
34. Linka K, Goriely A, Kuhl E (2021) Global and local mobility as a barometer for COVID-19 dynamics. Biomechanics and Modeling in Mechanobiology 20:651–669.
35. Linka K, Peirlinck M, Schafer A, Ziya Tikenogullari O, Goriely A, Kuhl E (2021) Effects of B.1.1.7 and B.1.351 on COVID-19 dynamics. A campus reopening study. Archives of Computational Methods in Engineering. doi:10.1007/s11831-021-09638-y.
36. Liu J, Shang, X (2020) Computational Epidemiology. Springer International Publishing.
37. Liu Y, Gayle AA, Wilder-Smith A, Rocklöv J (2020) The reproductive number of COVID-19 is higher compared to SARS coronavirus. Journal of Travel Medicine (2020) 27:taaa021.
38. Lu H, Weintz C, Pace J, Indana D, Linka K, Kuhl E (2021) Are college campuses superspreaders? A data-driven modeling study. Computer Methods in Biomechanics and Biomedical Engineering doi:10.1080/10255842.2020.1869221.
39. Martcheva M (2015) An Introduction to Mathematical Epidemiology. Springer Science + Business Media New York.
40. New York Times (2020) Coronavirus COVID-19 Data in the United States. https://github.com/nytimes/covid-19-data/blob/master/us-states.csv assessed: June 1, 2021.
41. Osvaldo M (2018) Bayesian Analysis with Python: Introduction to Statistical Modeling and Probabilistic Programming Using PyMC3 and ArviZ. Packt Publishing, 2nd edition.
42. Paul JR (1966) Clinical Epidemiology. University of Chicago Press, Chicago.
43. Peirlinck M, Linka K, Sahli Costabal F, Bendavid E, Bhattacharya J, Ioannidis J, Kuhl E (2020) Visualizing the invisible: The effect of asymptomatic transmission on the outbreak dynamics of COVID-19. Computer Methods in Applied Mechanics and Engineering 372:113410.
44. Peng GCY, Alber M, Buganza Tepole A, Cannon W, De S, Dura-Bernal S, Garikipati K, Karniadakis G, Lytton WW, Perdikaris P, Petzold L, Kuhl E (2021) Multiscale modeling meets machine learning: What can we learn? Archive of Computational Methods in Engineering 28:1017-1037.
45. Siegenfeld AF, Taleb NN, Bar-Yam Y (2020) What models can and cannot tell us about COVID-19. Proceedings of the National Academy of Sciences 117:16092-16095.
46. Snow J (1855) On the Mode of Communication of Cholera (2nd edition). London, John Churchill.
47. Soper GA (1919) The lessons of the pandemic. Science XLIX 501-506.
48. Viceconte G, Petrosillo N (2020) COVID-19 R0: Magic number or conundrum? Infectious Disease Reports 12:8516.
49. World Health Organization. WHO Virtual Press Conference on COVID-19, March 11, 2020 https://www.who.int/docs/default-source/coronaviruse/transcripts/who-audio-emergencies-coronavirus-press-conference-full-and-final-11mar2020.pdf?sfvrsn=cb432bb3$_2$ accessed: June 1, 2021.

Chapter 2
The classical SIS model

Abstract The SIS model is the simplest compartment model with only two populations, the susceptible and infectious groups S and I. It characterizes infectious diseases like the common cold or influenza that do not provide immunity upon infection. While the SIS model is too simplistic to explain the outbreak dynamics of complex infectious diseases, it is the only compartment model with an explicit analytical solution for the time course of its populations. This makes it a popular model to explain and illustrate the basic principles of compartment modeling. The learning objectives of this chapter on classical SIS modeling are to

- explain concepts of mass action incidence and constant rate recovery
- interpret contact and infectious rates and periods
- solve the analytical solutions for the susceptible and infectious populations
- distinguish disease free and endemic equilibria
- analyze the final size relation
- demonstrate how the infectious rate, reproduction number, and initial conditions modulate outbreak dynamics
- discuss limitations of classical SIS modeling

By the end of the chapter, you will be able to analyze, simulate, and predict the outbreak dynamics of simple infectious diseases like the common cold or influenza that do not provide immunity to reinfection.

2.1 Introduction of the SIS model

Some infectious diseases, for example from the common cold or influenza, do not provide lifelong immunity upon recovery from infection [2, 2], and previously infected individuals become susceptible again [10]. We can model their behavior through the simplest of all compartment models, the SIS model [5]. Figure 11.4 shows that the SIS model consists of only two populations, the susceptible group S and the infectious group I [3]. As such, it provides valuable insight into the concept

E. Kuhl, *Computational Epidemiology*, https://doi.org/10.1007/978-3-030-82890-5_2

Fig. 2.1 Classical SIS model. The classical SIS model contains two compartments for the susceptible and infectious populations, S and I. The transition rates between the compartments, the contact and infectious rates, β and γ, are inverses of the contact and infectious periods, $B = 1/\beta$ and $C = 1/\gamma$.

of compartment modeling [8]. The SIS model is often used to illustrate the basic features of compartment models because we can solve the dynamics of its two populations, $S(t)$ and $I(t)$, analytically at any point t in time [7]. The SIS model also has simple analytical solutions for the converged final sizes, S_∞ and I_∞, as $t \to \infty$ [13]. For the transition from the susceptible to the infectious group, we assume a *mass action* incidence [4], which implies that the rate of new infections is proportional to the size of the susceptible and infectious groups S and I weighted by the contact rate β, $\dot{I} = \beta\,SI$. For the transition from the infectious to the susceptible group, we

Fig. 2.2 The transition from the susceptible to the infectious group is based on the assumption of *mass action* incidence for which the rate of new infections is proportional to the size of the susceptible and infectious groups S and I, weighted by the contact rate β, $\dot{I} = \beta\,SI$. The transition from the infectious to the susceptible group is based on the assumption of *constant rate recovery* for which the rate of recovered infections is proportional to the size of the infectious group I, weighted by the infectious rate γ, $\dot{I} = -\gamma I$.

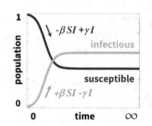

assume a *constant rate recovery*, which implies that the rate of recovered infections is proportional to the size of the infectious group I weighted by the infectious rate γ, $\dot{I} = \gamma I$. Figure 2.2 illustrates the dynamics of these two assumptions that result in the following system of two coupled ordinary differential equations,

$$\begin{aligned} \dot{S} &= -\ \beta\,SI\ +\ \gamma I \\ \dot{I} &= +\ \beta\,SI\ -\ \gamma I. \end{aligned} \tag{2.1}$$

The transition rates between both compartments are the contact rate β and the infectious rate γ in units [1/days], which are the inverses of the contact period $B = 1/\beta$ and the infectious period $C = 1/\gamma$ in units [days]. The ratio between the contact and infectious rates, or similarly, between the infectious and contact periods, defines the basic reproduction number R_0,

$$R_0 = \frac{\beta}{\gamma} = \frac{C}{B}. \tag{2.2}$$

In this simple format, the SIS model (2.1) neglects all vital dynamics, $\dot{S} + \dot{I} \doteq 0$ and $S + I = \text{const.} = 1$. It does not account for births, natural deaths, or death from the disease.

2.2 Analytical solution of the SIS model

From the condition of non-vital dynamics, $S = 1 - I$, we can rephrase the system of equations of the SIS model (2.1) in terms of only one independent variable, the size of the infectious group I, governed by the following nonlinear ordinary differential equation,

$$\dot{I} = \beta [1 - I] I - \gamma I = [\beta - \gamma] I \left[1 - \frac{I}{1 - \gamma/\beta} \right]. \tag{2.3}$$

Equation (2.3) is a *logistic differential equation* of the form,

$$\dot{I} = r I [1 - I/K] \quad \text{with} \quad r = \beta - \gamma \quad \text{and} \quad K = 1 - \gamma/\beta, \tag{2.4}$$

where K is the carrying capacity. This type of equation has an explicit analytical solution,

$$I(t) = \frac{K I_0}{I_0 + [K - I_0] \exp(-r t)}, \tag{2.5}$$

where $I_0 = I(0)$ is the initial infectious population [10]. It proves convenient to reparameterize the equation for the infectious population (2.5) in terms of the basic reproduction number $R_0 = \beta/\gamma$ and the infectious period $C = 1/\gamma$ with $r = [R_0 - 1]/C$ and $K = 1 - 1/R_0$. This provides the analytical solution for the SIS model in terms of the infectious period C, the basic reproduction number R_0, and the initial infectious population I_0,

$$
\begin{aligned}
S(t) &= 1 - \frac{[1 - 1/R_0] I_0}{I_0 + [1 - 1/R_0 - I_0] \exp([1 - R_0] t/C)} \\
I(t) &= \frac{[1 - 1/R_0] I_0}{I_0 + [1 - 1/R_0 - I_0] \exp([1 - R_0] t/C)}.
\end{aligned}
\tag{2.6}
$$

Figures 2.3, 2.4, and 2.5 highlight the outbreak dynamics of the SIS model, solved analytically using equations (2.6), for the time period of one year. The three figures demonstrate the sensitivity of the SIS model for varying infectious periods C, and varying basic reproduction numbers R_0, and initial infectious populations I_0. Increasing the infectious period C delays convergence to the endemic equilibrium, but the final sizes S_∞ and I_∞ remain unchanged. Increasing the basic reproduction number R_0 accelerates convergence to the endemic equilibrium, decreases S_∞, and increases I_∞. Increasing the initial infectious population I_0 accelerates the onset of the outbreak, but the final sizes S_∞ and I_∞ remain unchanged. Interestingly, increasing the initial exposed population I_0 by an order of magnitude shifts the population dynamics by a constant time increment, for the current parameterization by 50 days. This highlights the exponential nature of the SIS model, which causes a constant acceleration of the outbreak for a logarithmic increase of the initial infectious population, while the overall outbreak dynamics remain the same.

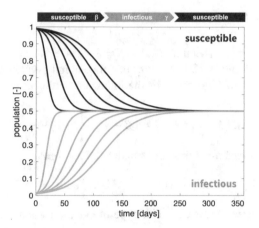

Fig. 2.3 Classical SIS model. Sensitivity with respect to the infectious period C. Increasing the infectious period C delays convergence to the endemic equilibrium, but the final sizes S_∞ and I_∞ remain unchanged. Basic reproduction number $R_0 = 2.0$, initial infectious population $I_0 = 0.01$, and infectious period $C = 5, 10, 15, 20, 25, 30$ days.

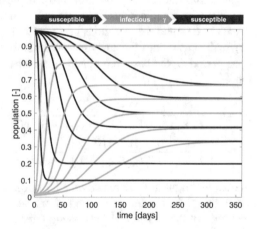

Fig. 2.4 Classical SIS model. Sensitivity with respect to the basic reproduction number R_0. Increasing the basic reproduction number R_0 accelerates convergence to the endemic equilibrium, decreases S_∞, and increases I_∞. Infectious period $C = 20$ days, initial infectious population $I_0 = 0.01$, and basic reproduction number $R_0 = 1.5, 1.7, 2.0, 2.4, 3.0, 5.0, 10.0$.

2.3 Final size relation of the SIS model

For practical purposes, it is interesting to estimate the final susceptible and infectious populations S_∞ and I_∞ [15]. From the analytical solution (2.5),

$$I_\infty = \frac{K \, I_0}{I_0 + [\, K - I_0 \,] \exp{(-r \, t_\infty)}} \quad \text{with} \quad r = \beta - \gamma \quad \text{and} \quad K = 1 - \gamma/\beta, \quad (2.7)$$

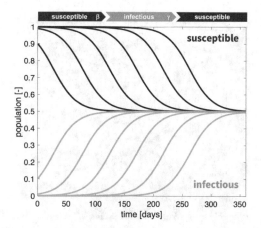

Fig. 2.5 Classical SIS model. Sensitivity with respect to the initial infectious population I_0.
Increasing the initial infectious population I_0 accelerates the onset of the outbreak, but the final
sizes S_∞ and I_∞ remain unchanged. Infectious period $C = 20$ days, basic reproduction number
$R_0 = 2.0$, and initial infectious population $I_0 = 10^{-1}, 10^{-2}, 10^{-3}, 10^{-4}, 10^{-5}, 10^{-6}$.

it is easy to show that we can distinguish two cases. If there is a finite initial infectious
population, $I_0 > 0$, the infectious population will converge to either zero, for $\beta < \gamma$,
$r < 0$, and $\exp(-r\, t_\infty) \to \infty$, or to K for $\beta > \gamma$, $r > 0$, and $\exp(-r\, t_\infty) \to 0$ [2, 9],
thus

$$I_\infty = \begin{cases} 0 & \text{if} \quad \beta < \gamma \\ K & \text{if} \quad \beta > \gamma. \end{cases} \tag{2.8}$$

We can rephrase these limit conditions (2.8) in terms of the basic reproduction
number R_0. In classical epidemiology, the two converged states are known as the
disease-free equilibrium and the *endemic equilibrium* [3]. The equations that define
the final converged populations of the endemic equilibrium state are the *final size
relation*.

$$\begin{array}{lll} R_0 < 1 & \text{disease-free equilibrium:} & S_\infty = 1 \quad \text{and} \ I_\infty = 0 \\ R_0 > 1 & \text{endemic equilibrium:} & S_\infty = 1/R_0 \ \text{and} \ I_\infty = 1 - 1/R_0. \end{array} \tag{2.9}$$

Figure 2.6 illustrates the final size relation and emphasizes the role of the basic
reproduction number R_0 as the distinguishing outbreak characteristic between the
disease-free and endemic equilibrium. For reproduction numbers smaller than one,
$R_0 < 1$, the SIS model converges to a disease-free equilibrium with $S_\infty = 1$ and
$I_\infty = 0$. For reproduction numbers larger than one, $R_0 > 1$, the SIS model converges
to an endemic equilibrium with $S_\infty = 1/R_0$ and $I_\infty = 1 - 1/R_0$. The larger the
reproduction number R_0, the smaller the final susceptible population S_∞ and the
larger the final infectious population I_∞ [3].

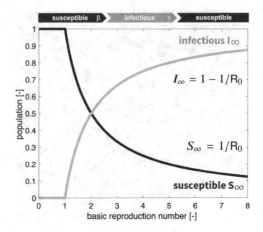

Fig. 2.6 Classical SIS model. Final size relation as a function of the basic reproduction number R_0. For $R_0 < 1$, the SIS model converges to a disease-free equilibrium with $S_\infty = 1$ and $I_\infty = 0$. For $R_0 > 1$, the SIS model converges to an endemic equilibrium with $S_\infty = 1/R_0$ and $I_\infty = 1 - 1/R_0$. Increasing R_0 reduces the susceptible population S_∞ and increases the infectious population I_∞.

Problems

2.1 Basic reproduction number. Throughout the winter of 1980, a Scottish boarding school reported an influenza that, on any day, affected on average 408 of its 1632 students. Estimate the basic reproduction number.

2.2 Basic reproduction number. Throughout the winter of 1980, a Scottish boarding school reported an influenza with an infectious period of $C = 4$ days and a contact period of $B = 2.8$ days. How many of its 1632 students are infectious at endemic equilibrium?

2.3 Basic reproduction number. Assume the common flu has an infectious period of $C = 7$ days and a contact rate of $\beta = 0.2/$ days. Determine the basic reproduction number R_0 and the final sizes of the susceptible and infectious populations S_∞ and I_∞.

2.4 Contact rate. Assume the common flu has an infectious period of $C = 7$ days. Determine the contact rate β for which the infectious population I never increases beyond 20% of the population. What is the basic reproduction number under these conditions?

2.5 Contact rate. Assume the common flu has an infectious period of $C = 7$ days. Determine the contact rate β for which the infectious population I never increases beyond 30% of the population. Comment on how the basic reproduction number for a maximum infectious population of 30% differs to the basic reproduction number for a maximum infectious population of 20%.

2.6 Contact period. For policy making, the contact period B is a critical parameter that we can manage through community mitigation strategies. Derive the general condition for the critical contact period B below which a disease with an infectious period C never exceeds a critical infectious population I_∞.

2.7 Timeline of infection. At the end of the winter quarter, 2590 of all 15540 students had a common flu with an infection rate of $\gamma = 0.25/$days. How long did it take from 20 students reporting symptoms until 200 students reporting symptoms on a single day?

2.8 Timeline of infection. In the Canadian province of Newfoundland with a population of 500,000 people, 50 individuals report symptoms of a common flu with known contact and infectious periods of $B = 3$ days and $C = 4$ days. Estimate how long it takes until one percent of the population is infectious. How many people will eventually be infectious?

2.9 Timeline of infection. In the Canadian province of Newfoundland with a population of 500,000 people, 100 individuals report symptoms of a common flu with known contact and infectious periods of $B = 3$ days and $C = 4$ days. Guess how much longer it takes until one percent of the population is infectious compared to only 50 individuals reporting symptoms. Now, calculate how long it take until one percent of the population is infectious. How many people will eventually be infectious? Compare your results to the previous problem and comment on what this implies for the health care system.

2.10 Timeline of infection. For the health care system, it could be important to estimate by when a certain fraction of the population is infectious. Derive the general equation for the time at which a population $I(t)$ is infectious for a disease with known infectious period C, reproduction number R_0, and initial infectious population I_0.

References

1. Anderson RM, May RM (1981) The population dynamics of microparasites and their invertebrate hosts. Philosophical Transactions of the Royal Society London B. 291:451–524.
2. Anderson RM, May RM (1991) Infectious Diseases of Humans. Oxford University Press, Oxford.
3. Bailey NTJ (1957) The mathematical theory of infectious diseases and its applications. Griffin, London UK.
4. Brauer F, van den Dreissche P, Wu J (2008) Mathematical Epidemiology. Springer-Verlag Berlin Heidelberg.
5. Brauer F (2017) Mathematical epidemiology: Past, present and future. Infectious Disease Modelling 2:113-127.
6. Brauer F, Castillo-Chavez C, Feng Z (2019) Mathematical Models in Epidemiology. Springer-Verlag New York.
7. Gray A, Greenhalgh D, Hu L, Mao X, Pan J. A stochastic differential model equation SIS epidemic model. SIAM Journal of Applied Mathematics 71:876-902.
8. Hethcote HW (1976) Quantitative analyses of communicable disease models. Mathematical Biosciences, 28:335-356.

 9. Heesterbeek JAP, Roberts MG. How mathematical biology became a field of biology. Philo-
     sophical Transactions of the Royal Society London B. 370:20140307.
10. Hethcote HW, Stech HW, van den Driessche P (1981) Stability analysis for models of diseases
     without immunity. Journal of Mathematical Biology, 13:185-198.
11. Hethcote HW (1989) Three basic epidemiological models. Biomathematics: Applied Mathe-
     matical Ecology, 18:119-144.
12. Hethcote HW (2000) The mathematics of infectious diseases. SIAM Review 42:599-653.
13. Kermack WO, McKendrick G (1932) Contributions to the mathematical theory of epidemics.
     Part II. The problem of endemicity. Proceedings of the Royal Society London Series A 138:55-
     83.

Chapter 3
The classical SIR model

Abstract The SIR model is the most popular compartment model with three populations, the susceptible, infectious, and recovered groups S, I, and R. It characterizes infectious diseases that provide immunity upon infection. While the SIR model does not have an analytical solution for the time course of its populations, it has explicit analytical solutions for its maximum infectious population and for the final sizes of its susceptible and recovered populations at endemic equilibrium. This makes it a widely used model to rapidly estimate the dimensions of an outbreak. The learning objectives of this chapter on classical SIR modeling are to

- interpret contact and infectious rates and periods
- explain the concept of the basic reproduction number
- estimate the growth rate of the initial outbreak
- solve the analytical solution of maximum infection
- analyze the final size relation
- develop your own SIR type models
- discuss limitations of classical SIR modeling

By the end of the chapter, you will be able to analyze, estimate, and predict the dimensions of infectious diseases like the measles or COVID-19 that provide immunity upon infection.

3.1 Introduction of the SIR model

A popular model in epidemiology is the SIR model [2]. Figure 3.1 shows that it consists of three populations, the susceptible group S, the infectious group I, and the recovered group R [10]. Unfortunately, we can no longer solve the dynamics of the SIR model analytically. However, in its simplest form, the SIR model has analytical solutions for the maximum infectious population I_{max} and for the converged final sizes, S_∞, I_∞, and R_∞ as $t \rightarrow \infty$ [13]. This allows us to rapidly estimate the dimensions of an outbreak [3]. In the SIR model, the transition between the

| susceptible | β | infectious | γ | recovered |

Fig. 3.1 Classical SIR model. The classical SIR model contains three compartments for the susceptible, infectious, and recovered populations, S, I, and R. The transition rates between the compartments, the contact and infectious rates, β and γ, are inverses of the contact and infectious periods, $B = 1/\beta$ and $C = 1/\gamma$.

Fig. 3.2 The transition from the susceptible to the infectious group is based on the assumption of *mass action* incidence. In a population of size N, an individual makes βN contacts per unit time, and all contacts are assumed to be effective. The probability of a random contact between an infectious and a susceptible individual is S/N, resulting in $[\beta N][S/N]$ new infections per infectious individual I per unit time and a rate of new infections of $\dot{I} = [\beta N][S/N]I = \beta SI$.

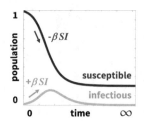

susceptible, infectious, and recovered populations is governed by three ordinary differential equations [1]. For the transition from the susceptible to the infectious group, we assume a we assume a *mass action* incidence [4] according to Figure 3.2, which implies that the rate of new infections is proportional to the size of the susceptible and infectious groups weighted by the contact rate β, $\dot{I} = \beta SI$. For the transition between the infectious and recovered groups, we assume a *constant rate recovery* according to Figure 3.3, which implies that the rate of recovered infections is proportional to the size of the infectious group weighted by the infectious rate γ, $\dot{I} = \gamma I$. This results in the following system of three coupled ordinary differential equations,

$$
\begin{aligned}
\dot{S} &= - \beta SI \\
\dot{I} &= + \beta SI - \gamma I \\
\dot{R} &= + \gamma I .
\end{aligned}
\tag{3.1}
$$

The transition rates between the three compartments are the contact rate β and the infectious rate γ in units [1/days], which are the inverses of the contact period $B = 1/\beta$ and the infectious period $C = 1/\gamma$ in units [days]. The ratio between the contact and infectious rates, or similarly, between the infectious and contact periods, defines the basic reproduction number R_0 [6],

Fig. 3.3 The transition from the infectious to the recovered group is based on the assumption of *constant rate recovery*. Infectious individuals become recovered at a constant rate, γI, and the rate of recovered infections is $\dot{I} = -\gamma I$. This rate equation has an explicit solution, $I(t) = I_0 \exp(-\gamma t)$, that states that a fraction of $\exp(-\gamma t)$ individuals remains infectious t time units after entering the infectious group I. This implies that the infectious period $C = 1/\gamma$ is distributed exponentially with a mean of $\int_0^\infty \exp(-\gamma t)dt = 1/\gamma$.

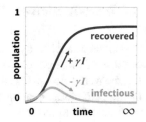

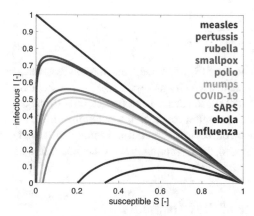

Fig. 3.4 Phase diagram of the SIR model in the susceptible-infectious plane. Each curve begins in he bottom right corner at t_0 with $S_0 \approx 1$ and $I_0 \approx 0$ and ends on the S-axis at t_∞ with $S_\infty > 0$ and $I_\infty = 0$. The curves represent the equation, $I(t) = S_0 + I_0 - S(t) + \log(S(t)/S_0)/R_0$, which we can no longer solve analytically. The color-code represents different basic reproduction numbers from $R_0 = 1.6$ for an influenza to $R_0 = 15$ for the measles as summarized in Table 1.3. Basic reproduction numbers $R_0 = 15, 14, 6.5, 6.5, 6.0, 5.5, 4.0, 3.5, 2.0, 1.6$ from Table 1.3.

Phase diagram of the SIR model. For the SIR model without vital dynamics (3.1), we can eliminate the third equation, $R = 1 - S - I$, and simply analyze the first two equations, $\dot{S} = -\beta\,SI$ and $\dot{I} = +\beta\,SI - \gamma\,I$. This results in a single equation for the susceptible-infectious relation,

$$I(t) = S_0 + I_0 - S(t) + \frac{1}{R_0}\,\log\frac{S(t)}{S_0}\,.$$

Unlike the SIS model, the SIR model does not have an explicit analytical solution. We solve it numerically to calculate the trajectory of points in the susceptible-infectious plane [13]. We discuss the time discretization and computational solution in detail in Chapter 5, but have already mapped the trajectories for the common infectious diseases from Table 1.3 in Figure 3.4 [9]. Each curve begins in the bottom right corner at t_0 with $S_0 \approx 1$ and $I_0 \approx 0$ and ends on the S-axis at t_∞ with $S_\infty > 0$ and $I_\infty = 0$ [8]. The larger the basic reproduction number, from blue to red, the larger the maximum infectious population I_{max} during the course of the disease, and the smaller the remaining susceptible population S_∞. While these phase diagrams provide valuable insight into the importance of the basic reproduction number R_0, they do not tell us anything about the advancement of the susceptible and infectious populations in time, nor do they tell us how fast the infectious population grows. In the following sections, we will explore estimates for the growth rate of the infectious population $G = \dot{I}/I$ and analytical solutions for the maximum infectious population I_{max} and the size of the final susceptible population S_∞.

$$R_0 = \frac{\beta}{\gamma} = \frac{C}{B}. \tag{3.2}$$

In this simple format, the SIR model neglects all vital dynamics, it does not account for births or natural deaths, which implies that $\dot{S}+\dot{I}+\dot{R} \doteq 0$ and $S+I+R = \text{const.} = 1$. The SIR model is widely used because of its conceptual simplicity. However, to study disease mechanisms, it is often considered too simplistic. Throughout this book, we will gradually build on the basic SIR model and generalize it to a family of SIR models that includes an exposed, an asymptomatic, a vaccinated, or a dead group [18].

3.2 Growth rate of the SIR model

Early in the pandemic, we can use the second equation of the SIR model (3.1),

$$\dot{I} = \beta SI - \gamma I = [\beta S - \gamma] I \tag{3.3}$$

to estimate the initial growth of the infectious population,

$$G = \frac{\dot{I}}{I} = \frac{[\beta S - \gamma] I}{I} = \beta S - \gamma. \tag{3.4}$$

At the beginning of an outbreak, the susceptible population is still close to one, $S \approx 1$, and we can approximate the *growth rate* G of the epidemic as

$$G \approx \beta - \gamma = \gamma [R_0 - 1] \quad \text{with} \quad R_0 = \beta/\gamma. \tag{3.5}$$

This implies that, during this early phase of the outbreak, the infectious population experiences *exponential growth* at a rate G [3],

$$I(t) \approx I_0 \, \exp(Gt) = I_0 \, \exp([\beta - \gamma] t) = I_0 \, \exp(\gamma [R_0 - 1] t). \tag{3.6}$$

In the logarithmic $\log(I(t))$ vs. time t diagram, ideal exponential growth results in a line with slope G. If we plot the infectious population of the early outbreak in a logarithmic plot, we can estimate the contact rate β and contact period B,

$$\beta = G + \gamma \quad \text{and} \quad B = C/[1 + GC], \tag{3.7}$$

and with them the basic reproduction number $R_0 = \beta/\gamma = C/B$

$$R_0 = 1 + G/\gamma = 1 + GC, \tag{3.8}$$

as functions of the infectious rate γ or infectious period C and the slope G in this logarithmic plot using equation (3.5).

Table 3.1 Early COVID-19 outbreak in Italy. Newly reported COVID-19 cases in Italy from Friday, February 21, the first day of reported cases, to Thursday, April, 2, 2020; case data, ΔN_I, are reported as seven-day moving averages.

	Fri	Sat	Sun	Mon	Tue	Wed	Thu
week 1	3	11	22	32	46	66	93
week 2	124	150	221	258	312	374	458
week 3	536	680	811	1020	1093	1340	1609
week 4	1861	2183	2484	2688	3052	3324	3705
week 5	4196	4633	4913	5135	5380	5521	5646
week 6	5635	5552	5502	5398	5228	5168	4950

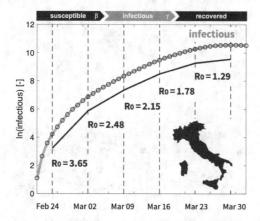

Fig. 3.5 Problem. Estimating the growth rate G, basic reproduction number, R_0 and contact period B. In this logarithmic plot of case numbers vs. time, circles represent the seven-day moving averages of the reported case data for Italy from Table 3.1, the orange curve highlights the cumulative infectious population for an infectious period of $C = 7$ days, and the red lines are the tangents to the curve that define the growth rates G, basic reproduction numbers R_0, and contact periods B for each interval.

Example. Estimating the growth rate G, basic reproduction number R_0, and contact period B. For the seven-day moving averages of the newly reported COVID-19 cases in Italy [9], from the first day of reporting, from Feb 21, 2020, to the peak of the first wave, April, 2, 2020 in Table 3.1, estimate the growth rate G, the basic reproduction number R_0, and the contact period B for an infectious period of $C = 7$ days.

Solution. We first calculate the total infectious population at each day t_n within each interval,

$$N_I(t_n) = \sum_{i=n+1-C}^{n} \Delta N_I(t_i),$$

by summing the newly reported cases ΔN_I for the $C = 7$ previous days from Table 3.1. Then we plot these seven-day cumulative cases in a logarithmic plot in Figure 3.5. We observe that the orange infectious curve is not a straight line, which suggests that the reproduction number R_0 varies within the interval.

We approximate the curve through five line segments between February 24, March 2, March 9, March 16, March 23, and March 30.

	Feb 24	Mar 2	Mar 9	Mar 16	Mar 23	Mar 30
day t	4	11	18	25	32	39
N_I	68	958	4191	13258	28958	38634
$\ln(N_I)$	4.22	6.86	8.34	9.49	10.27	10.56

We calculate the growth rates G as the slopes of these lines in all five intervals,

$$ G = \frac{\ln(N_I(t_{n+1})) - \ln(N_I(t_n))}{t_{n+1} - t_n} = \frac{1}{B} - \frac{1}{C} = \frac{R_0 - 1}{C}, $$

the contact periods B,

$$ B = C/[\,1 + GC\,], $$

and the basic reproduction numbers R_0,

$$ R_0 = 1 + GC. $$

The table below summarizes the growth rates G, the contact periods B, and the basic reproduction numbers R_0 in the five week-long intervals.

	Feb 24 - Mar 2	Mar 2 - Mar 9	Mar 9 - Mar 16	Mar 16 - Mar 23	Mar 23 - Mar 30
G	0.3779	0.2108	0.1645	0.1116	0.0412
B	1.9203	2.8273	3.2533	3.9298	5.4336
R_0	3.6453	2.4758	2.1517	1.7812	1.2883

After some early local lockdowns that effectively reduced the initial extreme basic reproduction number from $R_0 = 3.65$ to $R_0 = 2.48$, the entire country of Italy was subjected to a nation-wide lockdown on March 9, 2020 to further reduce the reproduction number to $R_0 = 2.15$, $R_0 = 1.78$, and $R_0 = 1.29$ within three weeks.

3.3 Maximum infection of the SIR model

The hallmark of a typical epidemic outbreak with lifelong immunity is that it begins with a small infectious population I_0. The infectious population $I(t)$ increases, reaches a peak I_{max}, and then decays to zero, $I_\infty = 0$. The size and timing of *maximum infection* are critical disease parameters since they determine the potential burden on the health care system [3]. To characterize the maximum infectious population I_{max}, we begin with the first equation of the SIR model (3.1),

$$ \dot{S} = -\beta SI, \tag{3.9} $$

perform a *separation of variables* by dividing it by the susceptible population S,

$$\dot{S}/S = -\beta I, \tag{3.10}$$

and integrate it in time from t_0 to t,

$$\int_{t_0}^{t} \frac{1}{S}\frac{dS}{dt} = -\int_{t_0}^{t} \beta I \, dt. \tag{3.11}$$

To evaluate the integral on the righthand side, $\int_{t_0}^{t} I \, dt$, we add the first and second equations of the SIR model (3.1),

$$\begin{aligned} \dot{S} &= -\beta SI \\ \dot{I} &= +\beta SI - \gamma I, \end{aligned} \tag{3.12}$$

to eliminates the explicit dependence on the susceptible population S,

$$\dot{S} + \dot{I} = -\gamma I. \tag{3.13}$$

We integrate this equation in time from t_0 to t,

$$\int_{t_0}^{t} \dot{S} + \dot{I} \, dt = -\int_{t_0}^{t} \gamma I \, dt, \tag{3.14}$$

and use the initial conditions, S_0 and I_0 with $S_0 + I_0 = 1$ at t_0, and the current conditions, $S(t)$ and $I(t)$ at t, to simplify the lefthand side, $S(t) - S_0 + I(t) - I_0 = S(t) + I(t) - 1$. With these considerations, the integral expression takes the following explicit representation,

$$\int_{t_0}^{t} I \, dt = [1 - S(t) - I(t)]/\gamma. \tag{3.15}$$

We return to equation (4.16) and substitute the righthand side by $\int_{t_0}^{t} 1/S \, dS/dt = \log(S(t)) - \log(S_0) = \log(S(t)/S_0)$, and the lefthand side using the above expression, $-\int_{t_0}^{t} \beta I \, dt = \beta/\gamma [1 - S(t) - I(t)]$,

$$\log \frac{S(t)}{S_0} = -R_0 [1 - S(t) - I(t)] \quad \text{with} \quad R_0 = \frac{\beta}{\gamma}. \tag{3.16}$$

To rephrase the susceptible population $S(t)$ at the time t of maximum infection I_{max}, we evaluate the stationarity of the second equation of the SIR model (3.1),

$$\dot{I} = \beta SI - \gamma I \doteq 0 \quad \text{thus} \quad S = \gamma/\beta. \tag{3.17}$$

By substituting $I(t) = I_{max}$ and $S(t) = \gamma/\beta$ in equation (3.16), we obtain an explicit expression for the maximum infectious population I_{max},

$$I_{max} = 1 - \frac{1}{R_0}\left[1 + \log(S_0) - \log\left(\frac{1}{R_0}\right)\right] \quad \text{with} \quad R_0 = \frac{\beta}{\gamma}. \tag{3.18}$$

Figure 3.6 illustrates the maximum infectious population I_{max} from equation (3.18)

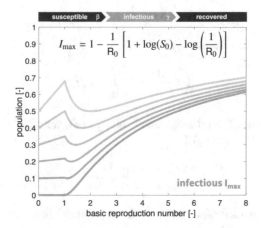

$$I_{max} = 1 - \frac{1}{R_0}\left[1 + \log(S_0) - \log\left(\frac{1}{R_0}\right)\right]$$

Fig. 3.6 Classical SIR model. Maximum infection as a function of the basic reproduction number R_0. For small initial infectious populations I_0, the maximum infectious population I_{max} increases with increasing R_0. For large initial infectious populations I_0, the maximum infectious population I_{max} increases for basic reproduction numbers up to $R_0 = 1$, then drops, and then increases with increasing R_0. Initial infectious population $I_0 = 0.0001, 0.1, 0.2, 0.3, 0.4, 0.5$, from dark to light.

and its dependence on the basic reproduction number R_0. Similar to the the SIS model in Chapter 2, we can distinguish two regimes for reproduction numbers smaller and larger than one. The I_{max} curves display a kink at $R_0 = 1$, where the $\log(1/R_0)$ term in equation (3.18) flips its sign. Interestingly, the maximum infectious population I_{max} is not only sensitive to the basic reproduction number R_0, but also to the initial infectious populations I_0, as we can see from the six different curves. For small initial infectious populations I_0, the maximum infectious population I_{max} increases with increasing R_0 for reproduction numbers larger than one, $R_0 > 0$. For large initial infectious populations I_0, the maximum infectious population I_{max} increases for basic reproduction numbers up to $R_0 = 1$, then drops, and then increases with increasing R_0. For many infectious diseases, the maximum infectious population, I_{max}, is a significant metric for planning, because it is closely related to the number of complications, hospitalizations, and required intensive care units.

3.4 Final size relation of the SIR model

For infectious diseases with lifelong immunity, the susceptible population $S(t)$ decreases continuously throughout the outbreak, but the final susceptible population S_∞ always remains larger than zero. This equation for the final converged populations is called the *final size relation* and the final converged state is called the *endemic equilibrium* [5]. To derive the final size relation of the SIR model, we begin with the first equation of the SIR model (3.1),

$$\dot{S} = -\beta\, SI, \tag{3.19}$$

perform a *separation of variables* by dividing this equation by the susceptible population S,

$$\dot{S}/S = -\beta\, I, \tag{3.20}$$

and integrate it in time from t_0 to t_∞,

$$\int_{t_0}^{t_\infty} \frac{1}{S}\frac{dS}{dt} = -\int_{t_0}^{t_\infty} \beta\, I\, dt. \tag{3.21}$$

Similar to the previous section, we evaluate the integral on the righthand side, $\int_{t_0}^{t_\infty} I\, dt$, by adding the first and second equations of the SIR model (3.1)

$$\begin{aligned} \dot{S} &= -\ \beta\, SI \\ \dot{I} &= +\ \beta\, SI\ -\ \gamma\, I, \end{aligned} \tag{3.22}$$

and add both equations to eliminates the explicit dependence on the susceptible population S,

$$\dot{S} + \dot{I} = -\gamma\, I. \tag{3.23}$$

We integrate this equation in time from t_0 to t_∞,

$$\int_{t_0}^{t_\infty} \dot{S} + \dot{I}\, dt = -\int_{t_0}^{t_\infty} \gamma\, I\, dt, \tag{3.24}$$

and use the initial conditions, S_0 and I_0 with $S_0 + I_0 = 1$ at t_0, and the final conditions, $S_\infty > 0$ and $I_\infty = 0$ at t_∞, to simplify the lefthand side, $S_\infty - S_0 + I_\infty - I_0 = S_\infty - 1$. With these considerations, the integral expression takes the following explicit representation,

$$\int_{t_0}^{t_\infty} I\, dt = [1 - S_\infty]/\gamma. \tag{3.25}$$

We return to equation (3.21) and substitute the righthand side by $\int_{t_0}^{t_\infty} 1/S\, dS/dt = \log(S_\infty) - \log(S_0) = \log(S_\infty/S_0)$, and the lefthand side using the above expression, $\int_{t_0}^{t_\infty} \beta I\, dt = \beta//\gamma[1 - S_\infty]$, similar to equation (3.16), to obtain an equation for the final susceptible population S_∞, which is known in epidemiology as the *final size relation*,

$$\log \frac{S_\infty}{S_0} = -\mathsf{R}_0\,[1 - S_\infty] \quad \text{with} \quad \mathsf{R}_0 = \frac{\beta}{\gamma}. \tag{3.26}$$

By applying the exponential on both sides of the equation, we obtain the following expression,

$$S_\infty = S_0 \exp(-\mathsf{R}_0\,[1 - S_\infty]) = 1 - R_\infty. \tag{3.27}$$

This transcendental equation has an explicit analytical solution for the final sizes of the susceptible and recovered populations S_∞ and R_∞ in terms of the Lambert function W,

$$\begin{aligned} S_\infty &= -\,W(-S_0\,\mathsf{R}_0 \exp(-\mathsf{R}_0[\,1 - \mathsf{R}_0\,]))/\mathsf{R}_0 \\ R_\infty &= 1 - W(-S_0\,\mathsf{R}_0 \exp(-\mathsf{R}_0[\,1 - \mathsf{R}_0\,]))/\mathsf{R}_0. \end{aligned} \tag{3.28}$$

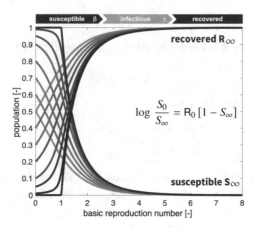

Fig. 3.7 Classical SIR model. Final size relation as a function of the basic reproduction number R_0. Increasing the basic reproduction number R_0 reduces the final susceptible population S_∞ and increases the final recovered population R_∞. The larger the initial infectious population I_0, the smaller the initial and final susceptible populations $S_0 = 1 - I_0$ and S_∞ and the larger the final recovered population R_∞. Initial infectious population $I_0 = 0.0001, 0.01, 0.05, 0.1, 0.2, 0.3, 0.4, 0.5$, from dark to light.

The final size relation (3.27), or evaluated explicitly as (3.28), confirms that, unless the initial susceptible population is zero, $S_0 = 0$, the final susceptible population will always be larger than zero, $S_\infty > 0$.

Final size relation and attack ratio. In epidemiology, the final size relation provides a relation between the basic reproduction number R_0 and the size of an outbreak,

$$\log \frac{S_0}{S_\infty} = R_0 \left[1 - S_\infty \right] .$$

It allows us to calculate the final sizes of the susceptible and recovered populations, S_∞ and $R_\infty = 1 - S_\infty$ using equations (3.28). The final recovered population R_∞ is often called the attack rate, although it is dimensionless and not truly a rate. A better term would be attack ratio. The attack ratio,

$$A = 1 - S_\infty = R_\infty ,$$

defines the fraction of the population that has become infected throughout the course of an outbreak. For example, according to the CDC, the attack ratio of the seasonal flu is $A = 9.3\%$ for children 0-17 years, $A = 8.8\%$ for adults 18-64 years, and $A = 3.9\%$ for adults 65 years and older.

Figure 3.7 illustrates the final size relation (3.28) and its dependence on the basic reproduction number R_0. Similar to the final size relation of the SIS model in Figure

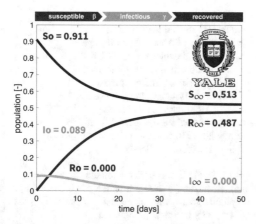

Fig. 3.8 Example. Final size relation for an influenza. Outbreak dynamics of an influenza among Yale university undergraduates with an initial and final susceptible population of $S_0 = 0.911$ and $S_\infty = 0.513$ and an infectious period of $C = 1/\gamma = 3$ days. Using the final size relation, we can estimate the basic reproduction number, $R_0 = 1.18$, and the contact rate, $\beta = 0.39/$days. The outbreak decays after approximately one month.

Example. Final size relation for an influenza. An influenza among Yale University undergraduates recorded an initial and final susceptible population of $S_0 = 0.911$ and $S_\infty = 0.513$ and an infectious period of $C = 1/\gamma = 3$ days. Estimate the contact rate β and the basic reproduction number R_0.

Solution. Using the final size relation, $R_0 = \log(S_0/S_\infty)/[1 - S_\infty]$ and the definition of the basic reproduction number, $R_0 = \beta/\gamma$, the contact rate and basic reproduction number are

$$\beta = \log \frac{S_0}{S_\infty} \frac{\gamma}{1 - S_\infty} \quad \text{and} \quad R_0 = \frac{\beta}{\gamma}$$

and, for the given values,

$$\beta = \log \frac{0.911}{0.513} \frac{1}{3[1 - 0.513]\text{days}} = 0.39/\text{days} \quad \text{and} \quad R_0 = 0.39 \cdot 3 = 1.18 \,.$$

This implies that every infected individual had on average 0.39 contacts per day throughout an infectious period of three days, and infected on average 1.18 new individuals, see Figure 3.8.

2.6, increasing the basic reproduction number R_0 reduces the final susceptible population S_∞ and increases the final recovered population R_∞. Even though the susceptible population S_∞ will be close to zero for large basic reproduction numbers R_0, it will always remain finite, $S_\infty > 0$. Unlike the SIS model in Figure 2.6, the final size

Fig. 3.9 William Ogilvy Kermack was born on April 26, 1898 in Kirriemuir, Scotland. He studied mathematics and natural philosophy at the University of Aberdeen, At the age of 26, he was blinded by a chemical explosion in his lab and never regained his sight. As a biochemist, he performed mathematical studies of epidemic spread and established links between environmental factors and specific diseases. Together with Anderson Gray McKendrick, he created the Kermack-McKendrick theory of infectious diseases. William Ogilvy Kermack died on July 20, 1970 while working at his desk inside Marischal College in Aberdeen, Scotland.

Fig. 3.10 Anderson Gray McKendrick was born on September 8, 1876 in Edinburgh, Scotland. He was a Scottish military physician and epidemiologist who pioneered the use of mathematical methods in epidemiology. He is known for the McKendrick–von Foerster equation, $\partial n/\partial t + \partial n/\partial a = -\mu(t, a)n$, a linear first-order partial differential equation in mathematical biology that is often used in demography and cell proliferation modeling. His work with William Ogilvy Kermack resulted in a general theory of infectious disease transmission. Anderson Gray McKendrick died on May 30, 1943 in Speyside, Scotland.

relation of the SIR model also depends on the initial susceptible population S_0, or, in other words, on the size of the initial outbreak I_0. This implies that, unlike for the SIS model in Chapter 2, the SIR model does not have a disease free equilibrium state. Even if the reproduction number is below one, $R_0 < 1$, there will be an outbreak, and the size of this outbreak depends on the initial infectious population I_0: The larger the initial infectious population I_0, the smaller the initial and final susceptible populuations $S_0 = 1 - I_0$ and S_∞ and the larger the final recovered population R_∞.

3.5 The Kermack-McKendrick theory

The classical SIR model (3.1) dates back to a set of three articles by William Ogilvy Kermack and Anderson Gray McKendrick published in 1927 [14], 1932 [15], and 1933 [16]. In this initial series of manuscripts, Kermack and McKendrick make three fundamental modeling assumptions that lay the foundation for compartment modeling and would later be adopted by many mathematical epidemiologists: (i) the population under investigation is *isolated* from the rest of the world; (ii) the contact rate between individuals is *homogeneous*; and (iii) the model itself is *deterministic* [13]. The initial Kermack-McKendrick theory uses a generalized SIR model in which the total infectious population I is further divided by the *age of infection a* into subpopulations i [14],

| susceptible $\beta(a)$ | ⟩⟩⟩ | infectious $\gamma(a)$ | ⟩⟩⟩ | recovered |

Fig. 3.11 The Kermack-McKendrick model. The Kermack-McKendrick model contains three compartments for the susceptible, infectious, and recovered populations, S, I, and R, where the infectious population I is further subdivided by the age of infection a into subgroups $i(a)$. The transition rates between the compartments, the contact and infectious rates, $\beta(a)$ and $\gamma(a)$ are not constant, but depend on the age of infection.

$$I(t) = \int_0^\infty i(a,t)\, da\, . \tag{3.29}$$

Figure 3.1 shows that this results in an age-of-infection dependent SIR model,

$$\begin{aligned}
\frac{dS}{dt} &= -\lambda S \\
\frac{\partial i}{\partial t} + \frac{\partial i}{\partial a} &= \delta(a)\,\lambda S - \gamma(a)\,i \\
\frac{dR}{dt} &= \int_0^\infty \gamma(a)\,i(a,t)\, da\, ,
\end{aligned} \tag{3.30}$$

in which the transition rates between the compartments, the contact rate $\beta(a)$ and the infectious rate $\gamma(a)$ are not constant, but depend on the age of infection a. In the Kermack-McKendrick model (3.30), $\delta(a)$ denotes the Dirac delta function and λ is the *infection pressure*,

$$\lambda = \int_0^\infty \beta(a)\,i(a,t)\, da\, . \tag{3.31}$$

For the special case with constant contact and recovery rates $\beta(a) = \beta$ and $\gamma(a) = \gamma$ across all infection ages a, the second equation of the Kermack-McKendrick model (3.30) simplifies as follows,

$$\int_0^\infty \frac{\partial i}{\partial t} + \frac{\partial i}{\partial a}\, da = \int_0^\infty \beta\, S\, i(a,t)\, da - \int_0^\infty \gamma\, i(a,t)\, da\, . \tag{3.32}$$

By substituting the integral (3.29), $I(t) = \int_0^\infty i(a,t)\, da$, we obtain the classical SIR model from equation (3.1),

$$\frac{dS}{dt} = -\beta\, SI \qquad \frac{dI}{dt} = +\beta\, SI - \gamma\, I \qquad \frac{dR}{dt} = +\gamma\, I\, . \tag{3.33}$$

In its general form, the Kermack-McKendrick model (3.30) allows us to study the effect of contact and infectious rates that vary during the age of the infection, for example with infectious rates that are initially low, increase towards the peak of infection, and decay gradually upon recovery.

Problems

3.1 Basic reproduction number. A survey at Stanford University reported that 92.5% of undergraduates were susceptible to an influenza with an infectious period of $C = 6$ days at the beginning of the year, and 52.5% were susceptible at the end. Determine the basic reproduction number R_0 and the contact period B.

3.2 Maximum infectiousness. A survey at Stanford University reported that 92.5% of undergraduates were susceptible to an influenza with an infectious period of $C = 6$ days at the beginning of the year, and 52.5% were susceptible at the end. Determine the maximum infectious population during the year.

3.3 Immunization. A survey at Stanford University reported that 92.5% of undergraduates were susceptible to an influenza with an infectious period of $C = 6$ days at the beginning of the year, and 52.5% were susceptible at the end. Determine the fraction of undergraduates that would have to be immunized to prevent the outbreak.

3.4 Basic reproduction number. A French boarding school reported an influenza that spread through 1224 of its 1632 students. Estimate the basic reproduction number.

3.5 Maximum infectiousness. A French boarding school reported an influenza that spread through 1224 of its 1632 students. Estimate how many students were infectious at the peak of the outbreak.

3.6 Immunization. A French boarding school reported an influenza that spread through 1224 of its 1632 students. Would it have been enough to immunize 200 students to prevent the outbreak?

3.7 Maximum infectiousness. Two visitors introduce COVID-19 to Stanford campus with a population of 5000 students. Each infectious individual meets on average 0.4286 students per day and is infectious for seven days. The university plans to isolate all infectious individuals. Estimate the maximum number of required isolation units.

3.8 Immunization. Two visitors introduce COVID-19 to Stanford campus with a population of 5000 students. Each infectious individual meets on average 0.4286 students per day and is infectious for seven days. How many students would have to be immunized to avoid a COVID-19 outbreak on campus?

3.9 Growth rate. The University of Notre Dame recorded the first case of COVID-19 on August 8, began in-person instruction on August 10, and transitioned to online instruction on August 18, 2020. For the newly reported COVID-19 cases [17], from the first day of reporting, Aug 8, 2020, to Aug, 28, 2020, in Table 3.2, estimate the growth rates G for the last week of in-person and first week of online instruction from the slopes within two time intervals between August 11, 18, and 25, 2020. Remember to first calculate the infectious population, $N_I(t_n) = \sum_{i=n+1-C}^{n} \Delta N_I(t_i)$, from the seven-day cumulative case numbers for each day. Assume an infectious period of $C = 7$ days.

Table 3.2 Early COVID-19 outbreak at the University of Notre Dame. Newly reported COVID-19 cases at the University of Notre Dame from Saturday, August 8, to Friday, August 28, 2020. Notre Dame recorded the first case of COVID-19 on August 8, began in-person instruction on August 10, and transitioned to online instruction on August 18, 2020.

	Sat	Sun	Mon	Tue	Wed	Thu	Fri
week 1	1	2	6	6	9	13	15
week 2	4	17	104	95	85	26	30
week 3	19	32	40	20	18	21	19

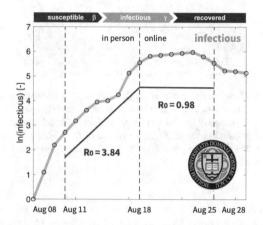

Fig. 3.12 Problem. Estimating the growth rate G, basic reproduction number, R_0 and contact period B. In this logarithmic plot of case numbers vs. time, circles represent the newly reported case data for the University of Notre Dame from Table 3.2, the orange curve highlights the cumulative infectious population for an infectious period of $C = 7$ days, and the red lines are the tangents to the curve that define the growth rates G, basic reproduction numbers R_0, and contact periods B for the last week of in-person and first week of online instruction between August 11, 18, and 25, 2020.

3.10 Basic reproduction number. For the newly reported COVID-19 cases at the University of Notre Dame [17] in Table 3.2, estimate the basic reproduction numbers R_0 for the last week of in-person and first week of online instruction from the slopes within two time intervals between August 11, 18, and 25, 2020. Compare you results against the values in Figure 3.12.

3.11 Contact period. For the newly reported COVID-19 cases at the University of Notre Dame [17] in Table 3.2, estimate the contact periods B for the last week of in-person and first week of online instruction from the slopes within two time intervals between August 11, 18, and 25, 2020. Interpret how the contact period B changes in relation to the infectious period C. Comment on whether the transition from in-person to online instruction on August 18, 2020 was effective.

3.12 Immunization. To prevent a campus outbreak at the University of Notre Dame [17], what fraction of the student population would have to be immunized without transitioning to online instruction? For an enrollment of 12607 students, how many

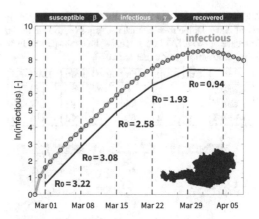

Fig. 3.13 Problem. Estimating the growth rate G, basic reproduction number, R_0 and contact period B. In this logarithmic plot of case numbers vs. time, circles represent the seven-day moving averages of the reported case data for Austria [9] from Table 3.3, the orange curve highlights the cumulative infectious population for an infectious period of $C = 7$ days, and the red lines are the tangents to the curve that define the growth rates G, basic reproduction numbers R_0 and contact periods B for each interval.

students would need to be immunized? You can use the basic reproduction number R_0 during in-person instruction from August 11 to August 18, 2020.

Table 3.3 Early COVID-19 outbreak in Austria. Newly reported COVID-19 cases, ΔN_I, in Austria from Friday, Feb 28, 2020, the first day of reported cases, to Thursday, Apr, 09, 2020; case data, ΔN_I, are reported as seven-day moving averages.

	Fri	Sat	Sun	Mon	Tue	Wed	Thu
week 1	1	2	2	2	3	4	6
week 2	8	10	13	16	23	31	45
week 3	63	82	108	127	164	200	260
week 4	306	334	389	493	565	563	676
week 5	721	754	744	735	699	732	603
week 6	547	501	466	383	351	319	302

3.13 Growth rate. For the seven-day moving averages of the newly reported COVID-19 cases in Austria, from the first day of reporting, Feb 28, to Apr, 03, 2020, in Table 3.3, estimate the growth rates G from the slopes within five time intervals between March 1, March 8, March 15, March 22, March 29, and April 5. Remember to first calculate the infectious population, $N_I(t_n) = \sum_{i=n+1-C}^{n} \Delta N_I(t_i)$, from the seven-day cumulative case numbers for each day. Assume an infectious period of $C = 7$ days.

3.14 Basic reproduction number. For the seven-day moving averages of the newly reported COVID-19 cases in Austria [9], from the first day of reporting, Feb 28, to

Apr, 03, 2020, in Table 3.3, estimate the basic reproduction numbers R from the slopes within five time intervals between March 1, March 8, March 15, March 22, March 29, and April 5. Remember to first calculate the infectious population from the seven-day cumulative case numbers for each day. Compare you results against the values in Figure 3.13.

3.15 Contact period. For the seven-day moving averages of the newly reported COVID-19 cases in Austria [9], from the first day of reporting, Feb 28, to Apr, 03, 2020, in Table 3.3, estimate the contact periods B from the slopes within five time intervals between March 1, March 8, March 15, March 22, March 29, and April 5. Remember to first calculate the infectious population from the seven-day cumulative case numbers for each day. Comment on whether the total lockdown on March 16, 2020 was effective.

Fig. 3.14 Classical SIRS model. The classical SIRS model contains three compartments for the susceptible, infectious, and recovered populations, S, I, and R. The transition rates between the compartments, the contact, infectious, and immune rates, β, γ, and δ, are inverses of the contact, infectious, and immune periods, $B = 1/\beta$, $C = 1/\gamma$, and $D = 1/\delta$.

3.16 Classical SIRS model. Derive the governing equations for the SIRS model in Figure 3.14 with three compartments for the susceptible, infectious, and recovered populations, S, I, and R. Assume that, similar to the classical SIR model (3.1), individuals transition between the compartments by the contact and infectious rates, β and γ, which are inverses of the contact and infectious periods, $B = 1/\beta$ and $C = 1/\gamma$. However, recovered individuals gradually lose immunity, at a rate δ with an immune period $D = 1/\delta$, and return to the susceptible group. Characterize the SIRS model through a system of three ordinary differential equations.

3.17 Classical SIRS model. Derive the governing equations for the SIRS model in Figure 3.14 with three compartments for the susceptible, infectious, and recovered populations, S, I, and R and transition rates β, γ, and δ. Neglect all vital dynamics, $\dot{S} + \dot{I} + \dot{R} = 0$ and $S + I + R = \text{const.} = 1$, and reduce the system of three ordinary differential equations to two.

3.18 Classical SIRD model. Derive the governing equations for the SIRD model in Figure 3.15 with four compartments for the susceptible, infectious, recovered, and dead populations, S, I, R, and D. Assume that the transition between the compartments is similar to the SIR model with contact and infectious rates, β and γ. However, only a fraction, $\nu_r = 1 - \nu_d$, recovers from the infection, and a fraction, ν_d, dies. Characterize the SIRD model through a system of four ordinary differential equations.

Fig. 3.15 Classical SIRD model. The classical SIRD model contains four compartments for the susceptible, infectious, recovered, and dead populations, S, I, R, and D. The transition rates between the compartments, the contact and infectious rates, β and γ, are inverses of the contact and infectious periods, $B = 1/\beta$ and $C = 1/\gamma$. A fraction, $\nu_r = 1 - \nu_d$, recovers from infection and a fraction, ν_d, dies.

3.19 Classical SIRD model. For the SIRD model in Figure 3.15, rationalize how the initial growth rate G, the basic reproduction number R_0, the maximum infectious population $I_{\max}$, and the final sizes of the susceptible and recovered groups, S_∞ and R_∞, differ from the classical SIR model.

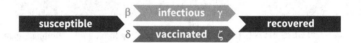

Fig. 3.16 Classical SIVR model. The classical SIVR model contains four compartments for the susceptible, infectious, vaccinated, and recovered populations, S, I, V, and R. The transition rates of the disease path, the contact and infectious rates, β and γ, are inverses of the contact and infectious periods, $B = 1/\beta$ and $C = 1/\gamma$. The transition rates of the vaccination path, the vaccination and immunization rates, δ and ζ, are inverses of the vaccination and immunization periods, $D = 1/\delta$ and $Z = 1/\zeta$.

3.20 Classical SIVR model. Derive the governing equations for the SIVR model in Figure 3.16 with four compartments for the susceptible, infectious, vaccinated, and recovered populations, S, I, V, and R. Assume that the disease path between the S, I, and R compartments is similar to the SIR model with contact and infectious rates, β and γ. However, it is also possible to recover through a vaccination path between S, V, and R with vaccination and immunization rates δ and ζ. Characterize the SIVR model through a system of four ordinary differential equations.

3.21 Classical SIVR model. The SIVR model in Figure 3.16 assumes two paths to immunity, either through disease or through vaccination. How does the additional vaccination path affect the basic reproduction number R_0 and the herd immunity threshold H?

3.22 Mathematical modeling in epidemiology. In his famous essay *Mathematical models of the spread of infection* [13], David G. Kendall writes "Almost certainly, these models are ridiculously over-simplified. Often they could be made more realistic by introducing a much larger number of additional mechanisms, each with its own array of arbitrary constants; in most cases the analysis of such variations does not introduce any new difficulties of principle, and does not necessarily lead to any better understanding of the basic problems." Discuss this assessment in view of the simplifications and limitations of the SIR model.

References

1. Anderson RM, May RM (1979) Population biology of infectious diseases: Part I. Nature 280:361-366.
2. Anderson RM, May RM (1991) Infectious Diseases of Humans. Oxford University Press, Oxford.
3. Brauer F, Castillo-Chavez C (2001) Mathematical Models in Population Biology and Epidemiology. Springer-Verlag New York.
4. Brauer F, van den Dreissche P, Wu J (2008) Mathematical Epidemiology. Springer-Verlag Berlin Heidelberg.
5. Brauer F (2017) Mathematical epidemiology: Past, present and future. Infectious Disease Modelling 2:113-127.
6. Brauer F, Castillo-Chavez C, Feng Z (2019) Mathematical Models in Epidemiology. Springer-Verlag New York.
7. Dietz K (1993) The estimation of the basic reproduction number for infectious diseases. Statistical Methods in Medical Research 2:23-41.
8. European Centre for Disease Prevention and Control. Situation update worldwide. `https://www.ecdc.europa.eu/en/geographical-distribution-2019-ncov-cases` accessed: June 1, 2021.
9. Fine PEM (1993) Herd immunity: history, theory, practice. Epidemiologic Reviews 15:265-302.
10. Hethcote HW (1976) Quantitative analyses of communicable disease models. Mathematical Biosciences, 28:335-356.
11. Hethcote HW (1989) Three basic epidemiological models. Biomathematics: Applied Mathematical Ecology, 18:119-144.
12. Hethcote HW (2000) The mathematics of infectious diseases. SIAM Review 42:599-653.
13. Kendall DG (1965) Mathematical models of the spread of infection. Mathematics and Computer Science in Biology and Medicine 213–225.
14. Kermack WO, McKendrick G (1927) Contributions to the mathematical theory of epidemics, Part I. Proceedings of the Royal Society London Series A 115:700-721.
15. Kermack WO, McKendrick G (1932) Contributions to the mathematical theory of epidemics. Part II. The problem of endemicity. Proceedings of the Royal Society London Series A 138:55-83.
16. Kermack WO, McKendrick G (1933) Contributions to the mathematical theory of epidemics. Part III. Further studies of the problem of endemicity. Proceedings of the Royal Society London Series A 141:94-122.
17. University of Notre Dame (2020) Notre Dame Covid-19 Dashboard. `https://here.nd.edu/our-approach/dashboard` accessed: June 1, 2021.
18. Zohdi TA (2020) An agent-based computational framework for simulation of global pandemic and social response on planet X. Computational Mechanics 66:1195-1209.

Chapter 4
The classical SEIR model

Abstract The SEIR model is a popular compartment model with four populations, the susceptible, exposed, infectious, and recovered groups S, E, I, and R. It characterizes infectious diseases with a significant incubation period during which individuals have been infected, but are not yet infectious themselves. While the SEIR model does not have an analytical solution for the time course of its populations, it has explicit analytical solutions for the maximum exposed infectious populations combined and for the final sizes of the susceptible and recovered populations at endemic equilibrium. This makes it a good candidate for computational modeling. The learning objectives of this chapter on classical SEIR modeling are to

- interpret contact, latent, and infectious rates and periods
- explain the concept of the basic reproduction number
- estimate the growth rate of the initial outbreak
- solve the analytical solution of maximum exposure and infection
- analyze the final size relation
- develop your own SEIR type models
- discuss limitations of classical SEIR modeling

By the end of the chapter, you will be able to analyze, estimate, and predict the dimensions of infectious diseases like the measles or COVID-19 that have an incubation period during which individuals have been infected, but are not yet infectious themselves.

4.1 Introduction of the SEIR model

In epidemiology, the period from the time of exposure to the appearance of the first symptoms is called the *incubation period* [3]. The incubation period of infectious diseases can vary significantly, from 24-72 hours for the common cold, to 10-12 days for measles, 14-16 days for chickenpox, 16-18 days for mumps, and 2-3 weeks for rubella [9]. For most infectious diseases, a patient is not infectious throughout

---

Fig. 4.1 Classical SEIR model. The classical SEIR model contains four compartments for the susceptible, exposed, infectious, and recovered populations, S, E, I, and R. The transition rates between the compartments, the contact, latent, and infectious rates β, α, and γ, are inverses of the contact, latent, and infectious periods, $B = 1/\beta$, $A = 1/\alpha$, and $C = 1/\gamma$.

most of the incubation period. For COVID-19, the mean incubation period has been estimated to 4-5 days [14, 15, 28], but exposed individuals can already infect others approximately 2-3 days before symptoms start [17]. The period from the time of exposure to the potential to infect others is calls the *latent period* [4].

A popular model in epidemiology that explicitly accounts for individuals who have been exposed to an infection but will only become infectious after a certain time is the SEIR model [1]. Figure 4.1 shows that it consists of four populations, the susceptible group S, the exposed group E, the infectious group I, and the recovered group R [20]. We cannot solve the dynamics of the SEIR model analytically, but we can explore the analytical solution for the converged final sizes, S_∞, E_∞, I_∞, and R_∞, as $t \to \infty$ [3]. In the SEIR model, the transition between the susceptible, exposed, infectious, and recovered populations is governed by four ordinary differential equations [2]. For the transition from the susceptible to the exposed group, we assume a *mass action incidence* [4], which implies that the rate of new exposure is proportional to the size of the susceptible and infectious groups weighted by the contact rate β, $\dot{E} = \beta SI$. For the transition between the infectious and recovered groups, we assume a *constant rate recovery*, which implies that the rate of recovered infections is proportional to the size of the infectious group weighted by the infectious rate γ, $\dot{I} = \gamma I$. This results in the following system of four coupled ordinary differential equations [8],

$$
\begin{aligned}
\dot{S} &= -\beta\,SI \\
\dot{E} &= +\beta\,SI - \alpha\,E \\
\dot{I} &= \qquad +\alpha\,E - \gamma\,I \\
\dot{R} &= \qquad\qquad +\gamma\,I\,.
\end{aligned}
\tag{4.1}
$$

The transition rates between the four compartments are the contact rate β, the latent rate α, and the infectious rate γ in units [1/days], which are the inverses of the contact period $B = 1/\beta$, the latent period $A = 1/\alpha$, and the infectious period $C = 1/\gamma$ in units [days]. The ratio between the contact and infectious rates, or similarly, between the infectious and contact periods, defines the basic reproduction number R_0,

$$
R_0 = \frac{\beta}{\gamma} = \frac{C}{B}\,.
\tag{4.2}
$$

In this simple format, the SEIR model neglects all vital dynamics, it does not account for births or natural deaths, which implies that $\dot{S} + \dot{E} + \dot{I} + \dot{R} \doteq 0$ and $S + E + I + R = \text{const.} = 1$ [12]. For the special case of large exposed rates $\alpha \to \infty$/days, or, equivalently, short exposed periods $A \to 0$ days, susceptible individuals become almost instantly infectious, and the SEIR model (4.17) converges to the SIR model (3.1).

4.2 Growth rate of the SEIR model

To characterize the growth rate of the SEIR model, a simple evaluation similar to the SIR model in Chapter 3 is no longer possible. Instead, we linearize the SEIR model (4.17),

$$
\mathbf{K} = \begin{bmatrix} d\dot{S}/dS & d\dot{S}/dE & d\dot{S}/dI & d\dot{S}/dR \\ d\dot{E}/dS & d\dot{E}/dE & d\dot{E}/dI & d\dot{E}/dR \\ d\dot{I}/dS & d\dot{I}/dE & d\dot{I}/dI & d\dot{I}/dR \\ d\dot{R}/dS & d\dot{R}/dE & d\dot{R}/dI & d\dot{R}/dR \end{bmatrix} = \begin{bmatrix} -\beta I & 0 & -\beta S & 0 \\ +\beta I & -\alpha & +\beta S & 0 \\ 0 & +\alpha & -\gamma & 0 \\ 0 & 0 & +\gamma & 0 \end{bmatrix}. \tag{4.3}
$$

and evaluate this linearization at the beginning of the outbreak, $S_0 = 1$, $E_0 = 0$, $I_0 = 0$, and $R_0 = 0$,

$$
\begin{bmatrix} -\beta I & 0 & -\beta S & 0 \\ +\beta I & -\alpha & +\beta S & 0 \\ 0 & +\alpha & -\gamma & 0 \\ 0 & 0 & +\gamma & 0 \end{bmatrix} = \begin{bmatrix} 0 & 0 & -\beta & 0 \\ 0 & -\alpha & +\beta & 0 \\ 0 & +\alpha & -\gamma & 0 \\ 0 & 0 & +\gamma & 0 \end{bmatrix}. \tag{4.4}
$$

It is easy to see that the first and fourth columns are associated with zero eigenvalues [3]. For the remaining 2×2 matrix $\mathbf{K}^*$ of the second and third columns and rows, we perform an *eigenvalue analysis*,

$$
\mathbf{K}^* \cdot \mathbf{n} = \lambda \mathbf{n} \quad \text{with} \quad \mathbf{K}^* = \begin{bmatrix} -\alpha & +\beta \\ +\alpha & -\gamma \end{bmatrix}. \tag{4.5}
$$

The above equation has a nonzero solution $\mathbf{n}$ if and only if the determinant of the matrix $(\mathbf{K}^* - \lambda \mathbf{I})$ is zero,

$$
\det(\mathbf{K}^* - \lambda \mathbf{I}) \doteq 0 \quad \text{with} \quad \mathbf{K}^* - \lambda \mathbf{I} = \begin{bmatrix} -(\alpha + \lambda) & +\beta \\ +\alpha & -(\gamma + \lambda) \end{bmatrix}. \tag{4.6}
$$

which results in the following *characteristic equation* for the two eigenvalues λ,

$$
\lambda^2 + (\alpha + \gamma)\lambda + (\gamma - \beta)\alpha \doteq 0. \tag{4.7}
$$

For positive basic reproduction numbers, $R_0 > 0$, one of the eigenvalues is positive and one is negative. If there is an epidemic, the positive eigenvalue λ, the largest eigenvalue of the matrix $\mathbf{K}^*$, determines the initial exponential *growth rate* of the outbreak,

$$
G \approx -\frac{1}{2}[\alpha + \gamma] + \frac{1}{2}\sqrt{(\alpha - \gamma)^2 + 4\alpha\beta} = \lambda_{\max}. \tag{4.8}
$$

Table 4.1 Early COVID-19 outbreak in Germany. Newly reported COVID-19 cases in Germany from Wednesday, February 26, the first day of reported cases, to Tuesday, April, 7, 2020; case data ΔN_I are reported as seven-day moving averages.

	Wed	Thu	Fri	Sat	Sun	Mon	Tue
week 1	2	4	9	9	16	21	27
week 2	33	71	85	103	130	152	194
week 3	244	314	429	543	682	864	1114
week 4	1481	1796	2310	2538	2723	3112	3375
week 5	3581	4088	4432	5047	5366	5404	5546
week 6	5808	5837	5755	5485	5384	5213	5122

This implies that, during this early phase of the outbreak, the infectious population experiences *exponential growth* at a rate G [3],

$$I(t) \approx I_0 \exp(G\,t). \tag{4.9}$$

We can estimate the contact rate β,

$$\beta = \frac{1}{4\,\alpha} [[\, 2G + \alpha + \gamma\,]^2 - [\,\alpha - \gamma\,]^2], \tag{4.10}$$

or the contact period B,

$$B = \frac{C}{1 + G\,[A + C] + G^2 AC}, \tag{4.11}$$

and with them the basic reproduction number $R_0 = \beta/\gamma$, as a function of the latent and infectious rates α and γ,

$$R_0 = \frac{1}{4\,\alpha\gamma} [[\, 2G + \alpha + \gamma\,]^2 - [\,\alpha - \gamma\,]^2] \tag{4.12}$$

or, equivalently, $R_0 = C/B$, as a function of the latent and infectious periods A and C,

$$R_0 = 1 + G\,[A + C] + G^2\,AC, \tag{4.13}$$

in terms of the slope G in the logarithmic plot. Interestingly, the exponential growth rate G for the SEIR model in equation (4.8) differs notably from the exponential growth rate, $G \approx \beta - \gamma$, for the SIR model in equation (3.5) of Chapter 3: Adding an exposed population E with an exposed period A to the timeline of the disease decreases the initial growth rate of the outbreak. Conversely, estimating the contact rate β and basic reproduction number R_0 from equations (4.10) and (4.13) for the the SEIR model results in larger values than the estimates, $\beta = G + \gamma$ and $R_0 = 1 + G/\gamma$ for equations (3.7) and (3.8) for the SIR model in Chapter 3. In the limit of $A \to 0$, the contact rate β and the basic reproduction number R_0 of the SEIR model converge to their SIR model limits.

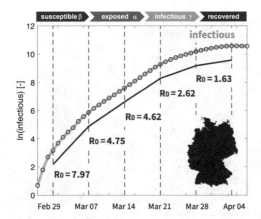

Fig. 4.2 Problem. Estimating the growth rate G, basic reproduction number R_0, and contact period B. In this logarithmic plot of case numbers vs. time, circles represent the seven-day moving averages of the reported case data for Germany from Table 4.1, the orange curve highlights the cumulative infectious population for latent and infectious periods of $A = 3$ days and $C = 7$ days, and the red lines are the tangents to the curve that define the growth rates G, basic reproduction numbers R_0, and contact periods B for each interval.

Example. Estimating the growth rate G, basic reproduction number R_0, and contact period B. For the seven-day moving averages of the newly reported COVID-19 cases in Germany [9], from the first day of reporting, from Feb 26, 2020, to the peak of the first wave, April, 7, 2020 in Table 4.1, estimate the growth rate G, the basic reproduction number R_0, and the contact period B for a latent period of $A = 3$ days and an infectious period of $C = 7$ days.

Solution. We first calculate the total infectious population at each day t_n within the interval,

$$N_I(t_n) = \sum_{i=n+1-C}^{n} \Delta N_I(t_i),$$

by summing the newly reported cases ΔN_I for the $C = 7$ previous days from Table 4.1. Then we plot these seven-day cumulative cases in a logarithmic plot in Figure 4.2. We observe that the orange infectious curve is not a straight line, which suggests that the reproduction number R_0 varies within the interval. We approximate the curve through five line segments between February 29, March 7, March 14, March 21, March 28, and April 4.

	Feb 29	Mar 7	Mar 14	Mar 21	Mar 28	Apr 4
day t	4	11	18	25	32	39
N_I	24	356	2006	10785	26358	39201
$\ln(N_I)$	3.18	5.87	7.60	9.29	10.18	10.58

We calculate the growth rates G as the slopes of these lines in all five intervals,

$$G = \frac{\ln(N_I(t_{n+1})) - \ln(N_I(t_n))}{t_{n+1} - t_n},$$

and from them the contact periods B,

$$B = \frac{C}{1 + G\,[\,A + C\,] + G^2\,AC}\,,$$

and the basic reproduction numbers R_0,

$$R_0 = 1 + G\,[\,A + C\,] + G^2\,AC\,.$$

The table below summarizes the growth rates G, the contact periods B, and the basic reproduction numbers R_0 in the five week-long intervals.

	Feb 29 - Mar 7	Mar 7 - Mar 14	Mar 14 - Mar 21	Mar 21 - Mar 28	Mar 28 - Apr 4
G	0.3853	0.2470	0.2403	0.1277	0.0567
B	0.8783	1.4733	1.5167	2.6730	4.2825
R_0	7.9697	4.7511	4.6154	2.6188	1.6346

On March 13, 2020, most German schools were closed and on March 17, 2020, the European Union closed all its external boarders. Along with several other local regulations, this reduced the initial high reproduction numbers of $R_0 = 7.97$, $R_0 = 4.75$, and $R_0 = 4.62$ to $R_0 = 2.62$ and $R_0 = 1.63$ within a time window of two weeks [17].

4.3 Maximum exposure and infection of the SEIR model

An epidemic outbreak with lifelong immunity typically begins with a small exposed population E_0 that gradually becomes infectious, which, turn, generates new exposure. With a time delay that depends on the exposed period A, the exposed and infectious populations $E(t)$ and $I(t)$ increase, reach a peak E_{max} and I_{max}, and then decays to zero, $E_\infty = 0$ and $I_\infty = 0$. Unlike the SIR model in Section 3.3, the SEIR model does not provide an explicit analytical solution for the maximum infectious population I_{max}. However, we can estimate the *maximum exposure and infection*, $[\,E + I\,]_{max}$. Similar to Section 3.3 for the SIR model, we begin with the first equation of the SEIR model (4.17),

$$\dot{S} = -\beta\,SI\,, \tag{4.14}$$

perform a *separation of variables* by dividing it by the susceptible population S,

$$\dot{S}/S = -\beta\,I\,, \tag{4.15}$$

and integrate it in time from t_0 to t,

$$\int_{t_0}^{t} \frac{1}{S}\frac{dS}{dt} = -\int_{t_0}^{t} \beta I\,dt\,. \tag{4.16}$$

To evaluate the integral on the righthand side, $\int_{t_0}^{t} I \, dt$, we add the first, second, and third equations of the SEIR model (4.17),

$$\begin{aligned}
\dot{S} &= -\beta \, SI \\
\dot{E} &= +\beta \, SI - \alpha \, E \\
\dot{I} &= \qquad +\alpha \, E - \gamma \, I
\end{aligned} \qquad (4.17)$$

to eliminates the explicit dependence on the susceptible population S,

$$\dot{S} + \dot{E} + \dot{I} = -\gamma \, I \, . \qquad (4.18)$$

We integrate this equation in time from t_0 to t,

$$\int_{t_0}^{t} \dot{S} + \dot{E} + \dot{I} \, dt = -\int_{t_0}^{t} \gamma \, I \, dt \, , \qquad (4.19)$$

and use the initial conditions, S_0, E_0, and I_0 with $S_0 + E_0 + I_0 = 1$ at t_0, and the current conditions, $S(t)$, $E(t)$, and $I(t)$ at t, to simplify the lefthand side, $S(t) - S_0 + E(t) - E_0 + I(t) - I_0 = S(t) + E(t) + I(t) - 1$. With these considerations, the integral expression takes the following explicit representation,

$$\int_{t_0}^{t} I \, dt = [\, 1 - S(t) - E(t) - I(t)]/\gamma \, . \qquad (4.20)$$

We return to equation (4.16) and substitute the righthand side by $\int_{t_0}^{t} 1/S \, dS/dt = \log(S(t)) - \log(S_0) = \log(S(t)/S_0)$, and the lefthand side using the above expression, $-\int_{t_0}^{t} \beta \, I \, dt = \beta/\gamma \, [\, 1 - S(t) - E(t) - I(t)]$,

$$\log \frac{S(t)}{S_0} = -\mathsf{R}_0 \, [\, 1 - S(t) - E(t) - I(t) \,] \quad \text{with} \quad \mathsf{R}_0 = \frac{\beta}{\gamma} \, . \qquad (4.21)$$

To rephrase the susceptible population $S(t)$ at the time t of maximum exposure and infection $[\, E + I \,]_{\max}$, we evaluate the stationarity of the sum of the second and third equations of the SEIR model (4.17),

$$\dot{E} + \dot{I} = \beta \, SI - \gamma I \doteq 0 \quad \text{thus} \quad S = \gamma/\beta \, . \qquad (4.22)$$

By substituting $E(t) + I(t) = [\, E + I \,]_{\max}$ and $S(t) = \gamma/\beta$ in equation (4.21), we obtain an explicit expression for maximum exposure and infection $[\, E + I \,]_{\max}$,

$$[\, E + I \,]_{\max} = 1 - \frac{1}{\mathsf{R}_0} \left[1 + \log(S_0) - \log\left(\frac{1}{\mathsf{R}_0}\right) \right] \quad \text{with} \quad \mathsf{R}_0 = \frac{\beta}{\gamma} \, . \qquad (4.23)$$

Figure 4.3 illustrates the maximum exposed and infectious population $[\, E + I \,]_{\max}$ from equation (4.23) and its dependence on the basic reproduction number R_0. Similar to the the SIS and SIR models in Chapters 2 and 3, we can distinguish two regimes for reproduction numbers smaller and larger than one. The $[\, E + I \,]_{\max}$ curves display a kink at $\mathsf{R}_0 = 1$, where the $\log(1/\mathsf{R}_0)$ term in equation (4.23) flips

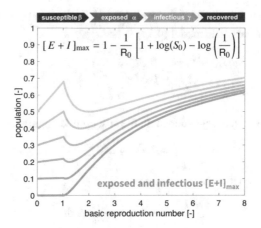

Fig. 4.3 Classical SEIR model. Maximum exposure and infection as a function of the basic reproduction number R_0. For a small initial exposed and infectious population $E_0 + I_0$, the maximum exposed and infectious population $[\,E + I\,]_{max}$ increases with increasing R_0. For large initial exposed and infectious populations $E_0 + I_0$, the maximum exposed and infectious population $[\,E + I\,]_{max}$ increases for basic reproduction numbers up to $R_0 = 1$, then drops, and then increases with increasing R_0. Initial exposed and infectious population $E_0 + I_0 = 0.0001, 0.1, 0.2, 0.3, 0.4, 0.5$, from dark to light.

its sign. Similar to the SIR case, the maximum exposed and infectious population $[\,E + I\,]_{max}$ is not only sensitive to the basic reproduction number R_0, but also to the initial infectious population I_0. For a small initial exposed and infectious population $E_0 + I_0$, the maximum exposed and infectious population $[\,E + I\,]_{max}$ increases with increasing R_0. For large initial exposed and infectious populations $E_0 + I_0$, the maximum exposed and infectious population $[\,E + I\,]_{max}$ increases for basic reproduction numbers up to $R_0 = 1$, then drops, and then increases with increasing R_0. When we solve the timeline of the outbreak computationally in Chapter 7, we will see that the $[\,E + I\,]_{max}$ estimate is typically smaller than both individual maxima, $[\,E + I\,]_{max} \leq E_{max} + I_{max}$, since a finite exposed period $A > 0$ triggers a time lag between the individual peaks, E_{max} and I_{max}.

4.4 Final size relation of the SEIR model

To derive the final size relation of the SEIR model, we use the first and fourth equations of the SEIR model 4.17,

$$\dot{S} = -\,\beta\,SI$$
$$\dot{R} = +\,\gamma I \tag{4.24}$$

and divide the equation for the susceptible population by the equation for the recovered population,

$$\frac{\dot{S}}{\dot{R}} = -\frac{\beta S I}{\gamma I} \tag{4.25}$$

Performing a *separation of variables* and using the definition of the basic reproduction number $R_0 = \beta/\gamma$ yields the following relation,

$$\dot{S}/S = -R_0 \dot{R}, \tag{4.26}$$

which we integrate in time,

$$\int_0^t \frac{1}{S}\frac{dS}{dt}\,dt = -\int_0^t R_0 \frac{dR}{dt}\,dt, \tag{4.27}$$

to obtain the following expression,

$$\ln\frac{S(t)}{S_0} = -R_0[\,R(t) - R_0\,], \tag{4.28}$$

where $\ln(S(t)/S_0) = \ln(S(t)) - \ln(S_0)$. Here S_0 and R_0 are the initial susceptible and recovered populations at time t_0 and $S(t)$ and $R(t)$ are these populations at time t. By applying the exponential function on both sides of the equation, we obtain the following explicit representation for the susceptible population at time t,

$$S(t) = S_0 \exp(-R_0[\,R(t) - R_0\,]). \tag{4.29}$$

We evaluate equation (4.29) at the limit $t \to \infty$ with S_∞, $E_\infty = 0$, $I_\infty = 0$, and $R_\infty = 1 - S_\infty$, to obtain the following expression for the susceptible population at *endemic equilibrium* [11],

$$S_\infty = S_0 \exp(-R_0[\,R_\infty - R_0\,]) = 1 - R_\infty. \tag{4.30}$$

This transcendental equation has an explicit solution in terms of the Lambert function W [17],

$$\begin{aligned} S_\infty &= -W(-S_0\,R_0 \exp(-R_0[\,1 - R_0\,]))/R_0 \\ R_\infty &= 1 + W(-S_0\,R_0 \exp(-R_0[\,1 - R_0\,]))/R_0. \end{aligned} \tag{4.31}$$

The *final size relation* (4.31) confirms that, unless the initial susceptible population is zero, $S_0 = 0$, the final susceptible population will always be larger than zero, $S_\infty > 0$. This implies that there will always be a group of individuals that will not become infected throughout the time course of the outbreak. Interestingly, the final size relation of the SEIR model (4.31) is similar to the final size relation of the SIR model (3.28) in Section 3.4. This suggests that, while the timeline of both models might be different, the final effect on a population will be similar.

Figure 7.9 illustrates the final size relation (4.31) and its dependence on the basic reproduction number R_0. Similar to the final size relation of the SIR model in Figure 3.7, increasing the basic reproduction number R_0 reduces the final susceptible population S_∞ and increases the final recovered population R_∞. Unlike the SIS model in Figure 2.6, but similar to the SIR model in Figure 3.7, the final size relation of

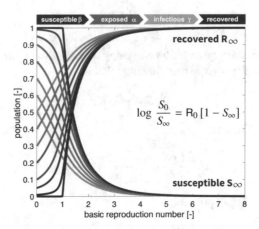

Fig. 4.4 Classical SEIR model. Final size relation as a function of the basic reproduction number R_0. Increasing the basic reproduction number R_0 reduces the final susceptible population S_∞ and increases the final recovered population R_∞. The larger the initial exposed and infectious populations E_0 and I_0, the smaller the initial and final susceptible populations $S_0 = 1 - E_0 - I_0$ and S_∞ and the larger the final recovered population R_∞. Initial exposed and infectious populations $E_0 + I_0 = 0.0001, 0.01, 0.05, 0.1, 0.2, 0.3, 0.4, 0.5$, from dark to light.

the SEIR model also depends on the initial susceptible population S_0, or, in other words, on the size of the initial exposed and infectious populations, E_0 and I_0. Even if the reproduction number is below one, $R_0 < 1$, the number of infections will rise initially. The larger the initial exposed and infectious populations E_0 and I_0, the smaller the initial and final susceptible populations $S_0 = 1 - E_0 - I_0$ and S_∞, and the larger the final recovered population R_∞.

Problems

4.1 Growth rate. Assume the COVID-19 pandemic has an exposed period of $A = 2.5$ days, an infectious period of $C = 6$ days, and a contact period of $B = 3.5$ days. Calculate the growth rate G for the SEIR model. Calculate the growth rate of the SIR model, compare, and comment on the results.

4.2 Growth rate. For the previous example, reduce the exposed period to $A = 0.01$ days, while keeping the infectious period and a contact period unchanged at $C = 6$ days and $B = 3.5$ days. Calculate the growth rate G for the SEIR model. Calculate the growth rate of the SIR model, compare, and comment on the results.

4.3 Immunization. For for a growth rate $G = 0.385$ and the initial basic reproduction number R_0 of the SEIR model in Figure 4.2, what percentage of the German population would have to be immunized to manage the outbreak? How would this estimate change for the initial basic reproduction number $R_0 = 1 + GC$ of the SIR model?

For a population of 82.900.000, how many Germans would have to be immunized in both cases?

4.4 Herd immunity. On March 10, 2020 the German chancellor Angela Merkel announced that eventually, 60% to 70% of Germans would get the COVID-19 virus. From Table 4.1 and Figure 4.2 we can approximate the growth rate in Germany on March 10, 2020 as $G = 0.247$. Did she use the SIR or SEIR model for this prediction?

4.5 Growth rate. The University of North Carolina Chapel Hill began in-person instruction on August 10, and transitioned to online instruction on August 19, 2020. For the newly reported COVID-19 cases [17], from August 11, 2020, to August, 31, 2020, in Table 4.2, estimate the growth rates G for the last week of in-person and first week of online instruction from the slopes within two time intervals between August 12, 19, and 26, 2020. Remember to first calculate the infectious population from the seven-day cumulative case numbers for each day.

4.6 Basic reproduction number. For the newly reported COVID-19 cases at the University of North Carolina Chapel Hill [17], in Table 4.2, estimate the basic reproduction numbers R_0 for the last week of in-person and first week of online instruction from the slopes within two time intervals between August 12, 19, and 26, 2020 using latent and infectious periods of $A = 3$ days and $C = 7$ days. Compare you results against the values in Figure 4.5.

4.7 Basic reproduction number. For the newly reported COVID-19 cases at the University of North Carolina Chapel Hill [17], in Table 4.2, estimate the basic reproduction numbers R_0 for the last week of in-person and first week of online instruction from the slopes within two time intervals between August 12, 19, and 26, 2020. In contrast to the previous problem, instead of using the SEIR model to calculate R_0 from the growth rate G, use the SIR model to calculate $R_0 = 1 + C + G$ with $C = 7$ days. Compare you results against the values in Figure 4.5. Discuss the impact of the exposed population E and the exposed period A.

4.8 Contact period. For the newly reported COVID-19 cases at the University of North Carolina Chapel Hill [17], in Table 4.2, estimate the contact periods B for the last week of in-person and first week of online instruction from the slopes within two time intervals between August 12, 19, and 26, 2020 using latent and infectious

Table 4.2 Early COVID-19 outbreak at the University of North Carolina Chapel Hill. Newly reported COVID-19 cases at UNC Chapel Hill from Tuesday, August 11, to Monday, August 31, 2020. The university began in-person instruction on August 10, and transitioned to online instruction on August 19, 2020.

	Tue	Wed	Thu	Fri	Sat	Sun	Mon
week 1	1	2	6	6	9	13	15
week 2	4	17	104	95	85	26	30
week 3	19	32	40	20	18	21	19

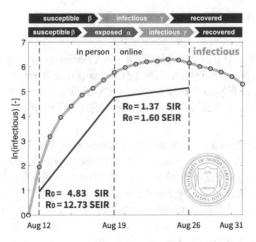

Fig. 4.5 Problem. Estimating the growth rate G, basic reproduction number, R_0, and contact period B. In this logarithmic plot of case numbers vs. time, circles represent the reported case data for the University of North Carolina Chapel Hill from Table 4.2, the orange curve highlights the cumulative infectious population for latent and infectious periods of $A = 3$ days and $C = 7$ days, and the red lines are the tangents to the curve that define the growth rates G, basic reproduction numbers R_0, and contact periods B for the last week of in-person and first week of online instruction between August 12, 19, and 26, 2020.

periods of $A = 3$ days and $C = 7$ days. Interpret how the contact period B changes in relation to the infectious period C. Comment on whether the transition from in-person to online instruction on August 19, 2020 was effective.

4.9 Immunization. To prevent a campus outbreak at the University of North Carolina Chapel Hill, what fraction of the student population would have to be immunized without transitioning to online instruction? For an enrollment of 30011 students, how many students would need to be immunized? Use the basic reproduction number R_0 for the SEIR model during in-person instruction from August 12 to August 19, 2020 from Figure 4.5.

4.10 Immunization. To prevent a campus outbreak at the University of North Carolina Chapel Hill, what fraction of the student population would have to be immunized without transitioning to online instruction? For an enrollment of 30011 students, how many students would need to be immunized? Use the basic reproduction number R_0 for the SIR model during in-person instruction from August 12 to August 19, 2020 from Figure 4.5. Compare your result to the SEIR model and comment on the impact of model selection in policy making.

4.11 Basic reproduction number. For the seven-day moving averages of the newly reported COVID-19 cases in Austria [9], in Table 4.3, estimate the growth rates G and basic reproduction numbers R_0 from the slopes within five time intervals between March 1, March 8, March 15, March 22, March 29, and April 5 using the SEIR

Table 4.3 Early COVID-19 outbreak in Austria. Newly reported COVID-19 cases in Austria from Friday, Feb 28, 2020, the first day of reported cases, to Thursday, Apr, 09, 2020; case data ΔN_I are reported as seven-day moving averages.

	Fri	Sat	Sun	Mon	Tue	Wed	Thu
week 1	1	2	2	2	3	4	6
week 2	8	10	13	16	23	31	45
week 3	63	82	108	127	164	200	260
week 4	306	334	389	493	565	563	676
week 5	721	754	744	735	699	732	603
week 6	547	501	466	383	351	319	302

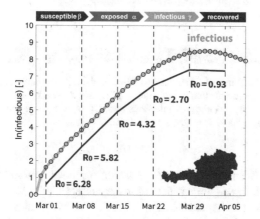

Fig. 4.6 Problem. Estimating the growth rate G, basic reproduction number, R_0 and contact period B. In this logarithmic plot of case numbers vs. time, circles represent the seven-day moving averages of the reported case data for Austria from Table 4.3, the orange curve highlights the cumulative infectious population for exposed and infectious periods of $A = 3$ days and $C = 7$ days, and the red lines are the tangents to the curve that define the growth rates G, basic reproduction numbers R_0 and contact periods B for each interval.

model with latent and infectious periods of $A = 3$ days and $C = 7$ days. Compare your results against the R_0 values in Figure 4.6 and against the $R_0 = 1 + GC$ values from the SIR model in Figure 3.13. Comment on your results.

4.12 Basic reproduction number. For the seven-day moving averages of the newly reported COVID-19 cases in Austria [9], in Table 4.3, estimate the growth rates G and basic reproduction numbers R_0 from the slopes within five time intervals between March 1, March 8, March 15, March 22, March 29, and April 5 using the SEIR model with latent and infectious periods of $A = 2$ days and $C = 7$ days. Compare your results against the R_0 values of the SEIR model with $A = 3$ days in Figure 4.6 and against the $R_0 = 1 + GC$ values from the SIR model with $A = 0$ days in Figure 3.13. What is the effect of reducing or increasing the latent period A?

4.13 Contact period. For the seven-day moving averages of the newly reported COVID-19 cases in Austria [9], in Table 4.3, estimate the contact periods B from the slopes within five time intervals between March 1, March 8, March 15, March

22, March 29, and April 5 using the SEIR model with latent and infectious periods of $A = 3$ days and $C = 7$. Compare your results against the $B = C/[1 + GC]$ values from the SIR model in Chapter 3 and comment on your results.

4.14 Immunization. For the initial basic reproduction number R_0 of the SEIR model in Figure 4.6, what percentage of the population would have to be immunized to manage the outbreak? How would this estimate change for the initial basic reproduction number R_0 of the SIR model in Figure 3.13? For a population of 8,840,000, how many Austrians would have to be immunized in both cases? Discuss the implication of model selection for public health care planning.

Fig. 4.7 Classical SEIS model. The classical SEIS model contains three compartments for the susceptible, exposed, and infectious populations, S, E, and I. The transition rates between the compartments, the contact, latent, and infectious rates, β, α, and γ, are inverses of the contact, latent, and infectious periods, $B = 1/\beta$, $A = 1/\alpha$, and $C = 1/\gamma$.

4.15 Classical SEIS model. Derive the governing equations for the SEIS model in Figure 4.7 with three compartments for the susceptible, exposed, and infectious populations, S, E, and I. Assume that, similar to the classical SEIR model (4.17), individuals transition between the compartments by the contact, latent, and infectious rates, β, α, and γ, which are inverses of the contact, latent, and infectious periods, $B = 1/\beta$, $A = 1/\alpha$, and $C = 1/\gamma$. However, individuals are not immune upon infection, but return directly to the susceptible group. Charcterize the SEIS model through a system of three ordinary differential equations.

4.16 Classical SEIS model. Derive the governing equations for the SEIS model in Figure 4.7 with three compartments for the susceptible, exposed, and infectious populations, S, E, and I and transition rates β, α, and γ. Neglect all vital dynamics, $\dot{S} + \dot{E} + \dot{I} = 0$ and $S + E + I = $ const. $= 1$, and reduce the system of three ordinary differential equations to two.

Fig. 4.8 Classical SEIRD model. The classical SEIRD model contains five compartments for the susceptible, exposed, infectious, recovered, and dead populations, S, E, I, R, and D. The transition rates between the compartments, the contact, latent, and infectious rates, β, α, and γ, are inverses of the contact, latent, and infectious periods, $B = 1/\beta$, $A = 1/\alpha$, and $C = 1/\gamma$. A fraction, $\nu_r = 1 - \nu_d$, recovers from infection and a fraction, ν_d, dies.

4.17 Classical SEIRD model. Derive the governing equations for the SEIRD model in Figure 4.8 with five compartments for the susceptible, exposed, infectious, recovered, and dead populations, S, E, I, R, and D. Assume that the transition between the compartments is similar to the SEIR model with contact, latent, and infectious rates, β, α, and γ. However, only a fraction, $v_r = 1 - v_d$, recovers from the infection, and a fraction, v_d, dies. Characterize the SEIRD model through a system of five ordinary differential equations.

4.18 Classical SEIRD model. For the SEIRD model in Figure 4.8, rationalize how the initial growth rate G, the basic reproduction number R_0, and the final sizes of the susceptible and recovered groups, S_∞ and R_∞, differ from the classical SEIR model.

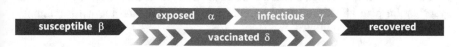

Fig. 4.9 Classical SEIR model with vaccination. The classical SEIR model contains four compartments for the susceptible, exposed, infectious, and recovered populations, S, E, I, and R. The transition rates between the compartments, the contact, latent, and infectious rates, β, α, γ are inverses of the contact, latent, and infectious periods, $B = 1/\beta, A = 1/\alpha, C = 1/\gamma$. There is also a direct path from susceptible S to recovered R via vaccination at a vaccination rate δ and vaccination period $D = 1/\delta$.

4.19 Classical SEIR model with vaccination. Derive the governing equations for the SEIR model with vaccination in Figure 4.9 with four compartments for the susceptible, exposed, infectious, and recovered populations, S, E, I, and R. Assume that the disease path between the S, E, I, and R compartments is similar to the SEIR model with contact, latent, and infectious rates, β, α, and γ. However, it is also possible to recover through a vaccination path between S and R with a vaccination rate δ. Characterize the SEIR model with vaccination through a system of four ordinary differential equations.

4.20 Classical SEIR model with vaccination. The SEIR model with vaccination in Figure 4.9 assumes that individuals from the susceptible group S are immediately immune upon receiving the first dose of their vaccine and transition directly into the recovered group R. From what you know about COVID-19 vaccines, identify and discuss at least three limitations of this model.

4.21 Classical SEIVR model. Derive the governing equations for the SEIVR model in Figure 4.10 with five compartments for the susceptible, exposed, infectious, vaccinated, and recovered populations, S, E, I, V, and R. Assume that the disease path between the S, E, I, and R compartments is similar to the SEIR model with contact, latent, and infectious rates, β, α, and γ. However, it is also possible to recover through a vaccination path between S, V, and R with vaccination and immunization rates δ and ζ. Characterize the SEIVR model through a system of five ordinary differential equations.

Fig. 4.10 Classical SEIVR model. The classical SEIVR model contains five compartments for the susceptible, exposed, infectious, vaccinated, and recovered populations, S, E, I, V, and R. The transition rates of the disease path, the contact, latent, and infectious rates, β, α, γ, are inverses of the contact, latent, and infectious periods, $B = 1/\beta$, $A = 1/\alpha$, $C = 1/\gamma$. The transition rates of the vaccination path, the vaccination and immunization rates, δ and ζ, are inverses of the vaccination and immunization periods, $D = 1/\delta$ and $Z = 1/\zeta$.

4.22 Classical SEIVR model. The SEIVR model in Figure 4.10 assumes two paths to immunity, either through disease or through vaccination. How does the additional vaccination path affect the basic reproduction number R_0 and the herd immunity threshold H?

References

1. Anderson RM, May RM (1982) Directly transmitted infectious diseases: control by vaccination. Science 215:1053-1060.
2. Anderson RM, May RM (1991) Infectious Diseases of Humans. Oxford University Press, Oxford.
3. Aron JL, Schwartz IB (1984) Seasonality and period-doubling bifurcations in an epidemic model. Journal of Theoretical Biology 110:665-679.
4. Bjornstad ON, Shea K, Krzywinski M, Altman N (2020) The SEIRS model for infectious disease dynamics. Nature Methods 17:557–558.
5. Brauer F, van den Dreissche P, Wu J (2008) Mathematical Epidemiology. Springer-Verlag Berlin Heidelberg.
6. Brauer F, Castillo-Chavez C, Feng Z (2019) Mathematical Models in Epidemiology. Springer-Verlag New York.
7. European Centre for Disease Prevention and Control. Situation update worldwide. https://www.ecdc.europa.eu/en/geographical-distribution-2019-ncov-cases accessed: June 1, 2021.
8. Evans AS (1976) Viral Infections of Humans. Epidemiology and Control. Plenum Medical Book Company, New York and London.
9. Hethcote HW (2000) The mathematics of infectious diseases. SIAM Review 42:599-653.
10. Lauer SA, Grantz KH, Bi Q, Jones FK, Zheng Q, Meredith HR, Azman AS, Reich NG, Lessler J (2020) The incubation period of coronavirus disease 2019 (COVID-19) from publicly reported confirmed cases: estimation and application. Annals of Internal Medicine 172:577-582.
11. Li MY, Muldownew JS (1995) Global stability for the SEIR model in epidemiology. Mathematical Biosciences 125:155-164.
12. Li MY, Graef JR, Wang L, Karsai J (1999) Global dynamics of a SEIR model with varying total population size. Mathematical Biosciences 160:191-213.
13. Li Q, Guan X, Wu P, Wang X, ... Feng Z (2020) Early transmission dynamics in Wuhan, China, of novel coronavirus-infected pneumonia. New England Journal of Medicine 382:1199-1207.
14. Linka K, Peirlinck M, Kuhl E (2020) The reproduction number of COVID-19 and its correlation with public heath interventions. Computational Mechanics 66:1035-1050.
15. Peirlinck M, Linka K, Sahli Costabal F, Kuhl E (2020) Outbreak dynamics of COVID-19 in China and the United States. Biomechanics and Modeling in Mechanobiology 19:2179-2193.

16. Sanche S, Lin Y, Xu C, Romero-Severson E, Hengartner N, Ke R (2020) High contagiousness and rapid spread of severe acute respiratory syndrome coronavirus 2. Emerging Infectious Diseases 2020;26(7):1470-1477.
17. University of North Carolina Chapel Hill (2020) UNC-Chapel Hill CV-19 Dashboard. https://carolinatogether.unc.edu/dashboard/ accessed: June 1, 2021.

Part II
Computational epidemiology

Chapter 5
Introduction to computational epidemiology

Abstract Most compartment models in epidemiology characterize the time evolution of their populations through a set of coupled nonlinear ordinary differential equations. Except for the SIS model, these equations have no analytical solution and we generally solve them numerically. Here we introduce the basic concepts of numerical methods for first order differential equations and illustrate explicit and implicit time integration schemes to solve them. To demonstrate the features of different classes of numerical methods, we derive and compare the explicit forward Euler, implicit backward Euler, and midpoint methods for the SIS model. We calculate their errors compared to the analytical solution and discuss concepts of convergence and accuracy. The learning objectives of this chapter on computational epidemiology are to

- discretize a nonlinear ordinary differential equation in time
- solve the discrete equation using the explicit forward Euler method
- derive the discrete residual equation using the implicit backward Euler method
- linearize and solve the residual equation using the Newton method
- derive the residual equation, linearize and solve it using the midpoint method
- understand the difference between explicit and implicit time integration schemes
- explain the concepts of error, convergence, and order of a numerical method
- identify discretization errors and rationalize time step sizes to minimize them

By the end of the chapter, you will be able to discretize, linearize, and solve nonlinear ordinary differential equations and analyze, simulate, and predict the outbreak dynamics of simple infectious diseases that do not provide immunity to reinfection.

5.1 Numerical methods for ordinary differential equations

Ordinary differential equations are abundant in many scientific disciplines including physics, chemistry, biology, and economics [1]. Ordinary differential equations contain differentials with respect to only one variable [8]. For most scientific problems, this variable is the time t [12]. Most ordinary differential equations in physics,

E. Kuhl, *Computational Epidemiology*, https://doi.org/10.1007/978-3-030-82890-5_5

chemistry, and biology are first-order equations in time of the general form,

$$\dot{c} = \frac{dc}{dt} = f(c(t)) \tag{5.1}$$

where c is the state of the system with initial condition $c(t_0) = c_0$, $\dot{c}$ is the change of the state in time, and $f(c(t))$ is a function that is often referred to as source or reaction [26]. The notion *first order* implies that only the first derivatives of c are present in the equation [11]. In the first part of this book, we have seen that only very few ordinary differential equations have an explicit analytical solution. In fact, of all compartment models, we can only solve the SIS model from Chapter 2 analytically [2]. All other compartment models require numerical methods [22] to approximate the solution of their ordinary differential equations computationally [3].

To solve equation (5.1) in time, we divide the time interval $\mathcal{T}$ into n_{step} discrete time steps, $\mathcal{T} = \bigcup_{n=1}^{n_{step}} [t_n, t_{n+1}]$, where the subscripts $(\circ)_n$ and $(\circ)_{n+1}$ are associated with the beginning and the end of the current time step. The objective of a numerical method is to advance the system in time, from the state c_n at the beginning of the time step t_n to the state c_{n+1} end of the time step t_{n+1}. In computational epidemiology, this translates into predicting how many people will be infectious tomorrow from the known susceptible and infectious people today.

Single-step methods vs. multi-step methods. Numerical methods for solving first-order differential equations differ by how they approximate the time derivative dc/dt and by where they evaluate the function $f(c(t))$. We can classify numerical methods into two categories: *single-step methods* and *multi-step methods* [22]. Single-step methods approximate the time derivative dc/dt and the function $f(c(t))$ only from knowledge of the previous step t_n, whereas multi-step methods use information from the previous s steps $t_n, t_{n-1}, t_{n-2}, \ldots t_{n-s+1}$ [15]. Multi-step methods are generally more efficient than single-step methods because they keep and use information from previous steps rather than discarding it; but they are also more complex to understand and require more storage of previous information.

Single-step methods are widely used because of their conceptual simplicity. They are sufficient for most practical purposes and we will only use single-step methods here. To approximate the time derivative dc/dt, most single-step methods use the *finite difference method* ,

$$\dot{c} = \frac{dc}{dt} = \frac{c_{n+1} - c_n}{\Delta t} \quad \text{with} \quad \Delta t = t_{n+1} - t_n > 0, \tag{5.2}$$

where Δt denotes time step size. To approximate the function $f(c(t))$, single-step methods typically use *Runge–Kutta methods*, a family of methods that evaluate the function $f(c(t))$ not only at the beginning, t_n, or end, t_{n+1}, of the interval of interest, but also at several points within the interval $[t_n, t_{n+1}]$ [22]. For example, the fourth-order Runge-Kutta method that approximates $f(c(t))$ through a weighted average of slopes at the beginning, midpoint, and end of the interval is especially popular and widely used.

Fig. 5.1 Leonhard Euler was born on April 15, 1707
in Basel, Switzerland. He was one of the most eminent
mathematicians of the 18th century. Amongst his many
contributions, one of his greatest successes was solving real-
world problems analytically. His monograph *Institutionum
calculi integralis* advanced the numerical approximation
of integrals and introduced what is now broadly known as
Euler's method. Euler's method is the most widely used
explicit method for the numerical integration of ordinary
differential equations. Leonhard Euler died on September
18, 1783 in Saint Petersburg, Russia.

Explicit methods vs. implicit methods. Numerical methods for solving first-order
differential equations differ by what information goes into evaluating the function
$f(c(t))$. We can classify numerical methods into two categories: *explicit methods*
and *implicit methods* [8]. Explicit methods approximate the function $f(c(t))$ only
from knowledge at the beginning of the time step t_n, whereas implicit methods
use information from both the beginning t_n and the end t_{n+1}. Implicit methods
are generally more stable and more efficient, but they are also more complex and
more expensive because we typically cannot solve them directly in a single solution
step [6].

The simplest explicit method is the *explicit forward Euler method*, or simply the
Euler method [22]. It evaluates the function $f(c_n)$ at the beginning t_n of the interval
using only known values c_n. Conveniently, this always results in a single explicit
equation for the new state $c(t_{n+1})$ that we can solve directly to advance the system
in time. The forward Euler method is a first-order method; its local error, the error
per time step, is proportional to the square of the time step size Δt^2 and its global
error, the error at a given time, is proportional to the time step size Δt. We discuss
the explicit forward Euler method and its application to epidemiology in detail in
Section 5.2 [26].

The simplest implicit method is the *implicit backward Euler method*. It evaluates
the function $f(c_{n+1})$ at the end t_{n+1} of the interval using unknown values c_{n+1}. This
implies that, in most cases, we can no longer solve the update equation explicitly.
Instead, we typically have to apply an iterative method, for example a fix-point
iteration or the Newton method, to advance the system to the new state $c(t_{n+1})$ [6].
The backward Euler method is unconditionally stable, but it remains a first-order
method; its local error, the error per time step, is proportional to the square of the
time step size Δt^2 and its global error, the error at a given time, is proportional to the
time step size Δt. We discuss the implicit backward Euler method and its application
to epidemiology in detail in Section 5.3.

The *implicit midpoint method* is a combination of both explicit and implicit
Euler methods [6]. It is the simplest method in the class of collocation methods, and,
applied to Hamiltonian dynamics, a symplectic integrator. It evaluates the function
$f(c_{n+1/2})$ at the midpoint $t_{n+1/2}$ of the interval using a combination of known and
unknown values $c_{n+1/2} = \frac{1}{2}[c_{n+1} + c_n]$. Similar to the implicit backward Euler
method, we can no longer solve the update equation explicitly, but have to apply

Fig. 5.2 Isaac Newton was born on January 4, 1643 in Lincolnshire, England. He is considered one of the most influential scientists of all time. After completing his bachelor's degree at Cambridge in 1665, the Great Plaque broke out in London and forced him to return to his family and study from home. Amongst his many seminal contributions, he discovered Newton's method, a root-finding algorithm that iteratively generates better approximations to the zeros of a real-valued function. Newton's method convergences quadratically; the accuracy of its approximation doubles with each step. Sir Isaac Newton died in his sleep on March 31, 1726 in Kensington, Great Britain.

Table 5.1 Classical SIS model. Convergence of error with decreasing time step size. Maximum error, err = $|I_n - I(t_n)|$, between the numerical solution I_n using a time integration with the explicit forward Euler, implicit backward Euler, and midpoint methods and the analytical solution $I(t_n)$. The order of each method is derived from the slopes in the log-log plot in Figure 5.3. Infectious period $C = 6$ days, basic reproduction number $R_0 = 4.0$, initial infectious population $I_0 = 0.0025$. ** method becomes unstable; *** method does not converge.

time step	explicit	implicit	midpoint
0.03125	0.0063	0.0063	<0.0001
0.06250	0.0126	0.0127	<0.0001
0.12500	0.0251	0.0255	0.0003
0.25000	0.0499	0.0512	0.0010
0.50000	0.0979	0.1027	0.0041
1.00000	0.1850	0.2070	0 0165
2.00000	0.3285	0.4093	0.0677
4.00000	**	***	0.3024
order	0.9576	1.0036	1.9995

an iterative method to advance the system to the new state $c(t_{n+1})$. The implicit midpoint method is a second-order method; its local error, the error per time step, is proportional to the time step size to the power of three, Δt^3, and its global error, the error at a given time, is proportional to the time step size squared, Δt^2. The midpoint method is computationally more intensive than the explicit forward Euler method and equally expensive as the implicit backward Euler method, but its error decreases faster than those of both Euler methods. We will derive the implicit midpoint rule and its application to epidemiology in the problem sets of this chapter.

The choice of the numerical method depends on the underlying problem and the nature of the ordinary differential equation: Explicit methods are faster to solve, but can require smaller time steps, especially when using stiff systems; implicit methods are more complex and require iterations to solve, but they are usually more stable and allow for larger time steps. In general, stiff ordinary differential equations often require implicit schemes, whereas non-stiff equations can be solved more efficiently with explicit schemes.

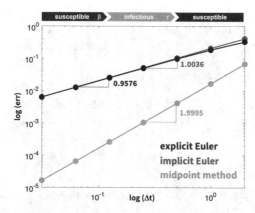

Fig. 5.3 Classical SIS model. Error vs time step size in log-log plot. Dots highlight the error, err $= |I_n - I(t_n)|$, between the numerical solution I_n using a time integration with the explicit forward Euler, implicit backward Euler, and midpoint methods for different time step sizes from Table 5.1 and the analytical solution $I(t_n)$. The slopes determine the order of convergence of each method. Infectious period $C = 6$ days, basic reproduction number $R_0 = 4.0$, initial infectious population $I_0 = 0.0025$.

Consistency, stability, and convergence are the three central concepts to analyze numerical methods for differential equations [5, 11, 8, 15]. An ideal method is consistent, stable, and convergent, with a high order of accuracy [22].

Consistency. A numerical method is consistent with the ordinary differential equation if the truncation error τ_{n+1} decreases as the time step size Δt decreases,

$$\tau_{n+1} = c(t_{n+1}) - c(t_n) - F(f, c_n, t_n, \Delta t)\Delta t \to 0 \quad \text{as} \quad \Delta t \to 0.$$

This definition shows that consistency depends critically on the increment function $F(f, c_n, t_n, \Delta t)$, which represents a numerical approximation of the slope between the beginning and end of the interval. The numerical methods we apply throughout this book, explicit forward Euler, implicit backward Euler, and midpoint, are consistent methods.

Stability. A numerical method is stable if small changes in the initial condition cause only small changes in the solution. Stability implies that the numerical method does not magnify errors that appear in the course of numerical solution process. The notion of stability is particularly important for stiff equations. When solving stiff equations, stability requirements, rather than accuracy requirements, determine the time step size.

Convergence. A numerical method is convergent if its maximum error,

$$\text{err} = |c_n - c(t_n)|,$$

the difference between the numerical solution c_n and the exact solution $c(t_n)$ within the time interval $\mathcal{T}$ decreases as the time step size Δt decreases,

$$\max_{\mathcal{T}} |c_n - c(t_n)| \to 0 \quad \text{as} \quad \Delta t \to 0.$$

The rate by which a method converges defines the global order of accuracy. A method is of order p, if its global error within the time interval $\mathcal{T}$ remains smaller than Δt^p,

$$\max_{\mathcal{T}} |c_n - c(t_n)| \le O(\Delta t^p) \quad \text{as} \quad \Delta t \to 0.$$

A method of higher order will converge more rapidly to the solution $c(t)$ than a method of lower order. The explicit forward Euler and implicit backward Euler methods are both first-order methods with $p = 1$, and dividing their time step by two reduces their error by a factor two. The implicit midpoint method is a second-order method with $p = 2$, and dividing its time step by two reduces its error by a factor four.

Table 5.1 illustrates the global error, $\max_{\mathcal{T}} |I_n - I(t_n)|$, between the numerical solution I_n for the forward Euler, backward Euler, and midpoint methods with different time step sizes Δt and the exact analytical solution for the infectious population $I(t)$ of the SIS model from equation (2.6) in Chapter 2. Comparing the global errors confirms that, for both Euler methods, dividing the time step by two reduces the error by a factor two, whereas for the midpoint method, dividing the time step by two reduces the error by a factor four. Figure 5.3 illustrates the error, err $= |I_n - I(t_n)|$, versus time step size Δt for all three methods in a log-log plot. The slopes of each graph define the order of each method and confirm that the explicit and implicit Euler methods with slopes of 0.9576 and 1.0036 are first-order methods, whereas the midpoint method with a slope of 1.9995 is a second-order method.

5.2 Explicit time integration

An explicit time integration scheme evaluates the state c of a system at a new time point t_{n+1} using only information from the previous time point t_n. In this section, we illustrate the concept of explicit time integration by means of the explicit forward Euler method, the most popular and most widely used explicit method to solve ordinary differential equations. We begin with the continuous ordinary differential equation for the state c with initial conditions c_0 at time t_0,

$$\dot{c} = \frac{dc}{dt} = f(c(t)) \quad \text{with} \quad c(t_0) = c_0. \tag{5.3}$$

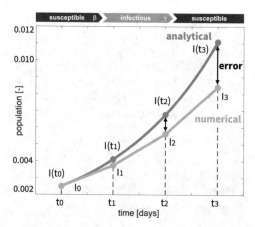

Fig. 5.4 Classical SIS model. Explicit forward Euler method vs analytical solution. The explicit time integration scheme delays infection and underestimates the outbreak. The difference between the numerical solution I_n and the analytical solution $I(t_n)$ introduces the error, err $= |I_n - I(t_n)|$. Infectious period $C = 6$ days, basic reproduction number $R_0 = 4.0$, initial infectious population $I_0 = 0.0025$, and time step size $\Delta t = 1.0$ day.

To solve this equation in time, we partition the time interval $\mathcal{T}$ into n_{step} discrete time steps, $\mathcal{T} = \bigcup_{n=1}^{n_{step}} [t_n, t_{n+1}]$, where the subscripts $(\circ)_n$ and $(\circ)_{n+1}$ are associated with the beginning and the end of the current time step. We assume that we know the state of the system c_n at the beginning of the time step t_n and advance it in time using an explicit time integration scheme. We approximate the first order time derivative using *finite differences* and adopt an *explicit forward Euler method* to evaluate the righthand side at the known time point t_n,

$$\dot{c} \approx \frac{c_{n+1} - c_n}{\Delta t} \quad \text{and} \quad \frac{c_{n+1} - c_n}{\Delta t} = f(c_n). \tag{5.4}$$

Here $\Delta t = t_{n+1} - t_n > 0$ denotes the time step size and n is the increment counter. This results in an *explicit* equation for the unknown c_{n+1} at the new time point t_{n+1},

$$c_{n+1} = c_n + f(c_n) \cdot \Delta t + O(\Delta t^2). \tag{5.5}$$

Starting with the initial condition $c(t_0) = c_0$, we incrementally advance the system in time by solving this explicit equation for each time step n_{step}. The explicit Euler method is a *first-order method* since its approximation of the time derivative $\dot{c} \approx [c_{n+1} - c_n]/\Delta t$ is first order in Δt. It has a local error per time step of order Δt^2 and a global error of order Δt.

Example. Explicit forward Euler method for the SIS model. To illustrate the explicit forward Euler method, we begin with the explicit SIS model from Chapter 2,

$$\dot{S} = - \beta SI + \gamma I$$
$$\dot{I} = + \beta SI - \gamma I.$$

This version of the SIS model neglects all births and deaths, such that $\dot{S} + \dot{I} \doteq 0$ and $S + I = \text{const.} = 1$. This implies that we can replace the susceptible population, $S = 1 - I$, and reduce the SIS model to a single non-linear ordinary differential for the infectious population,

$$\dot{I} = \beta [1 - I] I - \gamma I.$$

We replace the contact and infectious rates, $\beta = R_0/C$ and $\gamma = 1/C$, and express the evolution of the infectious population in terms of the infectious period C and basic reproduction number R_0,

$$\dot{I} = -\frac{R_0}{C} I^2 + \frac{R_0 - 1}{C} I.$$

To solve this equation in increments $\Delta t = t_{n+1} - t_n$, where n is the increment counter, we approximate the first order time derivative using finite differences and adopt an explicit forward Euler method to evaluate the righthand side at the known time point t_n,

$$\dot{I} = \frac{I_{n+1} - I_n}{\Delta t} = f(I_n) \quad \text{with} \quad f(I_n) = -\frac{R_0}{C} I_n^2 + \frac{R_0 - 1}{C} I_n.$$

This results in an explicit equation for the unknown infectious population I_{n+1} at the new time point t_{n+1},

$$I_{n+1} = -\frac{R_0}{C} \Delta t \, I_n^2 + \left[1 + \frac{R_0 - 1}{C} \Delta t \right] I_n$$

We start with the initial condition $I(t_0) = I_0$ and advance the system in time by solving this explicit equation in increments Δt. Since disease data are typically reported on a daily basis, solving the equation in increments of one day, $\Delta t = 1$ day, seems like a natural selection for the time step size.

Figure 5.4 contrasts the numerical solution I_n and the analytical solution $I(t_n)$ from equation (2.6) in Chapter (2),

$$I(t) = \frac{[1 - 1/R_0] I_0}{I_0 + [1 - 1/R_0 - I_0] \exp([1 - R_0] t/C)},$$

for the first three time steps of the explicit backward Euler method with an infectious period $C = 6$ days, a basic reproduction number $R_0 = 4.0$, an initial infectious population $I_0 = 0.0025$, and a time step size $\Delta t = 1.0$ day [17]. Compared to the analytical solution, the explicit time integration scheme delays infection and underestimates the outbreak.

5.3 Implicit time integration

An implicit time integration scheme evaluates the state c of a system at a new time point t_{n+1} using information from the current time point t_n and from the new time point t_{n+1}. In this section, we illustrate the concept of implicit time integration by means of the implicit backward Euler method, a popular explicit method to solve ordinary differential equations. We begin with the continuous ordinary differential equation for the state c with initial condition c_0 at time t_0,

$$\dot{c} = \frac{dc}{dt} = f(c(t)) \quad \text{with} \quad c(t_0) = c_0. \tag{5.6}$$

To solve this equation in time, we partition the time interval $\mathcal{T}$ into n_{step} discrete time steps, $\mathcal{T} = \bigcup_{n=1}^{n_{step}} [t_n, t_{n+1}]$, where the subscripts $(\circ)_n$ and $(\circ)_{n+1}$ are associated with the beginning and the end of the current time step. We assume that we know the state of the system c_n at the beginning of the time step t_n and advance it in time using an implicit time integration scheme. We approximate the first order time derivative using *finite differences* and adopt an *implicit backward Euler method* to evaluate the righthand side at the new unknown time point t_{n+1},

$$\dot{c} \approx \frac{c_{n+1} - c_n}{\Delta t} \quad \text{and} \quad \frac{c_{n+1} - c_n}{\Delta t} = f(c_{n+1}), \tag{5.7}$$

Here, $\Delta t = t_{n+1} - t_n > 0$ is the time step size and n is the increment counter. This results in an *implicit* equation for the unknown c_{n+1} at the new time point t_{n+1}. We rephrase this equation in its residual form,

$$\mathsf{R} = [c_{n+1} - c_n]/\Delta t - f(c_{n+1}) + O(\Delta t^2) \doteq 0, \tag{5.8}$$

where R is called the *residual*. Similar to the explicit Euler method in Section 5.2, the implicit Euler method is a *first-order method* since its approximation of the time derivative $\dot{c} \approx [c_{n+1} - c_n]/\Delta t$ is first order in Δt. It has a local error per time step of order Δt^2 and a global error of order Δt. Unlike the explicit Euler method, in most cases, the implicit Euler method does not have a direct explicit update equation and we have to solve it iteratively, for example, using Newton's method [6]. To solve the residual equation (5.8), we adopt Newton's method and linearize the residual R around the solution c using a first order Taylor expansion,

$$\mathsf{R}^{k+1} = \mathsf{R}^k + \frac{d\mathsf{R}}{dc} \cdot dc + O(dc^2) \doteq 0 \quad \text{with} \quad \mathsf{K} = \frac{d\mathsf{R}}{dc}, \tag{5.9}$$

where K is the tangent of the Newton method. We solve this equation to calculate the *incremental, iterative update* dc of the unknown c^{k+1},

$$c^{k+1} = c^k + dc \quad \text{with} \quad dc = \mathsf{R}^k/\mathsf{K}, \tag{5.10}$$

and iterate until the norm of the residual, $||\mathsf{R}|| < \text{tol}$, is smaller than a user defined tolerance tol. Here k is the iteration counter. An incremental iterative solution within

Table 5.2 Convergence of the Newton method. The Newton method is an iterative solution method to find the roots of a function, in this case of the residual R_n^k. During the first n = 4 time steps $t_1, ..., t_4$ the implicit backward Euler SIS model converges within four iterations $k = 1, .., 4$, meaning the residual drops below a user defined tolerance, R < tol = 1.0e-12. In the proximity of the solution, the Newton method converges quadratically, meaning the order of the residual is roughly divided by two in each iteration.

iteration k	time step t1	time step t2	time step t3	time step t4
1	1.7619e-03	3.4890e-03	6.8432e-03	1.3178e-02
2	5.7761e-06	2.2358e-05	8.3860e-05	2.9645e-04
3	6.1273e-11	8.9500e-10	1.1998e-08	1.3738e-07
4	3.5237e-17	5.5960e-17	2.6061e-16	2.9520e-14

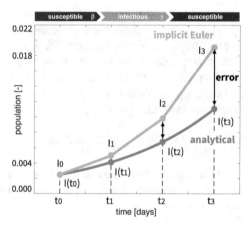

Fig. 5.5 Classical SIS model. Implicit backward Euler method vs analytical solution. The implicit time integration scheme accelerates infection and overestimates the outbreak. The difference between the numerical solution I_n and the analytical solution $I(t_n)$ introduces the error, err = $| I_n - I(t_n) |$. Infectious period C = 6 days, basic reproduction number R_0 = 4.0, initial infectious population I_0 = 0.0025, and time step size Δt = 1.0 day.

an implicit time integration scheme results in two nested loops, an outer loop for all time steps n and an inner loop one for all iterations k.

Example. Implicit backward Euler method for the SIS model. To illustrate the implicit backward Euler method, we derive the explicit SIS model from Chapter 2,

$$\dot{S} = - \beta SI + \gamma I$$
$$\dot{I} = + \beta SI - \gamma I.$$

This version of the SIS model neglects all births and deaths, such that $\dot{S} + \dot{I} \doteq 0$ and $S + I$ = const. = 1. By replacing the susceptible population, $S = 1 - I$, in the equation for the infectious population we can rephrase the SIS model in terms of a single non-linear ordinary differential equation,

$$\dot{I} = \beta[1 - I]I - \gamma I.$$

For convenience, we replace $\beta = R_0/C$ and $\gamma = 1/C$, and obtain the following equation,

$$\dot{I} = -\frac{R_0}{C} I^2 + \frac{R_0 - 1}{C} I \quad \text{with} \quad I(t_0) = I_0. \tag{5.11}$$

To solve this equation in increments $\Delta t = t_{n+1} - t_n$, we adopt a finite difference approximation with an implicit backward Euler method,

$$\dot{I} = \frac{I_{n+1} - I_n}{\Delta t} = f(I_{n+1}) \quad \text{with} \quad f(I_{n+1}) = -\frac{R_0}{C} I_{n+1}^2 + \frac{R_0 - 1}{C} I_{n+1},$$

which results in an implicit equation for the unknown infectious population I_{n+1} at the new time point t_{n+1}. We rephrase this equation in its residual form,

$$R = \frac{1}{\Delta t}[I_{n+1} - I_n] + \frac{R_0}{C} I_{n+1}^2 - \frac{R_0 - 1}{C} I_{n+1} \doteq 0.$$

Since there is no explicit solution for this equation, we solve it iteratively using Newton's method. We linearize the residual R around the solution I_{n+1},

$$R^{k+1} = R^k + \frac{dR}{dI_{n+1}} \cdot dI_{n+1} \doteq 0,$$

where K is the tangent of the Newton method,

$$K = \frac{dR}{dI_{n+1}} = \frac{1}{\Delta t} + 2\frac{R_0}{C} I_{n+1} - \frac{R_0 - 1}{C}.$$

We solve the linearized residual equation to calculate the incremental iterative update dI_{n+1} of the unknown I_{n+1}^{k+1},

$$I_{n+1}^{k+1} = I_{n+1}^k + dI_{n+1} \quad \text{with} \quad dI_{n+1} = R^k/K,$$

where k is the iteration counter. We start with the initial condition, $I(t_0) = I_0$, and advance the system in time by iteratively solving this implicit equation for all n time increments Δt, for example, using $\Delta t = 1$ day.

Figure 5.5 contrasts the numerical solution I_n and the analytical solution $I(t_n)$ from equation (2.6) in Chapter (2),

$$I(t) = \frac{[1 - 1/R_0] I_0}{I_0 + [1 - 1/R_0 - I_0] \exp([1 - R_0] t/C)},$$

for the first three time steps of the implicit backward Euler method with an infectious period of $C = 6$ days, a basic reproduction number of $R_0 = 4.0$, an initial infectious population $I_0 = 0.0025$, and a time step size $\Delta t = 1.0$ day. We observe that the implicit time integration scheme accelerates infection and overestimates the outbreak. Table 5.2 illustrates the convergence of the Newton

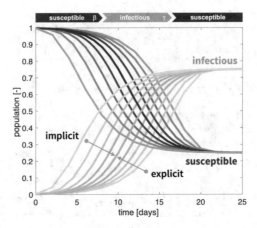

Fig. 5.6 Classical SIS model. Explicit and implicit time integration for varying time step sizes
Δt. The explicit time integration scheme delays infection and underestimates the outbreak. The
implicit time integration scheme accelerates infection and overestimates the outbreak. Infectious
period $C = 6$ days, basic reproduction number $R_0 = 4.0$, initial infectious population $I_0 = 0.0025$,
and time step size $\Delta t = 2.0, 1.5, 1.0, 0.5, 0.1$ days, from light to dark. The darkest curves represent
the analytical solution.

method for the first four time steps. The convergence table confirms that the
residual drops below a user defined tolerance, $R < \text{tol} = 1.0e-12$, within four
iterations k. The order of the residual is divided by two in each iteration, which
confirms that the Newton method converges quadratically.

5.4 Comparison of explicit and implicit time integration

We have already discussed that, in the proximity of the solution, both explicit and
implicit time integration schemes converge to the exact analytical solution as the time
step size decreases. Table 5.1 and Figure 5.3 illustrate this convergence by means
of the decreasing error, err $= |I_n - I(t_n)|$, between the numerical solution I_n and
the exact analytical solution $I(t)$ of the infectious population of the SIS model, for
decreasing time step sizes Δt. The slopes of 0.9576 and 1.0036 in the log-log plot in
Figure 5.3 confirm that the explicit and implicit Euler methods are both first-order
methods.

Figure 5.6 provides a side-by-side comparison of the explicit and implicit time
integration schemes across the entire time window, from the beginning of the disease
to endemic equilibrium, for varying time step sizes. Its $\Delta t = 1.0$ days curves contain
the first three time steps of Figures 5.4 and 5.5. The curves confirm that the explicit
time integration scheme delays infection and underestimates the outbreak while
the implicit time integration scheme accelerates infection and overestimates the

outbreak. For decreasing time step sizes, the infectious populations $I(t_n)$ of both schemes converge against the darkest orange curve of the analytical solution $I(t_n)$: Convergence is from below for the explicit scheme and from above for them implicit scheme.

In contrast to the explicit scheme, the implicit scheme requires an additional Newton iteration for each time increment [6]. Since the Newton iteration converges quadratically close to the solution as we have seen in Table 5.2, this implies that we need to solve four to five additional iteration steps k for each time step n. For the SIS model in this Chapter, each Newton iteration updates the unknowns as $I_{n+1}^{k+1} = I_{n+1}^k + R^k/K$, which does not add notably to the overall computational cost. For the SEIR model in Chapter 7, each Newton iteration involves inverting the 4×4 tangent matrix K to calculate the updates in all four compartments, which adds slightly more to the overall cost. For the network epidemiology models in Chapter 11, each Newton iteration involves inverting a tangent matrix K of the number of compartments times the number of network nodes [16], which could potentially add significant computational cost.

In general, the advantage of an implicit time integration is that it allows for larger time step sizes, which could potentially compensate for the cost of its additional Newton iterations. While this is clearly true for many engineering applications, for the epidemiology applications considered in this book, we have empirically observed that both schemes tend to fail approximately at the same time step size. The explicit method fails by becoming unstable, which is easy to overlook, especially within a parameter identification or machine learning setting. The implicit method fails by a loss of convergence, which is easy to detect by monitoring the residual or the number of Newton iterations. More importantly, we can use the convergence of the Newton method for an ad hoc time adaptation and simply increase or decrease the time step size if the Newton method converges in fewer or more iterations than a user defined target number. Ultimately, the choice of time integration depends on many different factors and there is no single one scheme that is best suited for all purposes. It is important to keep in mind the limitations and advantages of the different methods and be aware of potential numerical errors, especially those that are not immediately visible to the simulation itself.

Problems

5.1 Errors of numerical methods. When we approximate the solution of an ordinary differential equation using numerical methods, we introduce two types or errors, rounding errors and truncation errors. Rounding errors result from floating-point arithmetics associated with the precision of our computer and we cannot control them. Truncation errors are errors of the numerical method that we can control. What are the origins of truncation errors for the explicit forward Euler method?

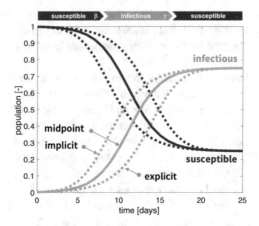

Fig. 5.7 Classical SIS model. Midpoint method vs explicit and implicit time integration. The simulated susceptible and infectious populations of the midpoint method lie in between the simulations of the explicit forward Euler and implicit backward Euler methods. Infectious period $C = 6$ days, basic reproduction number $R_0 = 4.0$, initial infectious population $I_0 = 0.0025$, and time step size $\Delta t = 1.0$ day.

- the approximation of dt/dc
- the approximation of c_{n+1}

- the approximation of c_n
- all of the above

Justify your answer.

5.2 Explicit forward Euler method for the SIS model. We have parameterized the update equation of the explicit forward Euler method, $I_{n+1} = -R_0 \Delta t \, I_n^2/C + [\, 1 + [\, R_0 - 1\,] \Delta t \,] I_n/C$, in terms of the infectious period C and the basic reproduction number R_0. Derive the update equation for the infectious population I_{n+1} from the continuous equation $\dot{I} = \beta[\, 1 - I\,] I - \gamma I$, but now paramterized in terms of the contact and infectious rates β and γ.

5.3 Explicit forward Euler method for the SIS model. With the update equation, $I_{n+1} = I_n + [-\beta I_n^2 + [\beta - \gamma] I_n] \Delta t$, that you have derived in the previous problems, implement the explicit forward Euler method. Simulate the infectious population of the SIS model with an infectious period of $C = 6$ days, a basic reproduction number of $R_0 = 4.0$, an initial infectious population of $I_0 = 0.0025$, and a time step size of $\Delta t = 1.0$ day. Compare your results against Figure 5.7.

5.4 Explicit forward Euler method and time step size. We have seen that the result of the numerical time integration depends on the time step size Δt. Solve the non-linear ordinary differential equation of the SIS model, $\dot{I} = \beta[\, 1 - I\,] I - \gamma I$, using an explicit forward Euler method. Simulate an outbreak with an infectious period of $C = 7$ days, an initial infectious population of $I_0 = 0.001$, and a varying reproduction number of $R_0 = 2, 3, 4, 5, 10, 15$. Gradually increase the time step size $\Delta t = 0.01, 0.1, 1, 2, 5, 10$ days. What happens? Make informed recommendations for a reasonable time step size Δt.

5.5 Explicit forward Euler method and convergence. Show convergence of the explicit forward Euler method. Solve the non-linear ordinary differential equation of the SIS model, $\dot{I} = \beta[1 - I]I - \gamma I$, using the explicit forward Euler method. Simulate an outbreak with an infectious period of $C = 7$ days, a reproduction number of $R_0 = 3$, and an initial infectious population of $I_0 = 0.001$. Calculate the maximum error, $\max_{\mathcal{T}} |I_n - I(t(n))|$, within your time interval $\mathcal{T}$ for varying time step sizes $\Delta t = 0.0315, 0.0625, 0.125.0.25, 0.5, 1.0, 2.0$. Plot your error vs time step size in a log-log plot. Determine the order p of the explicit forward Euler method from the slope in the graph. Compare your results qualitatively to Table 5.1 and Figure 5.3.

5.6 Implicit backward Euler method for the SIS model. We have parameterized the residual of the implicit backward Euler method, $R = [I_{n+1} - I_n]/\Delta t + R_0 I_{n+1}^2/C - [R_0 - 1]I_{n+1}/C \doteq 0$, in terms of the infectious period C and the basic reproduction number R_0. Derive the residual of the implicit backward Euler method R from the continuous equation $\dot{I} = \beta[1 - I]I - \gamma I$, but now in terms of the contact and infectious rates β and γ.

5.7 Implicit backward Euler method for the SIS model. We have parameterized the tangent of the Newton method of the implicit backward Euler method, $K = 1/\Delta t + 2 R_0 I_{n+1}^2/C - [R_0 - 1]/C$, in terms of the infectious period C and the basic reproduction number R_0. Show that the tangent of the Newton method of the implicit backward Euler method is $K = dR/dI_{n+1}$ is $K = 1/\Delta t + 2\beta I_{n+1}^2 - [\beta - \gamma]$, when parameterized in the contact and infectious rates β and γ.

5.8 Implicit backward Euler method for the SIS model. With the residual, $R = [I_{n+1} - I_n]/\Delta t + \beta I_{n+1}^2 - [\beta - \gamma] I_{n+1}$, and the tangent, $K = 1/\Delta t + 2\beta I_{n+1}^2 - [\beta - \gamma]$, that you have derived in the previous problems, implement the implicit backward Euler method. Simulate the infectious population of the SIS model with an infectious period of $C = 6$ days, a basic reproduction number of $R_0 = 4.0$, an initial infectious population of $I_0 = 0.0025$, and a time step size of $\Delta t = 1.0$ day. Compare your results against Figure 5.7.

5.9 Implicit backward Euler method for the SIS model. For the implicit backward Euler method that you have implemented in the previous problem, evaluate the residual R_n^k for the first four time steps, $n = 1, .., 4$, using a time step size of $\Delta t = 1$ day. Select a tolerance of $\text{tol} = 1.0e{-}12$. How many iterations k does the method take in each time step to converge to $||R|| < \text{tol} = 1.0e{-}12$? Compare your results qualitatively to Table 5.2. Does your Newton method converge quadratically?

5.10 Implicit backward Euler method and convergence. Show convergence of the implicit backward Euler method. Solve the non-linear ordinary differential equation of the SIS model, $\dot{I} = \beta[1 - I]I - \gamma I$, using the implicit backward Euler method. Simulate an outbreak with an infectious period of $C = 7$ days, a reproduction number of $R_0 = 3$, and an initial infectious population of $I_0 = 0.001$. Calculate the maximum error, $\max_{\mathcal{T}} |I_n - I(t(n))|$, within your time interval $\mathcal{T}$ for varying time step sizes $\Delta t = 0.0315, 0.0625, 0.125.0.25, 0.5, 1.0, 2.0$. Plot your error vs time step size in a log-log plot. Determine the order p of the implicit backward Euler method from the slope in the graph. Compare your results qualitatively to Table 5.1 and Figure 5.3.

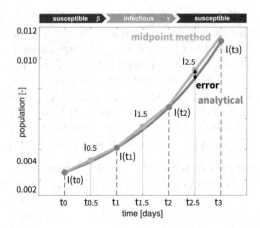

Fig. 5.8 Classical SIS model. Midpoint method vs analytical solution. The midpoint method combines the explicit forward Euler and implicit backward Euler methods. It evaluates the righthand side at the midpoint $t_{n+1/2}$. The difference between the numerical solution I_n and the analytical solution $I(t_n)$ introduces the error, err $= |I_n - I(t_n)|$. Infectious period $C = 6$ days, basic reproduction number $R_0 = 4.0$, initial infectious population $I_0 = 0.0025$, and time step size $\Delta t = 1.0$ day.

5.11 Midpoint method for the SIS model. We have seen that the explicit forward Euler method underestimates the outbreak and the implicit backward Euler method overestimates the outbreak. Figure 5.8 illustrates a method that combines both, forward Euler and backward Euler, the midpoint method. For the SIS model, $\dot{I} = \beta[1 - I]I - \gamma I$, adopt the midpoint method to evaluate the righthand side $f(I_{n+1/2})$ at the midpoint, $t_{n+1/2} = \frac{1}{2}[t_{n+1} + t_n]$, by replacing all I-terms on the righthand side by $I_{n+1/2} = \frac{1}{2}[I_{n+1} + I_n]$.

5.12 Midpoint method for the SIS model. For the SIS model, $\dot{I} = \beta[1 - I]I - \gamma I$, derive the residual R for the midpoint method. Approximate the time derivative $\dot{I}$ using finite differences and evaluate the righthand side at the midpoint with $I_{n+1/2} = \frac{1}{2}[I_{n+1} + I_n]$. Show that the residual of the midpoint method is R $= [I_{n+1} - I_n]/\Delta t + \frac{1}{4}\beta[I_{n+1} + I_n]^2 - \frac{1}{2}[\beta - \gamma][I_{n+1} + I_n]$.

5.13 Midpoint method for the SIS model. For the SIS model, $\dot{I} = \beta[1 - I]I - \gamma I$, derive the tangent K for the midpoint method. Approximate the time derivative $\dot{I}$ using finite differences and evaluate the righthand side at the midpoint with $I_{n+1/2} = \frac{1}{2}[I_{n+1} + I_n]$. Show that the tangent of the Newton method of the midpoint method K $= dR/dI_{n+1}$ is K $= 1/\Delta t + \frac{1}{2}\beta[I_{n+1} + I_n] - \frac{1}{2}[\beta - \gamma]$.

5.14 Midpoint method for the SIS model. With the residual, R $= [I_{n+1} - I_n]/\Delta t + \frac{1}{4}\beta[I_{n+1} + I_n]^2 - \frac{1}{2}[\beta - \gamma][I_{n+1} + I_n]$, and the tangent, K $= 1/\Delta t + \beta[I_{n+1} + I_n]/2 - [\beta - \gamma]/2$, that you have derived in the previous problems, implement the midpoint method. Simulate the infectious population of the SIS model with an infectious period of $C = 6$ days, a basic reproduction number of $R_0 = 4.0$, an initial infectious population of $I_0 = 0.0025$, and a time step size of $\Delta t = 1.0$ day. Compare your results against Figure 5.7.

5.15 Midpoint method for the SIS model. For the midpoint method that you have implemented in the previous problem, evaluate the residual R_n^k for the first five time steps, $n = 1, .., 4$, using a time step size of $\Delta t = 1$ day. Select a tolerance of tol $= 1.0$e-12. How many iterations k does the method take in each time step to converge to $||R|| <$ tol $= 1.0$e-12? Compare your results qualitatively to Table 5.2. Does your Newton method converge quadratically?

5.16 Midpoint method and convergence. Show convergence of the midpoint method. Solve the non-linear ordinary differential equation of the SIS model, $\dot{I} = \beta [1 - I] I - \gamma I$, using the midpoint method. Simulate an outbreak with an infectious period of $C = 7$ days, a reproduction number of $R_0 = 3$, and an initial infectious population of $I_0 = 0.001$. Calculate the maximum error, $\max_{\mathcal{T}} | I_n - I(t(n)) |$, within your time interval $\mathcal{T}$ for varying time step sizes $\Delta t = 0.0315, 0.0625, 0.125. 0.25, 0.5, 1.0, 2.0$. Plot your error vs time step size in a log-log plot. Determine the order p of the midpoint method from the slope in the graph. Compare your results qualitatively to Table 5.1 and Figure 5.3.

5.17 Time integration of the SIS model. Simulate the infectious population of the SIS model with an infectious period of $C = 6$ days, a basic reproduction number of $R_0 = 4.0$, an initial infectious population of $I_0 = 0.0025$, and a time step size of $\Delta t = 1.0$ day. Compare a time integration with the explicit forward Euler, implicit backward Euler, and midpoint methods. Comment on your observations. Compare your results against Figures 5.3 and 5.7. Which method would you recommend?

References

1. Alber M, Buganza Tepole A, Cannon W, De S, Dura-Bernal S, Garikipati K, Karniadakis G, Lytton WW, Perdikaris P, Petzold L, Kuhl E (2019) Integrating machine learning and multiscale modeling: Perspectives, challenges, and opportunities in the biological, biomedical, and behavioral sciences. npj Digital Medicine 2:115.
2. Anderson RM, May RM (1981) The population dynamics of microparasites and their invertebrate hosts. Philosophical Transactions of the Royal Society London B. 291:451–524.
3. Brauer F, Castillo-Chavez C, Feng Z (2019) Mathematical Models in Epidemiology. Springer-Verlag New York.
4. Deuflhard P (2011) Newton Methods for Nonlinear Problems. Springer-Verlag Berlin Heidelberg.
5. Eriksson K, Estep D, Hansbo P, Johnson C (1996) Computational Differential Equations. Cambridge University Press.
6. Goodwine B (2011) Engineering Differential Equations. Springer-Verlag New York.
7. Griffiths D, Higham DJ (2010) Numerical Methods for Ordinary Differential Equations. Springer-Verlag London.
8. Hermann M, Saravi M (2014) A First Course in Ordinary Differential Equations. Springer New Delhi Heidelberg New York.
9. Hundsdorfer W, Verwer J (2003) Numerical Solution of Time-Dependent Advection-Diffusion-Reaction Equations. Springer-Verlag Berlin Heidelberg.
10. Linka K, Peirlinck M, Sahli Costabal F, Kuhl E (2020) Outbreak dynamics of COVID-19 in Europe and the effect of travel restrictions. Computer Methods in Biomechanics and Biomedical Engineering 23: 710-717.

11. Linka K, Peirlinck M, Kuhl E (2020) The reproduction number of COVID-19 and its correlation with public heath interventions. Computational Mechanics 66:1035-1050.
12. Liu J, Shang, X (2020) Computational Epidemiology. Springer International Publishing.
13. Moin P (2000) Fundamentals of Engineering Numerical Analysis. Cambridge University Press.
14. Peirlinck M, Linka K, Sahli Costabal F, Bendavid E, Bhattacharya J, Ioannidis J, Kuhl E (2020) Visualizing the invisible: The effect of asymptomatic transmission on the outbreak dynamics of COVID-19. Computer Methods in Applied Mechanics and Engineering 372:113410.
15. Peng GCY, Alber M, Buganza Tepole A, Cannon W, De S, Dura-Bernal S, Garikipati K, Karniadakis G, Lytton WW, Perdikaris P, Petzold L, Kuhl E (2021) Multiscale modeling meets machine learning: What can we learn? Archive of Computational Methods in Engineering 28:1017-1037.

Chapter 6
The computational SIR model

Abstract The SIR model is popular compartment model with three populations, the susceptible, infectious, and recovered groups S, I, and R. It characterizes infectious diseases that provide immunity upon infection. Since the SIR model has no analytical solution for the time course of its populations, we discretize it in time using finite differences and adopt explicit and implicit time integration schemes to solve it. We compare the timeline of the SIR model to the analytical solution of the SIS model and show its sensitivity to the infectious period, reproduction number, and initial conditions. To illustrate the features of the SIR model, we simulate the early COVID-19 outbreak in Austria using reported case data. The learning objectives of this chapter on computational SIR modeling are to

- discretize the classical SIR model in time
- solve the SIR equations using the explicit forward Euler method
- derive the residual SIR equations using the implicit backward Euler method
- linearize and solve the residual SIR equations using the Newton method
- demonstrate how the infectious period, reproduction number, and initial conditions modulate outbreak dynamics
- simulate the first wave of the COVID-19 outbreak using reported case data
- discuss limitations of computational SIR modeling

By the end of the chapter, you will be able to computationally analyze, simulate, and predict the early outbreak dynamics of infectious diseases like COVID-19 using reported case data.

6.1 Explicit time integration of the SIR model

The SIR model consists of three populations, the susceptible, infectious, and recovered groups S, I, and R [2]. Figure 6.1 illustrates the three populations and the contact and infectious rates β and γ that define the transition between them [3]. The contact and infectious rates are inverses of the contact and infectious periods $B = 1/\beta$ and

| susceptible | β | infectious | γ | recovered |

Fig. 6.1 Classical SIR model. The classical SIR model contains three compartments for the susceptible, infectious, and recovered populations, S, I, and R. The transition rates between the compartments, the contact and infectious rates, β and γ, are inverses of the contact and infectious periods, $B = 1/\beta$ and $C = 1/\gamma$.

$C = 1/\gamma$, and their ratio defines the basic reproduction number, $R_0 = \beta/\gamma = C/B$. Unlike the SIS model [2], the SIR model does not have a straightforward explicit analytical solution [22]. To solve the set of coupled nonlinear ordinary differential equations of the SIR model [8],

$$
\begin{aligned}
\dot{S} &= -\ \beta\, SI \\
\dot{I} &= +\ \beta\, SI\ -\ \gamma\, I \\
\dot{R} &= \qquad\qquad +\ \gamma\, I,
\end{aligned}
\tag{6.1}
$$

we apply a finite difference approximation and replace the lefthand sides of the system of equations (6.1) by the discrete difference, $(\dot{\circ}) = [\,(\circ)_{n+1} - (\circ)_n\,]/\Delta t$, in terms of the unknown new populations, $(\circ)_{n+1}$, the known previous populations, $(\circ)_n$, and the discrete time step size, $\Delta t = t_{n+1} - t_n$, for all three populations, $(\circ) = S, I, R$,

$$
\dot{S} = \frac{S_{n+1} - S_n}{\Delta t} \qquad \dot{I} = \frac{I_{n+1} - I_n}{\Delta t} \qquad \dot{R} = \frac{R_{n+1} - R_n}{\Delta t}\,.
\tag{6.2}
$$

In the spirit of an explicit time integration, we adopt the classical explicit *forward Euler method* and evaluate the righthand sides of the system of equations (6.1) at the known previous time point t_n [22],

$$
\dot{S} = -\beta\, S_n I_n \qquad \dot{I} = +\beta\, S_n I_n - \gamma\, I_n \qquad \dot{R} = +\gamma\, I_n\,.
\tag{6.3}
$$

This explicit time integration results in the following discrete system of equations for the SIR model,

$$
\begin{aligned}
[\ S_{n+1} - S_n\ ]/\Delta t &= -\beta\, S_n I_n \\
[\ I_{n+1} - I_n\ ]/\Delta t &= +\beta\, S_n I_n - \gamma\, I_n \\
[\ R_{n+1} - R_n\ ]/\Delta t &= \qquad\qquad +\gamma\, I_n,
\end{aligned}
\tag{6.4}
$$

which we can solve directly for the unknown populations S_{n+1}, I_{n+1}, and R_{n+1} at the new time point t_{n+1},

$$
\begin{aligned}
S_{n+1} &= [\,1 - \beta\, I_n\, \Delta t\,]\, S_n \\
I_{n+1} &= +\ \beta\, I_n\, \Delta t\ \ S_n + [\,1 - \gamma\, \Delta t\,]\, I_n \\
R_{n+1} &= \qquad\qquad\quad +\qquad \gamma\, \Delta t\ \ I_n + R_n,
\end{aligned}
\tag{6.5}
$$

using the initial conditions $S_0 = 1 - I_0$, I_0, and $R_0 = 0$. Explicit time integration schemes are relatively straightforward, but can become oscillatory and numerically unstable, especially for stiff systems and large time steps Δt. In the next section, we suggest an alternative approach and discretize the SIR model in time using an implicit time integration scheme.

6.2 Implicit time integration of the SIR model

Similar to the explicit time integration in Section 6.1, we begin with the set of coupled nonlinear ordinary differential equations of the SIR model (3.1) from Chapter 3,

$$
\begin{aligned}
\dot{S} &= -\beta SI \\
\dot{I} &= +\beta SI - \gamma I \\
\dot{R} &= +\gamma I,
\end{aligned}
\tag{6.6}
$$

and apply a finite difference approximation to replace the lefthand sides of the system of equations (6.6), by the discrete differences, $(\dot{\circ}) = [(\circ)_{n+1} - (\circ)_n]/\Delta t$, in terms of the unknown new populations, $(\circ)_{n+1}$, the known previous populations, $(\circ)_n$, and the discrete time step size, $\Delta t = t_{n+1} - t_n$, for all three populations, $(\circ) = S, I, R$,

$$
\dot{S} = \frac{S_{n+1} - S_n}{\Delta t} \qquad \dot{I} = \frac{I_{n+1} - I_n}{\Delta t} \qquad \dot{R} = \frac{R_{n+1} - R_n}{\Delta t}.
\tag{6.7}
$$

However, now, we apply an implicit time integration and adopt the classical *backward Euler method* to evaluate the righthand sides of the system of equations (6.6) at the unknown current time point t_{n+1} [22],

$$
\dot{S} = -\beta S_{n+1} I_{n+1} \qquad \dot{I} = +\beta S_{n+1} I_{n+1} - \gamma I_{n+1} \qquad \dot{R} = +\gamma I_{n+1}.
\tag{6.8}
$$

This implicit time integration results in the following discrete system of equations for the SIR model,

$$
\begin{aligned}
[\ S_{n+1} - S_n\ ]/\Delta t &= -\beta S_{n+1} I_{n+1} \\
[\ I_{n+1} - I_n\ ]/\Delta t &= +\beta S_{n+1} I_{n+1} - \gamma I_{n+1} \\
[\ R_{n+1} - R_n\ ]/\Delta t &= +\gamma I_{n+1}.
\end{aligned}
\tag{6.9}
$$

To solve this system, we rephrase it in its residual form, $\mathbf{R} = [R_S, R_I, R_R] \doteq \mathbf{0}$, by bringing all terms to the righthand side,

$$
\begin{aligned}
R_S &= [\ S_{n+1} - S_n\ ]/\Delta t + \beta S_{n+1} I_{n+1} &&\doteq 0 \\
R_I &= [\ I_{n+1} - I_n\ ]/\Delta t - \beta S_{n+1} I_{n+1} + \gamma I_{n+1} &&\doteq 0 \\
R_R &= [\ R_{n+1} - R_n\ ]/\Delta t \qquad\qquad - \gamma I_{n+1} &&\doteq 0.
\end{aligned}
\tag{6.10}
$$

We apply the Newton method [6] and linearize the residual $\mathbf{R}$ with respect to the unknown populations $\mathbf{P}_{n+1} = [S_{n+1}, I_{n+1}, R_{n+1}]$, such that, $\mathbf{R}^{k+1} = \mathbf{R}^k + \mathbf{K} \cdot d\mathbf{P} \doteq \mathbf{0}$, with

$$
\begin{bmatrix} R_S^{k+1} \\ R_I^{k+1} \\ R_R^{k+1} \end{bmatrix} = \begin{bmatrix} R_S^k \\ R_I^k \\ R_R^k \end{bmatrix} + \begin{bmatrix} \frac{1}{\Delta t} + \beta\, I_{n+1} & \beta\, S_{n+1} & 0 \\ -\beta\, I_{n+1} & \frac{1}{\Delta t} - \beta\, S_{n+1} + \gamma & 0 \\ 0 & -\gamma & \frac{1}{\Delta t} \end{bmatrix} \cdot \begin{bmatrix} dS \\ dI \\ dR \end{bmatrix} \doteq \begin{bmatrix} 0 \\ 0 \\ 0 \end{bmatrix}, \quad (6.11)
$$

where the tangent matrix $\mathbf{K}$ of the Newton method contains the derivatives of the three residuals $\mathbf{R} = [\, R_S, R_I, R_R \,]$ with respect to the three populations $\mathbf{P}_{n+1} = [\, S_{n+1}, I_{n+1}, R_{n+1} \,]$,

$$
\mathbf{K} = \begin{bmatrix} \dfrac{dR_S}{dS_{n+1}} & \dfrac{dR_S}{dI_{n+1}} & \dfrac{dR_S}{dR_{n+1}} \\[2mm] \dfrac{dR_I}{dS_{n+1}} & \dfrac{dR_I}{dI_{n+1}} & \dfrac{dR_I}{dR_{n+1}} \\[2mm] \dfrac{dR_R}{dS_{n+1}} & \dfrac{dR_R}{dI_{n+1}} & \dfrac{dR_R}{dR_{n+1}} \end{bmatrix} = \begin{bmatrix} \frac{1}{\Delta t} + \beta\, I_{n+1} & \beta\, S_{n+1} & 0 \\ -\beta\, I_{n+1} & \frac{1}{\Delta t} - \beta\, S_{n+1} + \gamma & 0 \\ 0 & -\gamma & \frac{1}{\Delta t} \end{bmatrix}.
$$

$$(6.12)$$

The Newton method uses the inverse of this tangent matrix, $\mathbf{K}^{-1}$, to calculate the incremental iterative update of the populations, $d\mathbf{P} = [\, dS, dI, dR \,]$, by solving the linearized residual equation, $d\mathbf{P} = -\mathbf{K}^{-1} \cdot \mathbf{R}^k$, with

$$
\begin{bmatrix} dS \\ dI \\ dR \end{bmatrix} = - \begin{bmatrix} \frac{1}{\Delta t} + \beta\, I_{n+1} & \beta\, S_{n+1} & 0 \\ -\beta\, I_{n+1} & \frac{1}{\Delta t} - \beta\, S_{n+1} + \gamma & 0 \\ 0 & -\gamma & \frac{1}{\Delta t} \end{bmatrix}^{-1} \begin{bmatrix} R_S^k \\ R_I^k \\ R_R^k \end{bmatrix}. \quad (6.13)
$$

The solution provides the incremental iterative of update of the populations, $d\mathbf{P} = [\, dS, dI, dR \,]$, to update the current populations, $\mathbf{P}^{k+1} = \mathbf{P}^k + d\mathbf{P}$, with

$$
\begin{bmatrix} S^{k+1} \\ I^{k+1} \\ R^{k+1} \end{bmatrix} = \begin{bmatrix} S^k \\ I^k \\ R^k \end{bmatrix} + \begin{bmatrix} dS \\ dI \\ dR \end{bmatrix}. \quad (6.14)
$$

until the method has converged and the norm of the residual, $||\mathbf{R}|| < $ tol, or the norm of the update, $||d\mathbf{P}|| < $ tol is smaller than a user defined tolerance tol. An incremental iterative solution within an implicit time integration scheme results in two nested loops, an outer loop for all time steps n and an inner loop one for all iterations k.

6.3 Comparison of explicit and implicit SIR models

To illustrate the effect of explicit and implicit time integration for the SIR model, we compare two cases, slow dynamics with a long infectious period and a low basic reproduction number, and fast dynamics with a short infectious period and a higher basic reproduction number. We highlight the outbreak dynamics of the discrete SIR model for varying time step sizes and discuss converges towards the analytical solutions for the final size relations of the susceptible and recovered groups (3.28)

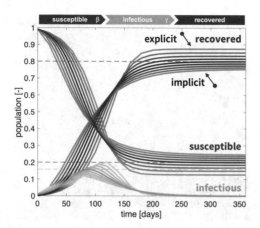

Fig. 6.2 Classical SIR model. Explicit and implicit time integration of slow dynamics with varying time step size Δt. Dashed lines highlight the analytical solutions $S_\infty = 0.1998$, $R_\infty = 0.8002$, and $I_{max} = 0.1585$. The explicit time integration scheme overestimates the outbreak; the final recovered population R_∞ converges from above. The implicit time integration scheme underestimates the outbreak; the final recovered population R_∞ converges from below. Infectious period $C = 20$ days, basic reproduction number $R_0 = 2.0$, initial infectious population $I_0 = 0.01$, and time step size $\Delta t = 20, 15, 10, 5, 1$ days, from light to dark.

in terms of the Lambert function W from Chapter 3,

$$S_\infty = - W(-S_0 \, R_0 \, \exp(-R_0[\,1 - R_0\,]))/R_0$$
$$R_\infty = 1 - W(-S_0 \, R_0 \, \exp(-R_0[\,1 - R_0\,]))/R_0,$$

and for the maximum infectious group (3.18),

$$I_{max} = 1 - \frac{1}{R_0}\left[1 + \log(S_0) - \log\left(\frac{1}{R_0}\right)\right]$$

for slow dynamics with a long infectious period of $C = 20$ days and a low basic reproduction number of $R_0 = 2.0$, and for fast dynamics with a short infectious period of $C = 6.5$ days and a higher basic reproduction number of $R_0 = 4.0$.

Figure 6.2 illustrates the effect of explicit and implicit time integration for the SIR model with *slow dynamics* with a long infectious period of $C = 20$ days and a low basic reproduction number of $R_0 = 2.0$. The differences between both methods are most visible in the red susceptible and blue recovered curves, while all orange infectious curves are relatively similar with a good approximation of the maximum infectious population $I_{max} = 0.1585$ from the analytical solution in equation (3.18). For the largest time step size, $\Delta t = 20$ days, the curves of both integration schemes are noticeably non-smooth. Decreasing the time step size as $\Delta t = 20, 15, 10, 5, 1, 0.1$ days results in smoother curves that converge towards the analytical solutions, $S_\infty = 0.1998$ and $R_\infty = 0.8002$, of the final size relation in equation (3.28). For the largest time step size of $\Delta t = 20$ days, the relative error, the difference between

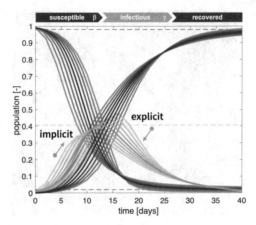

Fig. 6.3 Classical SIR model. Explicit and implicit time integration of fast dynamics with varying time step size Δt. Dashed lines highlight the analytical solutions $S_\infty = 0.0196$ and $R_\infty = 0.9804$, and $I_{max} = 0.4059$. The explicit time integration scheme overestimates the outbreak; the maximum infectious population I_{max} converges from above. The implicit time integration scheme underestimates the outbreak; the maximum infectious population I_{max} converges from below. Infectious period $C = 6.5$ days, basic reproduction number $R_0 = 4.0$, initial infectious population $I_0 = 0.01$, and time step size $\Delta t = 2.0, 1.5, 1.0, 0.5, 0.1$ days, from light to dark.

the analytical and numerical final size R_∞ scaled by the analytical final size, is 7.36% for the explicit and -4.64% for the implicit time integration scheme. For the time step size of $\Delta t = 1$ day, both errors have already decayed below ±0.3%, and for $\Delta t = 0.1$ day, both are well below ±0.05%. Interestingly, the explicit time integration overestimates the outbreak and its final recovered population R_∞ converges from above. Conversely, the implicit time integration underestimates the outbreak and its final recovered population R_∞ converges from below.

Figure 6.3 illustrates the effect of explicit and implicit time integration for the SIR model with *fast dynamics* with a short infectious period of $C = 6.5$ days and a higher basic reproduction number of $R_0 = 4.0$, values that are comparable to the first wave of the COVID-19 outbreak. Compared to the slow dynamics in Figure 6.2, the fast dynamics simulation in Figures 6.3 does not produce meaningful results for time step sizes larger than $\Delta t = 2$ days: The explicit method results in oscillating population sizes outside the zero-to-one interval; the implicit method simply does not converge. Differences between both methods are most visible in the orange infectious curves, while the red susceptible and blue recovered curves are relatively similar with a good converge to the analytical solutions, $S_\infty = 0.0196$ and $R_\infty = 0.9804$, of the final size relation in equation (3.28). Again, the curves or both methods are noticeably non-smooth for the largest time step size, $\Delta t = 2.0$ days. Decreasing the time step size as $\Delta t = 2.0, 1.5, 1.0, 0.5, 0.1$ days results in smoother curves that converge towards the analytical solution of the maximum infectious population $I_{max} = 0.4059$ from the analytical solution in equation (3.18). For the largest time step size of $\Delta t = 2$ days, the relative error, the difference between the analytical and numerical maximum infectious population I_{max} scaled by the analytical population, is 17.41% for the explicit and -13.29% for the implicit time integration schemes. For the time step size of $\Delta t = 1$ day, both errors are still notable as 8.19% for the explicit and -7.08% for the schemes. Only for $\Delta t = 0.1$ day, both errors are smaller than $\pm 1.00\%$. Interestingly, the explicit time integration overestimates the outbreak and its maximum infectious population I_{max} converges from above. Conversely, the implicit time integration underestimates the outbreak and its maximum infectious population I_{max} converges from below. The explicit scheme delays infection and predicts a later peak, whereas the implicit scheme accelerates infection and predicts an earlier peak.

This simple example illustrates that both explicit and implicit time integration can be quite sensitive to the selected time step size Δt [11, 12, 15], and that this sensitivity increases with increasing basic reproduction number R_0 [6]. For faster dynamics with basic reproduction numbers on the order of $R_0 = 4$ or larger, as have been reported for the early stages of the COVID-19 pandemic, this example demonstrates that time step sizes on the order of $\Delta t = 1$ day can produce errors in the maximum infectious population I_{max} on the order of 10%. This is particularly important because the reporting frequency on public dashboards is typically $\Delta t = 1$ day, and the data available to calibrate and validate the model come at a resolution where simple numerical integration schemes can become numerically unstable. For example, when fixing the time step size to $\Delta t = 1$ day, the critical basic reproduction number for which both methods either oscillated or failed to converge was $R_0 = 8.0$. For basic reproduction numbers above this value, and time step sizes of one day more sophisticated time integration schemes like the midpoint method discussed in Chapter 5 could become necessary.

6.4 Comparison of SIR and SIS models

Figure 6.4 shows a side-by-side comparison of the SIR and SIS models [13]. The early outbreak dynamics are identical for both models with the same initial slopes of the susceptible and infectious populations S and I. Compared to the SIS model, dashed lines, the SIR model, solid lines, has a vanishing infectious population I_∞ at endemic equilibrium [3]. The final size of the susceptible population S_∞ for the SIR model is smaller than for the SIS model, for which individuals return to the susceptible population after leaving the infectious population.

6.5 Sensitivity analysis of the SIR model

Figures 6.5, 6.6, and 6.7 highlight the outbreak dynamics of the SIR model, similar to Figures 2.3, 2.4, and 2.5 for the SIS model, but now solved numerically with an explicit time integration using equations (6.5) for the time period of one year. The three figures demonstrate the sensitivity of the of the SIR model for varying infectious periods C, basic reproduction numbers R_0, and initial infectious populations I_0.

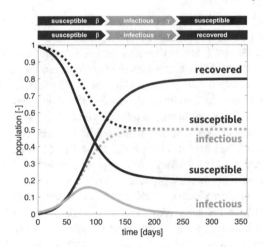

Fig. 6.4 Classical SIS and SIR models. The early outbreak dynamics are identical for both models with the same initial slopes of the susceptible and infectious populations S and I. Compared to the SIS model, dashed lines, the SIR model, solid lines, has a vanishing infectious population I_∞ at endemic equilibrium, and the final size of its susceptible population S_∞ is smaller. Infectious period $C = 20$ days, basic reproduction number $R_0 = 2.0$, and initial infectious population $I_0 = 0.01$.

Increasing the infectious period C increases the maximum infectious population I_{max} and delays convergence to the endemic equilibrium, but the final sizes S_∞ and R_∞ remain unchanged. The computational finite sizes $S_\infty = 0.1971$ and $R_\infty = 0.8029$ agree well with the analytical finite sizes from to Chapter 3, $S_\infty = 0.1998$ and $R_\infty = 0.8002$. Increasing the basic reproduction number R_0 increases the maximum infectious population I_{max}, accelerates convergence to the endemic equilibrium, decreases S_∞, and increases R_∞. Increasing the initial infectious population I_0 accelerates the onset of the outbreak, but the the maximum infectious population I_{max} and the final sizes S_∞ and I_∞ remain unchanged. The computational maximum infectious population $I_{maxf} = 0.1604$ agrees well with the the analytical maximum infectious population from Chapter 3, $I_{max} = 0.1585$. The shift in the population dynamics, similar to the SIS model in Figure 2.5, highlights the exponential nature of the SIR model, which causes a constant acceleration of the outbreak for a logarithmic increase of the infectious population, while the overall outbreak dynamics remain the same.

6.6 Example: Early COVID-19 outbreak in Austria

Austria was one was one of the first countries in Europe to see cases of COVID-19. The country implemented a strict lockdown and observed a rapid drop of new infections [16]. Table 6.1 summarizes the seven-day moving average of the newly reported COVID-19 cases in Austria [9], from Friday, Feb 28, 2020, the first day

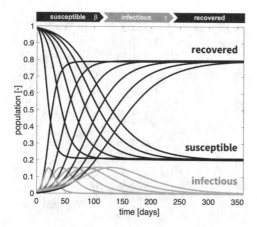

Fig. 6.5 Classical SIR model. Sensitivity with respect to the infectious period C**.** Increasing the infectious period C increases the maximum infectious population I_{max} and delays convergence to the endemic equilibrium, but the final sizes S_∞ and R_∞ remain unchanged. Basic reproduction number $R_0 = 2.0$, initial infectious population $I_0 = 0.01$, and infectious period $C = 5, 10, 15, 20, 25, 30$ days.

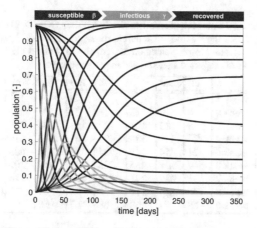

Fig. 6.6 Classical SIR model. Sensitivity with respect to the basic reproduction number R_0**.** Increasing the basic reproduction number R_0 increases the maximum infectious population I_{max}, accelerates convergence to the endemic equilibrium, decreases S_∞, and increases R_∞. Infectious period $C = 20$ days, initial infectious population $I_0 = 0.01$, and basic reproduction number $R_0 = 1.5, 1.7, 2.0, 2.4, 3.0, 5.0, 10.0$.

of reported cases, to Thursday, Apr, 30, 2020. This example builds on a problem in Chapter 3 with case data in Table 3.3, but now displayed for a window of the first nine weeks of the outbreak. Similar to Chapter 3, we first calculate the total infectious population at each day t_n within each interval,

$$N_I(t_n) = \sum_{i=n+1-C}^{n} \Delta N_I(t_i),$$

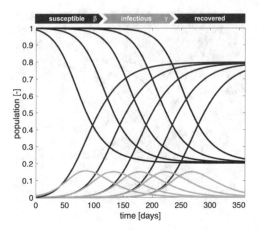

Fig. 6.7 Classical SIR model. Sensitivity with respect to the initial infectious population I_0. Increasing the initial infectious population I_0 accelerates the onset of the outbreak, but the the maximum infectious population I_{max} and the final sizes S_∞ and I_∞ remain unchanged. Infectious period $C = 20$ days, basic reproduction number $R_0 = 2.0$, and initial infectious population $I_0 = 10^{-1}, 10^{-2}, 10^{-3}, 10^{-4}, 10^{-5}, 10^{-6}$.

Table 6.1 Early COVID-19 outbreak in Austria. Newly reported COVID-19 cases in Austria from Friday, February 28, 2020, the first day of reported cases, to Thursday, April, 30, 2020; with a nation-wide stay-at-home order implemented on March, 16, 2020 and a mask mandate on March 30, 2020; case data ΔN_I are reported as seven-day moving averages.

	Fri	Sat	Sun	Mon	Tue	Wed	Thu
week 1	1	2	2	2	3	4	6
week 2	8	10	13	16	23	31	45
week 3	63	82	108	127	164	200	260
week 4	306	334	389	493	565	563	676
week 5	721	754	744	735	699	732	603
week 6	547	501	466	383	351	319	302
week 7	291	289	271	249	227	201	176
week 8	148	123	115	108	92	82	75
week 9	68	69	68	68	69	68	65

by summing the newly reported cases ΔN_I for the $C = 7$ previous days. Table 6.2 summarize the infectious population N_I assuming an infectious period of $C = 7$ days. Figure 3.13 illustrates the time evolution of the logarithmic infectious population, from which we can estimate the growth rates, $G = [\ln(N_I(t_{n+1})) - \ln(N_I(t_n))]/[t_{n+1} - t_n]$, as $G = [0.317, 0.297, 0.225, 0.133, -0.008]$ days, for five week-long intervals between March 1 and April 5, 2020. From these growth rates, we can estimate the weekly basic reproduction numbers,

$$R_0 = 1 + G C \quad \text{thus} \quad R_0 = [3.22, 3.08, 2.58, 1.93, 0.94].$$

We approximate the early basic reproduction number as $R_0 = 3$, and use the analytical solution (3.18) to estimate the maximum infectious population I_{max},

Table 6.2 Early COVID-19 outbreak in Austria. Infectious population of COVID-19 cases in Austria from Friday, February 28, 2020, the first day of reported cases, to Thursday, April, 30, 2020; with a nation-wide stay-at-home order implemented on March, 16, 2020 and a mask mandate on March 30, 2020; infectious population N_I estimated from case data ΔN_I assuming an infectious period of $C = 7$ days.

	Fri	Sat	Sun	Mon	Tue	Wed	Thu
week 1	1	3	5	7	10	14	20
week 2	27	35	46	60	80	107	146
week 3	201	273	368	479	620	789	1004
week 4	1247	1499	1780	2146	2547	2910	3326
week 5	3741	4161	4516	4758	4892	5061	4988
week 6	4814	4561	4283	3931	3583	3170	2869
week 7	2613	2401	2206	2072	1948	1830	1704
week 8	1561	1395	1239	1098	963	844	743
week 9	663	609	562	522	499	485	475

$$I_{max} = 1 - \frac{1}{R_0}\left[1 + \log(S_0) - \log\left(\frac{1}{R_0}\right)\right] \quad \text{thus } I_{max} = 1 - \frac{1}{3}\left[1 + 0 - \log\frac{1}{3}\right] = 0.3005.$$

Austria has a total population of $N = 8.84$ million people, but, clearly, because of the strict lockdown, only a fraction N^* of the entire population was potentially susceptible to the virus. We map the SIR model with a total population of one onto the potentially susceptible population of N^*, which we calculate from mapping the analytical maximum infectious population $I_{max} = 0.3005$ into the maximum number of infectious people, $\max\{N_I\} = 5,061$,

$$I_{max} = \frac{\max\{N_I\}}{N^*} \quad \text{thus} \quad N^* = \frac{\max\{N_I\}}{I_{max}} = \frac{5,061}{0.3005} = 16,842.$$

Theoretically, this mapping also defines the initial condition for the infectious population, $I_0 = N_I(0)/N^*$ to $I_0 = 1/16,842 = 0.000059 = 0.0059\%$, on the first day of the outbreak, February 28, 2020. To characterize the level of containment and compare it to other countries, we can calculate the *affected population* η, the ratio between the potentially susceptible population N^* and the total population N,

$$\eta = \frac{N^*}{N} \quad \text{thus} \quad \eta = \frac{16,842}{8,860,000} = 0.0019052 = 0.19\%.$$

Figure 6.8 summarizes the outbreak dynamics of COVID-19 in Austria in early 2020, during the first wave of the pandemic. The dots indicate the reported infectious population N_I from Table 6.2 [9], the curves highlight the simulated susceptible, infectious, and recovered populations. The SIR model uses an infectious period of $C = 7$ days, a basic reproduction number of $R_0 = 3$, and an initial infectious population of $I_0 = 2/N^*$, with a potentially susceptible population of $N^* = 16,842$. The infectious population increases gradually during the first two weeks of the outbreak, then grows rapidly for another two weeks, reaches a peak, and steadily declines. On 16 March, a nationwide stay-at-home order went into effect and the

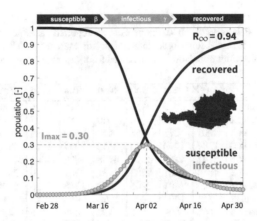

Fig. 6.8 Early COVID-19 outbreak dynamics in Austria. Reported infectious population and simulated susceptible, infectious, and recovered populations. Simulations are based on an infectious period of $C = 7$ days, a basic reproduction number of $R_0 = 3$, and an initial infectious population of $I_0 = 2/N^*$, with a potentially susceptible population of $N^* = 16,842$.

reproduction number dropped notably to $R_0 = 1.48$ in the following week. On 30 March, three days before the peak of infections, with $\max\{N_I\} = 5,061$, the government introduced a mask mandate that made Austria one of the first countries to adopt a rule, which has since then become standard around the globe. Figure 6.8 nicely highlights the success of these measures. The curves of the SIR model display an excellent agreement with the dots of the recorded case data. By design, the maximum infectious population $I_{max} = 0.30$ of the SIR model matches the maximum infectious population of the reported data $\max\{N_I\}/N^* = 0.30$. Both peaks occur on April 2. For the model, to peak on this day, we assumed that two individuals were infected at the beginning of the outbreak, $I_0 = 2/N^*$. For only one infectious individual, $I_0 = 1/N^*$, the peak would have occurred two days later, on April 4. The SIR model predicts a final recovered population of $R_\infty = 0.94$, both analytically and numerically. For a potential susceptible population of $N^* = 16,842$, this would imply that 15,831 individuals would have recovered at endemic equilibrium. Summing the reported case numbers ΔN_I in Table 6.1 yields 15,340 recovered cases, resulting in a model error of 3.2%. We can see this overestimate of the model in early April, slightly after the peak of the infectious population. The orange infectious curve is slightly wider than the reported infectious case data, which suggests that the true basic reproduction number could be slightly larger than $R_0 = 3$, maybe closer to $R_0 = 3.22$ as Figure 3.13 in Chapter 3.1 suggests. During the first wave of the outbreak, the affected population in Austria was $\eta = N^*/N = 0.19\%$, which shows that the strict and quick political measures in the country were quite effective in containing the spread of COVID-19 and reducing the number of new infections [16, 17].

Problems

6.1 Explicit vs. implicit time integration. We have seen that the explicit time integration overestimates the maximum infectious and recovered populations I_{max} and R_∞ and that implicit time integration underestimates I_{max} and R_∞. If we fit our SIR model to reported infectious and recovered data, what does this imply for the fitted basic reproduction number R_0 if we use an explicit versus implicit time integration?

6.2 Explicit vs. implicit time integration. We have seen that the explicit time integration predicts a later peak of the maximum infectious populations I_{max} and the implicit time integration predicts and earlier peak of I_{max}. If fit our SIR model to reported infectious data, what does this imply for the fitted infectious period C if we use an explicit versus implicit time integration?

6.3 Time step size. We have seen that the result of the numerical time integration depends on the time step size Δt. Implement the explicit SIR model. Simulate an outbreak with an infectious period of $C = 7$ days, an initial infectious population of $I_0 = 0.001$ and varying reproduction numbers $R_0 = 2, 3, 4, 5, 10, 15$. Gradually increase the time step size $\Delta t = 0.01, 0.1, 1, 2, 5, 10$ days. Make informed recommendation for a reasonable time step size Δt.

6.4 Phase diagram of the SIR model. Create a phase diagram of the SIR model in the susceptible-infectious plane similar to Figure 3.4 in Chapter 3. Use an explicit time integration to solve the susceptible and infectious populations of the SIR model in time, starting with $I_0 = 0.0001$ and $S_0 = 0.9999$ in the lower right corner. Assume a basic reproduction number of $R_0 = 3.0$. Does your model converge towards the final sizes $I_\infty = 0$ and S_∞? Compare your phase diagram against the curves in Figure 3.4. Does your curve lie between the $R_0 = 2.0$ and $R_0 = 3.5$ curves?

6.5 Early exponential growth. Simulate the time evolution of the susceptible, infectious, and recovered populations for an SIR model with an infectious period of $C = 7$ days, a basic reproduction number of $R_0 = 3$, and an initial infectious population of $I_0 = 0.001$. Plot the populations in time. Add the infectious population of the early exponential growth model, $I(t) = I_0 \exp(Gt)$ with $G = [R_0 - 1]C$. Compare your results against Figure 6.9. When do the infectious populations start to deviate? When does the error between the SIR model and the exponential growth model exceed 0.1% and 1%? What is the size of the susceptible population at the time of these two errors? Make recommendations when we can confidently use the exponential growth model as an approximation.

6.6 Maximum infectiousness. A survey at Stanford University reported that 92.5% of undergraduates were susceptible to an influenza with an infectious period of $C = 6$ days at the beginning of the year, and 52.5% were susceptible at the end. Determine the basic reproduction number R_0, the contact period B, and the maximum infectious population I_{max} analytically using the SIR equations from Chapter 3.

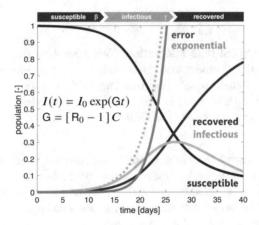

Fig. 6.9 Classical SIR model vs. exponential growth model. Solid red, orange, and blue curves represent the numerical solutions of the susceptible, infectious, and recovered populations of the SIR model; the dotted orange line represents the analytical solution of early exponential growth, $I(t) = I_0 \exp(Gt)$ with $G = [R_0 - 1] C$; dark grey line highlights the error between both models. Infectious period $C = 7$ days, basic reproduction number $R_0 = 3$, initial infectious population $I_0 = 0.001$.

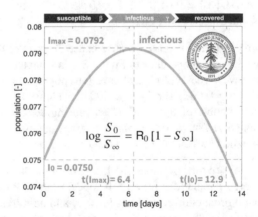

Fig. 6.10 Classical SIR model. Infectious population during an influenza. At the beginning of reporting, $S_0 = 92.5\%$ of the population were susceptible to an influenza with an infectious period of $C = 6$ days, with $S_\infty = 52.5\%$ remaining susceptible.

Determine the maximum infectious population numerically using either an explicit or implicit SIR model. What is a reasonable time step size to reproduce the analytical maximum infectious population I_{max}?

6.7 Maximum infectiousness. A survey at Stanford University reported that 92.5% of undergraduates were susceptible to an influenza with an infectious period of $C = 6$ days at the beginning of the year, and 52.5% were susceptible at the end. Determine the basic reproduction number R_0, the contact period B, and the maximum

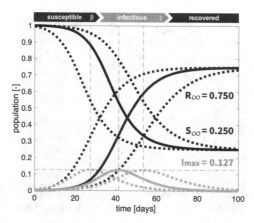

Fig. 6.11 Classical SIR model. Infectious population during an influenza. The influenza spread through 1224 of 1632 individuals assuming one individual was initially infectious with an infectious period of $C = 5$ days.

infectious population I_{max} analytically using the SIR equations from Chapter 3. Simulate the outbreak of the influenza numerically using either an explicit or implicit SIR model. Map the evolution of the infectious population in time $I(t)$. On which day $t(I_{max})$ does the infectious population reach its maximum I_{max}? On which day $t(I_0)$ does the infectious population first fall below its initial value I_0? Compare your results against Figure 6.10.

6.8 Maximum infectiousness. A French boarding school reported an influenza that spread through 1224 of its 1632 students. Estimate how many students were infectious at the peak of the outbreak. Simulate and map the evolution of the infectious population $I(t)$ in time if, on the first day of class, a single student was infectious and the infectious period was $C = 5$ days. On which day does the outbreak peak? Compare your results against Figure 6.11.

6.9 Maximum infectiousness. A French boarding school reported an influenza that spread through 1224 of its 1632 students. Estimate how many students were infectious at the peak of the outbreak. Simulate and map the evolution of the infectious population $I(t)$ in time if, on the first day of class, a single student was infectious and the infectious period was $C = 6.5$ days. Simulate and map the evolution of the infectious population in time $I(t)$ if, on the first day of class, ten students were infectious and the infectious period was $C = 5$ days. For both cases, on which day does the outbreak peak? Compare your results against Figure 6.11. How does changing the initial infectious population affect the results? How does this observation compare to Figure 6.5?

6.10 Recovery. A French boarding school reported an influenza that spread through 1224 of its 1632 students. Simulate and map the evolution of the susceptible and

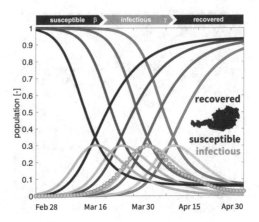

Fig. 6.12 Early COVID-19 outbreak dynamics in Austria. Reported infectious population and simulated susceptible, infectious, and recovered populations. Infectious period $C = 7$ days, basic reproduction number $R_0 = 3$, and initial infectious population $I_0 = 0.001\%, 0.01\%, 0.1\%, 1.0\%$, from light to dark.

recovered populations $S(t)$ and $R(t)$ in time for three different cases: a single infectious student with an infectious period of $C = 5$ days, a single infectious student with an infectious period of $C = 6.5$ days, and ten infectious students with an infectious period of $C = 5$ days. For all three cases, on which day have 816 students recovered from the infection? Compare your results against Figure 6.11.

6.11 Maximum infectiousness. Two visitors introduce COVID-19 to Stanford campus with a population of 5000 students. Each infectious individual meets on average 0.4286 students per day and is infectious for seven days. The university plans to isolate all infectious individuals. Simulate the outbreak. Find the numerical solution for the maximum number of required isolation units and compare it against the analytical solution. On what day are the most isolation units needed?

6.12 Initial conditions and community spreading. For the early COVID-19 outbreak in Austria with the seven-day moving average case data in Table 6.1, implement the SIR model and recreate Figure 6.8 for a simulation with an infectious period of $C = 7$ days and a basic reproduction number of $R_0 = 3$. Vary the initial infectious population $I_0 = 0.001\%, 0.01\%, 0.1\%, 1.0\%$, compare your graphs to Figure 6.12, and comment on the results. Discuss the implications of initial undetected community spreading.

6.13 Affected population and outbreak control. For the early COVID-19 outbreak in Austria with the seven-day moving average case data in Table 6.1, redo the Austria example from Section 6.6, but now with a basic reproduction number of $R_0 = 4$. Estimate the maximum infectious population I_{max}, the potentially susceptible population N^*, and the affected population η. Discuss how increasing the basic reproduction number affects the potentially susceptible and affected populations.

6.14 Outbreak dynamics. For the early COVID-19 outbreak in Austria with the seven-day moving average case data in Table 6.1, implement the SIR model and recreate Figure 6.8 for a simulation with an infectious period of $C = 7$ days, but now with a basic reproduction number of $R_0 = 4$. Estimate the maximum infectious population I_{max}, the potentially susceptible population N^*, and the initial infectious population I_0. How does the graph change, compared to Figure 6.8? How can you adjust the initial infectious population I_0 to improve the fit?

6.15 Parameter sensitivity. The fit of the SIR model to reported case data depends on four parameters, the infectious period C, the basic reproduction number R_0, the initial infectious population I_0, and the potentially susceptible population N^*. Discuss how increasing R_0, I_0, and N^* affects the fit of the SIR model to reported case data.

6.16 Outbreak control. In the Austria example in Section 6.6, we have modeled outbreak control by introducing a potentially susceptible population N^* that is a fraction η of the total population N. Take a closer look at Figure 3.13 and discuss other possibilities to model outbreak control within the SIR model.

6.17 Model your own country, state, or city. Draw the COVID-19 case data from the first wave of the outbreak of your own country, state, or city. From the daily new cases ΔN_I, calculate the infectious population N_I assuming an infectious periods of $C = 7$ days. Implement the SIR model and fit the basic reproduction number R_0, the initial infectious population I_0, and the potentially susceptible population N^*. Try to fit the model such that the peak I_{max} and the timing of the peak $t(I_{max})$ match the reported infectious population N_I/N^*. Comment on the fit of the model.

References

1. Anderson RM, May RM (1981) The population dynamics of microparasites and their invertebrate hosts. Philosophical Transactions of the Royal Society London B. 291:451–524.
2. Anderson RM, May RM (1991) Infectious Diseases of Humans. Oxford University Press, Oxford.
3. Brauer F, Castillo-Chavez C (2001) Mathematical Models in Population Biology and Epidemiology. Springer-Verlag New York.
4. Brauer F, Castillo-Chavez C, Feng Z (2019) Mathematical Models in Epidemiology. Springer-Verlag New York.
5. Deuflhard P (2011) Newton Methods for Nonlinear Problems. Springer-Verlag Berlin Heidelberg.
6. Dietz K (1993) The estimation of the basic reproduction number for infectious diseases. Statistical Methods in Medical Research 2:23-41.
7. European Centre for Disease Prevention and Control. Situation update worldwide. https://www.ecdc.europa.eu/en/geographical-distribution-2019-ncov-cases accessed: June 1, 2021.
8. Goodwine B (2011) Engineering Differential Equations. Springer-Verlag New York.
9. Griffiths D, Higham DJ (2010) Numerical Methods for Ordinary Differential Equations. Springer-Verlag London.

10. Hethcote HW (1989) Three basic epidemiological models. Biomathematics: Applied Mathematical Ecology, 18:119-144.
11. Hethcote HW (2000) The mathematics of infectious diseases. SIAM Review 42:599-653.
12. Hundsdorfer W, Verwer J (2003) Numerical Solution of Time-Dependent Advection-Diffusion-Reaction Equations. Springer-Verlag Berlin Heidelberg.
13. Linka K, Peirlinck M, Sahli Costabal F, Kuhl E (2020) Outbreak dynamics of COVID-19 in Europe and the effect of travel restrictions. Computer Methods in Biomechanics and Biomedical Engineering 23: 710-717.
14. Linka K, Peirlinck M, Kuhl E (2020) The reproduction number of COVID-19 and its correlation with public heath interventions. Computational Mechanics 66:1035-1050.
15. Liu J, Shang, X (2020) Computational Epidemiology. Springer International Publishing.
16. Moin P (2000) Fundamentals of Engineering Numerical Analysis. Cambridge University Press.

Chapter 7
The computational SEIR model

Abstract The SEIR model is the most popular compartment model with four populations, the susceptible, exposed, infectious, and recovered groups S, E, I, and R. It characterizes infectious diseases that have a significant incubation period and provide immunity upon infection. Since the SEIR model has no analytical solution for the time course of its populations, we discretize it in time using finite differences and apply explicit and implicit time integration schemes to solve it. To illustrate the features of the SEIR model, we simulate the early COVID-19 outbreak in Germany and China using reported case data. We compare two strategies to model outbreak control, the potentially susceptible population approach and the dynamic SEIR model. The learning objectives of this chapter on computational SEIR modeling are to

- discretize the classical SEIR model in time
- solve the SEIR equations using the explicit forward Euler method
- derive the residual SEIR equations using the implicit backward Euler method
- linearize and solve the residual SEIR equations using the Newton method
- rationalize and design a dynamic SEIR model with time-varying reproduction
- simulate the first wave of the COVID-19 outbreak using reported case data
- identify latent, contact, and infectious periods from COVID-19 case data
- discuss limitations of computational SEIR modeling

By the end of the chapter, you will be able to computationally analyze, simulate, and predict the early outbreak dynamics of infectious diseases like COVID-19 using reported case data and identify their disease parameters.

7.1 Explicit time integration of the SEIR model

The SEIR model consists of four populations, the susceptible, exposed, infectious, and recovered groups S, E, I, and R [1]. Figure 7.1 illustrates the four populations and the latent, contact, and infectious rates α, β, and γ that define the transition between them [8]. The latent, contact, and infectious rates are inverses of the latent,

© The Author(s), under exclusive license to Springer Nature Switzerland AG 2021
E. Kuhl, *Computational Epidemiology*, https://doi.org/10.1007/978-3-030-82890-5_7

| susceptible | β | exposed | α | infectious | γ | recovered |

Fig. 7.1 Classical SEIR model. The classical SEIR model contains four compartments for the susceptible, exposed, infectious, and recovered populations, S, E, I, and R. The transition rates between the compartments, the contact, latent, and infectious rates, β, α, and γ, are inverses of the contact, latent, and infectious periods, $B = 1/\beta$, $A = 1/\alpha$, and $C = 1/\gamma$.

contact, and infectious periods $A = 1/\alpha$, $B = 1/\beta$, and $C = 1/\gamma$, and their ratio defines the basic reproduction number, $R_0 = \beta/\gamma = C/B$ [3]. Similar to the SIR model, the SEIR model does not have a straightforward explicit analytical solution [22]. To solve the set of coupled nonlinear ordinary differential equations of the SEIR model,

$$
\begin{aligned}
\dot{S} &= -\beta\,SI \\
\dot{E} &= +\beta\,SI - \alpha\,E \\
\dot{I} &= \qquad\quad +\alpha\,E - \gamma\,I \\
\dot{R} &= \qquad\qquad\quad + \gamma\,I\,.
\end{aligned}
\tag{7.1}
$$

we apply a finite difference approximation and replace the lefthand sides of the system of equations (7.1) by the discrete difference, $(\dot{\circ}) = [\,(\circ)_{n+1} - (\circ)_n\,]/\Delta t$ in terms of the unknown new populations, $(\circ)_{n+1}$, the known previous populations, $(\circ)_n$, and the discrete time step size, $\Delta t = t_{n+1} - t_n$, for all four populations, $(\circ) = S, E, I, R$,

$$
\dot{S} = \frac{S_{n+1} - S_n}{\Delta t} \quad \dot{E} = \frac{E_{n+1} - E_n}{\Delta t} \quad \dot{I} = \frac{I_{n+1} - I_n}{\Delta t} \quad \dot{R} = \frac{R_{n+1} - R_n}{\Delta t}\,.
\tag{7.2}
$$

In the spirit of an explicit time integration, we adopt a classical *forward Euler method* and evaluate the righthand sides of the system of equations (7.1) at the known previous time point t_n [22],

$$
\dot{S} = -\beta\,S_n I_n \quad \dot{E} = +\beta\,S_n I_n - \alpha\,E_n \quad \dot{I} = +\alpha\,E_n - \gamma\,I_n \quad \dot{R} = +\gamma\,I_n\,.
\tag{7.3}
$$

This explicit time integration results in the following discrete system of equations for the SEIR model,

$$
\begin{aligned}
[\;S_{n+1} - S_n\;]/\Delta t &= -\beta\,S_n I_n \\
[\;E_{n+1} - E_n\;]/\Delta t &= +\beta\,S_n I_n - \alpha\,E_n \\
[\;I_{n+1} - I_n\;]/\Delta t &= \qquad\quad + \alpha\,E_n - \gamma\,I_n \\
[\;R_{n+1} - R_n\;]/\Delta t &= \qquad\qquad\quad + \gamma\,I_n\,,
\end{aligned}
\tag{7.4}
$$

which we can solve directly for the unknown populations S_{n+1}, E_{n+1}, I_{n+1}, and R_{n+1} at the new time point t_{n+1},

$$
\begin{aligned}
S_{n+1} &= [1 - \beta\,I_n\,\Delta t]\,S_n \\
E_{n+1} &= \quad + \beta\,I_n\,\Delta t\;S_n + [1 - \alpha\,\Delta t]\,E_n \\
I_{n+1} &= \qquad\qquad\qquad + \quad \alpha\,\Delta t\;E_n + [1 - \gamma\,\Delta t]\,I_n \\
R_{n+1} &= \qquad\qquad\qquad\qquad\quad + \quad \gamma\,\Delta t\;I_n + R_n\,,
\end{aligned}
\tag{7.5}
$$

using the initial conditions $S_0 = 1 - E_0 - I_0$, E_0, I_0, and $R_0 = 0$. Explicit time integration schemes are relatively straightforward, but can become oscillatory and numerically unstable, especially for stiff systems, i.e., systems with large basic reproduction numbers, and large time steps Δt. In the next section, we suggest an alternative approach and discretize the SEIR model in time using an implicit time integration scheme.

7.2 Implicit time integration of the SEIR model

Similar to the explicit time integration in Section 7.1, we begin with the set of coupled nonlinear ordinary differential equations of the SEIR model (4.17) from Chapter 4,

$$
\begin{aligned}
\dot{S} &= -\beta \, SI \\
\dot{E} &= +\beta \, SI - \alpha \, E \\
\dot{I} &= \qquad\;\; +\alpha \, E - \gamma \, I \\
\dot{R} &= \qquad\qquad\quad\; +\gamma \, I,
\end{aligned}
\tag{7.6}
$$

and apply a finite difference approximation to replace the lefthand sides of the system of equations (7.6) by the discrete difference, $(\dot{\circ}) = [\,(\circ)_{n+1} - (\circ)_n\,]/\Delta t$, in terms of the unknown new populations, $(\circ)_{n+1}$, the known previous populations, $(\circ)_n$, and the discrete time step size, $\Delta t = t_{n+1} - t_n$, for all four populations, $(\circ) = S, E, I, R$,

$$
\dot{S} = \frac{S_{n+1} - S_n}{\Delta t} \quad \dot{E} = \frac{E_{n+1} - E_n}{\Delta t} \quad \dot{I} = \frac{I_{n+1} - I_n}{\Delta t} \quad \dot{R} = \frac{R_{n+1} - R_n}{\Delta t} \; .
\tag{7.7}
$$

However, now, similar to Section 6.2 for the SIR model, we apply an implicit time integration and adopt a classical *backward Euler method* to evaluate the righthand sides of the system of equations (7.6) at the unknown current time point t_{n+1} [22],

$$
\dot{S} = -\beta S_{n+1} I_{n+1} \quad \dot{E} = +\beta S_{n+1} I_{n+1} - \alpha E_{n+1} \quad \dot{I} = +\alpha E_{n+1} - \gamma I_{n+1} \quad \dot{R} = +\gamma I_{n+1} \; .
\tag{7.8}
$$

This implicit time integration results in the following discrete system of equations for the SEIR model,

$$
\begin{aligned}
[\, S_{n+1} - S_n \,]/\Delta t &= -\beta \, S_{n+1} I_{n+1} \\
[\, E_{n+1} - E_n \,]/\Delta t &= +\beta \, S_{n+1} I_{n+1} - \alpha \, E_{n+1} \\
[\, I_{n+1} - I_n \,]/\Delta t &= \qquad\qquad\quad +\alpha \, E_{n+1} - \gamma \, I_{n+1} \\
[\, R_{n+1} - R_n \,]/\Delta t &= \qquad\qquad\qquad\qquad\quad +\gamma \, I_{n+1} \; .
\end{aligned}
\tag{7.9}
$$

We rephrase this system in its residual form, $\mathbf{R} = [\, \mathsf{R}_S, \mathsf{R}_E, \mathsf{R}_I, \mathsf{R}_R \,] \doteq \mathbf{0}$, by bringing all terms to the righthand side,

$$
\begin{aligned}
R_S &= [\ S_{n+1} - S_n\ ]/\Delta t + \beta\, S_{n+1}\, I_{n+1} && = 0 \\
R_E &= [\ E_{n+1} - E_n\ ]/\Delta t - \beta\, S_{n+1}\, I_{n+1} + \alpha\, E_{n+1} && = 0 \\
R_I &= [\ I_{n+1} - I_n\ ]/\Delta t && \quad - \alpha\, E_{n+1} + \gamma\, I_{n+1} = 0 \\
R_R &= [\ R_{n+1} - R_n\ ]/\Delta t && \quad\quad\quad\quad\quad - \gamma\, I_{n+1} = 0.
\end{aligned}
\tag{7.10}
$$

We apply the Newton method [6] and linearize the residual $\mathbf{R}$ with respect to the unknown populations $\mathbf{P}_{n+1} = [\ S_{n+1}, E_{n+1}, I_{n+1}, R_{n+1}\ ]$, such that, $\mathbf{R}^{k+1} = \mathbf{R}^k + \mathbf{K} \cdot d\mathbf{P} = \mathbf{0}$, with

$$
\begin{bmatrix} R_S^{k+1} \\ R_E^{k+1} \\ R_I^{k+1} \\ R_R^{k+1} \end{bmatrix}
= \begin{bmatrix} R_S^k \\ R_E^k \\ R_I^k \\ R_R^k \end{bmatrix}
+ \begin{bmatrix}
\frac{1}{\Delta t} + \beta\, I_{n+1} & 0 & \beta\, S_{n+1} & 0 \\
-\beta\, I_{n+1} & \frac{1}{\Delta t} + \alpha & -\beta\, S_{n+1} & 0 \\
0 & -\alpha & \frac{1}{\Delta t} + \gamma & 0 \\
0 & 0 & -\gamma & \frac{1}{\Delta t}
\end{bmatrix}
\cdot \begin{bmatrix} dS \\ dE \\ dI \\ dR \end{bmatrix}
= \begin{bmatrix} 0 \\ 0 \\ 0 \\ 0 \end{bmatrix},
\tag{7.11}
$$

where the tangent matrix $\mathbf{K}$ of the Newton method contains the derivatives of the four residuals $\mathbf{R} = [\ R_S, R_E, R_I, R_R\ ]$ with respect to the four populations $\mathbf{P}_{n+1} = [\ S_{n+1}, E_{n+1}, I_{n+1}, R_{n+1}\ ]$,

$$
\mathbf{K} =
\begin{bmatrix}
\dfrac{dR_S}{dS_{n+1}} & \dfrac{dR_S}{dE_{n+1}} & \dfrac{dR_S}{dI_{n+1}} & \dfrac{dR_S}{dR_{n+1}} \\[2mm]
\dfrac{dR_E}{dS_{n+1}} & \dfrac{dR_E}{dE_{n+1}} & \dfrac{dR_E}{dI_{n+1}} & \dfrac{dR_E}{dR_{n+1}} \\[2mm]
\dfrac{dR_I}{dS_{n+1}} & \dfrac{dR_I}{dE_{n+1}} & \dfrac{dR_I}{dI_{n+1}} & \dfrac{dR_I}{dR_{n+1}} \\[2mm]
\dfrac{dR_R}{dS_{n+1}} & \dfrac{dR_R}{dE_{n+1}} & \dfrac{dR_R}{dI_{n+1}} & \dfrac{dR_R}{dR_{n+1}}
\end{bmatrix}
=
\begin{bmatrix}
\frac{1}{\Delta t} + \beta\, I_{n+1} & 0 & \beta\, S_{n+1} & 0 \\
-\beta\, I_{n+1} & \frac{1}{\Delta t} + \alpha & -\beta\, S_{n+1} & 0 \\
0 & -\alpha & \frac{1}{\Delta t} + \gamma & 0 \\
0 & 0 & -\gamma & \frac{1}{\Delta t}
\end{bmatrix}.
\tag{7.12}
$$

The Newton method uses the inverse of this matrix $\mathbf{K}^{-1}$ to calculate the incremental iterative update of the populations, $d\mathbf{P} = [\ dS, dE, dI, dR\ ]$, by solving the linearized residual equation, $d\mathbf{P} = -\mathbf{K}^{-1} \cdot \mathbf{R}^k$, with

$$
\begin{bmatrix} dS \\ dE \\ dI \\ dR \end{bmatrix}
= - \begin{bmatrix}
\frac{1}{\Delta t} + \beta\, I_{n+1} & 0 & \beta\, S_{n+1} & 0 \\
-\beta\, I_{n+1} & \frac{1}{\Delta t} + \alpha & -\beta\, S_{n+1} & 0 \\
0 & -\alpha & \frac{1}{\Delta t} + \gamma & 0 \\
0 & 0 & -\gamma & \frac{1}{\Delta t}
\end{bmatrix}^{-1}
\cdot \begin{bmatrix} R_S^k \\ R_E^k \\ R_I^k \\ R_R^k \end{bmatrix}.
\tag{7.13}
$$

The solution provides the incremental update of the populations, $d\mathbf{P} = [\ dS, dE, dI, dR\ ]$ to update the populations, $\mathbf{P}^{k+1} = \mathbf{P}^k + d\mathbf{P}$, with

$$
\begin{bmatrix} S^{k+1} \\ E^{k+1} \\ I^{k+1} \\ R^{k+1} \end{bmatrix}
\leftarrow
\begin{bmatrix} S^k \\ E^k \\ I^k \\ R^k \end{bmatrix}
+
\begin{bmatrix} dS \\ dE \\ dI \\ dR \end{bmatrix}
\tag{7.14}
$$

until the method has converged and the norm of the residual, $||\mathbf{R}|| <$ tol, or the norm of the update, $||d\mathbf{P}|| <$ tol is smaller than a user defined tolerance tol. Similar to the SIR model in Chapter 6, the incremental iterative solution within an implicit time integration scheme results in two nested loops, an outer loop for all time steps n and an inner loop one for all iterations k.

7.3 Comparison of explicit and implicit SEIR models

To illustrate the effect of explicit and implicit time integration for the SEIR model, we compare two cases, slow dynamics with a long infectious period and a low basic reproduction number, and fast dynamics with a short infectious period and a higher basic reproduction number. We highlight the outbreak dynamics of the discrete SEIR model for varying time step sizes and discuss converges towards the analytical solutions from Chapter 4 for the final size relations of the susceptible and recovered groups (4.31) in terms of the Lambert function W,

$$
\begin{aligned}
S_\infty &= - W(-S_0 R_0 \exp(-R_0[1 - R_0]))/R_0 \\
R_\infty &= 1 - W(-S_0 R_0 \exp(-R_0[1 - R_0]))/R_0,
\end{aligned}
$$

for slow dynamics with long exposed and infectious periods of $A = 5$ days and $C = 20$ days and a low basic reproduction number of $R_0 = 2.0$, and for fast dynamics with short exposed and infectious periods of $A = 2.5$ days and $C = 6.5$ days and a higher basic reproduction number of $R_0 = 4.0$.

Figure 7.2 illustrates the effect of explicit and implicit time integration for the SEIR model with *slow dynamics* with long exposed and infectious periods of $A = 5$ days and $C = 20$ days and a low basic reproduction number of $R_0 = 2.0$. The differences between both methods are most visible in the red susceptible and blue recovered curves, while all red exposed and orange infectious curves are relatively similar. The explicit time integration scheme generates physically meaningless oscillatory populations outside the zero-to-one interval for larger time step sizes of $\Delta t = 20, 15$, and 10 days, but works well for smaller time step sizes of $\Delta t = 5$ and 1 days. The implicit time integration scheme works well for all time step sizes, $\Delta t = 20, 15, 10, 5$ and 1 days. Both schemes converge towards the analytical solutions, $S_\infty = 0.1998$ and $R_\infty = 0.8002$, of the final size relation in equation (4.31). Similar to the SIR discretization in Section 6.3, the final recovered population R_∞ converges from above for the explicit scheme and from below for the implicit scheme.

Figure 7.3 illustrates the effect of explicit and implicit time integration for the SEIR model with *fast dynamics* with short exposed and infectious periods of $A = 2.5$ days and $C = 6.5$ days and a higher basic reproduction number of $R_0 = 4.0$, values that are comparable to the first wave of the COVID-19 outbreak. Similar to the discrete SIR model in Section 6.3, for the fast dynamics simulation, the discrete SEIR model does not produce meaningful results for time step sizes larger than $\Delta t = 2$ days: The explicit method results in oscillating population sizes outside the

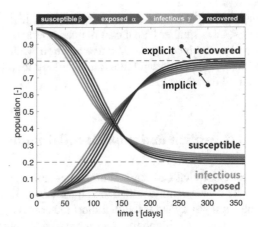

Fig. 7.2 Classical SEIR model. Explicit and implicit time integration of slow dynamics with varying time step size Δt. Dashed lines highlight the analytical solutions $S_\infty = 0.1998$ and $R_\infty = 0.8002$. The explicit time integration scheme overestimates the outbreak; the final recovered population R_∞ converges from above. The implicit time integration scheme underestimates the outbreak; the final recovered population R_∞ converges from below. Exposed and infectious periods $A = 5$ days and $C = 20$ days, basic reproduction number $R_0 = 2.0$, initial exposed population $E_0 = 0.01$, time step size $\Delta t = 20, 15, 10, 5, 1$ days, from light to dark.

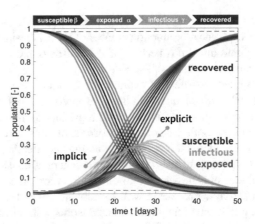

Fig. 7.3 Classical SEIR model. Explicit and implicit time integration of fast dynamics with varying time step size Δt. Dashed lines highlight the analytical solutions $S_\infty = 0.0196$ and $R_\infty = 0.9804$. The explicit time integration scheme overestimates the outbreak; the maximum infectious population $I_{\max}$ converges from above. The implicit time integration scheme underestimates the outbreak; the maximum infectious population $I_{\max}$ converges from below. Exposed and infectious periods $A = 2.5$ days and $C = 6.5$ days, basic reproduction number $R_0 = 4.0$, initial exposed population $E_0 = 0.01$, and time step size $\Delta t = 2.0, 1.5, 1.0, 0.5, 0.1$ days, from light to dark.

zero-to-one interval; the implicit method simply fails to converge. For decreasing time step sizes, $\Delta t = 2.0, 1.5, 1.0, 0.5, 0.1$ days, both methods converge towards the same solution with finite sizes of $S_\infty = 0.0196$ and $R_\infty = 0.9804$. Again, the explicit

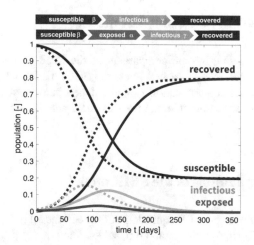

Fig. 7.4 Classical SEIR and SIR models. Compared to the SIR model, dashed lines, the SEIR model, solid lines, reduces the peak of the infectious population I_{max} and delays convergence to the endemic equilibrium, but the final sizes S_∞ and R_∞ are identical for both models. Latent period $A = 5$ days, infectious period $C = 20$ days, basic reproduction number $R_0 = 2.0$, and initial exposed or infectious population $E_0 = 0.01$ or $I_0 = 0.01$.

time integration overestimates the outbreak and its maximum exposed and infectious populations E_{max} and I_{max} converge from above. Conversely, the implicit time integration underestimates the outbreak and its maximum exposed and infectious populations E_{max} and I_{max} converge from below. The explicit scheme delays the outbreak and predicts a later peak, whereas the implicit scheme accelerates the outbreak and predicts an earlier peak.

Similar to the discrete SIR model in Section 6.3, we find that both time integration schemes can be quite sensitive to the selected time step size Δt [11, 12, 15], and that this sensitivity increases with increasing basic reproduction number R_0 [6]. The larger the basic reproduction number R_0, the smaller the required time step size Δt to obtain physically meaningful solutions. In view of the COVID-19 pandemic, for which basic reproduction numbers can be on the order of $R_0 = 4$ or larger, this example shows that time step sizes on the order of $\Delta t = 1$ day can generate noticeable discretization errors. This can become a problem, because the reporting frequency on public dashboards is typically $\Delta t = 1$ day. The data available to calibrate and validate the model come at a resolution where simple numerical integration schemes are numerically unstable and higher order time integration schemes would become necessary.

7.4 Comparison of SEIR and SIR models

Figure 7.4 shows a side-by-side comparison of the SEIR and SIR models [13]. Compared to the SIR model, dashed lines, the SEIR model, solid lines, reduces the peak of the infectious population I_{max} and delays convergence to the endemic equilibrium [3]. At the same time, the final sizes S_∞ and R_∞ are identical for both models.

7.5 Sensitivity analysis of the SEIR model

Figures 7.5 and 7.6 are the SEIR equivalents of the SIS model in Figures 2.3 and 2.4 and the SIR model in Figures 6.5 and 6.6, solved numerically [20]. Increasing the infectious period C, in Figures 7.5, reduces the maximum exposed population E_{max}, increases the maximum infectious population I_{max}, and delays converges towards the endemic equilibrium, but the final sizes S_∞ and R_∞ remain unchanged. In view of the COVID-19 outbreak, knowing the infectious time C is important to correctly estimate the timing and peak of the infectious population I_{max}, and with it the number of required hospital beds, ventilators, and intensive care units required to insure appropriate medical care. Increasing the basic reproduction number R_0, Figures 7.6, increases the maximum exposed and infectious populations E_{max} and I_{max}, accelerates convergence to the endemic equilibrium, decreases S_∞, and increases R_∞. In view of the COVID-19 outbreak, the basic reproduction number R_0 is a major parameter that we can influence by behavioral changes and political mitigation strategies. Reducing the basic reproduction number beyond its natural value by increasing the contact rate β through physical distancing or total lock down allows us to reduce the maximum infectious population I_{max} and delay the outbreak, a measure that has become known in the public media as flatting the curve.

Figures 7.7 and 7.8 highlight the sensitivity of the SEIR model with respect to the latent period A and the initial exposed population E_0 [20]. Increasing the latent period A, in Figure 7.7, increases the exposed population E_{max}, decreases the infectious population I_{max}, and delays converges towards the endemic equilibrium, but the final sizes S_∞ and R_∞ remain unchanged. Notably, the steepest set of curves for $A = 0$ days corresponds to the SIR model without a separate exposed population E in Figures 6.5, 6.6, and 7.4. In view of the COVID-19 outbreak, this implies that knowledge of the latent period is important to correctly estimate the timing and peak of the infectious population, which ultimately determines the absolute number of hospital beds, ventilators, and intensive care units. Increasing the initial exposed population E_0, in Figure 7.8, accelerates the onset of the outbreak, but the maximum exposed and infectious populations E_{max} and I_{max} and the final sizes S_∞ and R_∞ remain unchanged. This is conceptually similar to increasing the initial infectious population I_0, analytically for the SIS model in Figure 2.5 and computationally for the SIR model in Figure 6.7. Here, increasing the initial exposed population E_0 by an order of magnitude accelerates the peak of the infectious population I_{max} by a constant

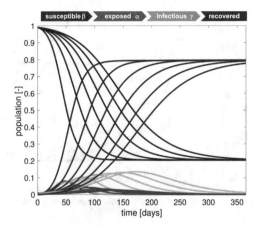

Fig. 7.5 Classical SEIR model. Sensitivity with respect to the infectious period C. Increasing the infectious period C reduces the maximum exposed population E_{max}, increases the maximum infectious population I_{max}, and delays converges towards the endemic equilibrium, but the final sizes S_∞ and R_∞ remain unchanged. Latent period $A = 5$ days, basic reproduction number $R_0 = 2.0$, initial exposed fraction $E_0 = 0.010$, and infectious period $C = 5, 10, 15, 20, 25, 30$ days [20].

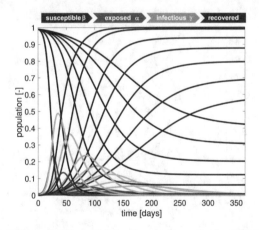

Fig. 7.6 Classical SEIR model. Sensitivity with respect to the basic reproduction number R_0. Increasing the basic reproduction number R_0 increases the maximum exposed and infectious populations E_{max} and I_{max}, accelerates convergence to the endemic equilibrium, decreases S_∞, and increases R_∞. Latent period $A = 5$ days, infectious period $C = 20$ days, initial exposed fraction $E_0 = 0.010$, and basic reproduction number $R_0 = 1.5, 1.7, 2.0, 2.4, 3.0, 5.0, 10.0$ [20].

time increment, for the current parameterization by 65 days [20]. This highlights the exponential nature of the SEIR model, which causes a constant acceleration for a logarithmic increase of the exponential population, while the overall outbreak dynamics remain the same [4]. Since the original date of the outbreak is often unknown, this correlation allows us to infer the beginning of the outbreak from known case data [26]. In view of the COVID-19 outbreak, this supports the general

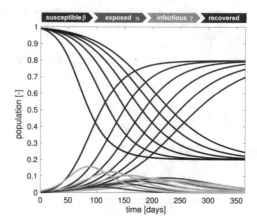

Fig. 7.7 Classical SEIR model. Sensitivity with respect to the latent period A. Increasing the latent period A increases the exposed population E_{max}, decreases the infectious populationand I_{max}, and delays converges towards the endemic equilibrium, but the final sizes S_∞ and R_∞ remain unchanged. Infectious period $C = 20$ days, basic reproduction number $R_0 = 2.0$, initial exposed fraction $E_0 = 0.010$, and latent period $A = 0, 5, 10, 15, 20, 25$ days [20].

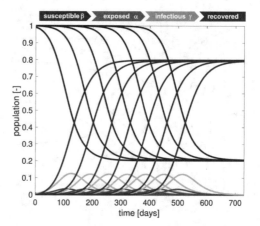

Fig. 7.8 Classical SEIR model. Sensitivity with respect to the initial exposed population E_0. Increasing the initial exposed population E_0 accelerates the onset of the outbreak , but the maximum exposed and infectious populations E_{max} and I_{max} and the final sizes S_∞ and R_∞ remain unchanged. Latent period $A = 5$ days, infectious period $C = 20$ days, basic reproduction number $R_0 = 2.0$, and initial exposed population $E_0 = 10^{-2}, 10^{-3}, 10^{-4}, 10^{-5}, 10^{-6}, 10^{-7}, 10^{-8}$ [20].

notion that even a single individual can cause an outbreak. If multiple individuals trigger the outbreak in a province, state, or country, the overall outbreak dynamics will remain the same, but the peak of the outbreak will happen earlier.

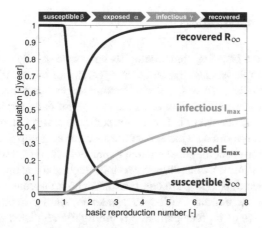

Fig. 7.9 Classical SEIR model. Final size relation and maximum exposure and infection as a function of the basic reproduction number R_0. Increasing the basic reproduction number beyond one, $R_0 > 1$, reduces the final susceptible population S_∞ and increases the final recovered population R_∞ and the maximum exposed and infectious populations E_{max} and I_{max}. Initial exposed population $E_0 = 0.0001$.

7.6 Maximum exposure and infection of the SEIR model

Figure 7.9 is the SEIR equivalent of the SIS model in Figure 2.6 and the SIR model in Figure 3.7; however, unlike the SIS and SIR models, the SEIR model has no explicit analytical solution for the individual maximum exposed and maximum infectious populations, and has to be solved numerically. Figure 7.9 once again highlights the importance of the basic reproduction number R_0: Increasing R_0 beyond one reduces the final susceptible population S_∞ and increases the final recovered population R_∞ and the maximum exposed and infectious populations E_{max} and I_{max}. From the SIR model in Figures 3.7 and 3.6, we know that these curves are sensitive to the size of the initial outbreak. Here we have assumed a small initial exposed population of $E_0 = 0.0001$. Increasing the initial exposed population population E_0 would reduce the initial and finite susceptible populations S_0 the S_∞ and increase the recovered population R_∞, similar to Figure 3.7, and increase the maximum exposed and infectious populations E_{max} and I_{max}, similar to Figure 3.6. For many infectious diseases, the peak of the infectious population, I_{max}, is of great epidemiological interest. It characterizes the dimension of the outbreak and is often viewed as an indicator for the number of hospitalizations and required intensive care units during an outbreak [15].

7.7 Dynamic SEIR model

For more than three decades, epidemiologists have successfully applied the classical SEIR model to understand the outbreak dynamics of the measles, chickenpox, mumps, polio, rubella, pertussis, and smallpox under unconstrained conditions [8]. In all these studies, the outbreak ends as the number of daily new cases, $\beta\,SI$, decreases. As such, the classical SEIR model is self-regulating: It naturally converges to an endemic equilibrium, at which either the susceptible group S, or the infectious group I, or both have become small enough to prevent new infections [3]. The SEIR model assumes that an infectious disease develops freely and that the contact, latency, and infectious rates β, α, and γ are constant throughout the course of the outbreak [2]. However, for the current COVID-19 pandemic, similar to SARS, MERS, or ebola, the dynamics of its four populations are tightly regulated by public health interventions [10]. This implies that model parameters like the contact rate β, the rate at which an infectious individual comes into contact and infects others, are not constant, but modulated strongly by social behavior and political actions. To address

Fig. 7.10 Dynamic SEIR model. The dynamic SEIR model contains four compartments for the susceptible, exposed, infectious, and recovered populations, S, E, I, and R. The transition rates between the compartments, the dynamic contact rate $\beta(t)$, the latent rate α, and the infectious rate γ, are inverses of the dynamic contact period, $B(t) = 1/\beta(t)$, the latent period, $A = 1/\alpha$, and the infectious period, $C = 1/\gamma$.

these interventions, we introduce a *dynamic* SEIR model that explicitly accounts for a time-varying dynamic contact rate $\beta(t)$ as indicated in Figure 7.10 and rewrite the system of equations (7.1) [17],

$$
\begin{aligned}
\dot{S} &= -\beta(t)\,S\,I \\
\dot{E} &= +\beta(t)\,S\,I - \alpha\,E \\
\dot{I} &= \qquad\qquad + \alpha\,E - \gamma\,I \\
\dot{R} &= \qquad\qquad\qquad + \gamma\,I\,.
\end{aligned}
\tag{7.15}
$$

To model the effects of interventions, we adopt a hyperbolic tangent type ansatz for the contact rate $\beta(t)$ [17],

$$
\beta(t) = \beta_0 - \tfrac{1}{2}[\,1 + \tanh([\,t - t^*\,]/T)\,][\,\beta_0 - \beta_t\,],
\tag{7.16}
$$

where β_0 is the initial contact rate at the onset of the pandemic, β_t is the contact rate in response to public health interventions, t^* is the adaptation time, and T is the transition time. For easier interpretation, we reparameterize the system (7.15) in term of the time-dependent dynamic reproduction number $R(t) = \beta(t)/\gamma$,

Fig. 7.11 The dynamic hyperbolic-tangent type reproduction
number, $R(t) = R_0 - \frac{1}{2}[1 + \tanh([t - t^*]/T)][R_0 - R_t]$,
ensures a smooth transition from the initial basic reproduction
number, $R_0 = \beta_0/\gamma$, at the beginning of the outbreak to
the reduced reproduction number, $R_t = \beta_t/\gamma$, in response
to public health interventions, where t^* and T are the
adaptation and transition times. The hyperbolic tangent
function, $\tanh(t) = [\exp(t) - \exp(-t)]/[\exp(t) + \exp(-t)]$,
smoothly transitions from -1 to +1. It is a rescaling of the
logistic sigmoid function, $f(t) = \exp(t)/[1 + \exp(t)]$, with
$\tanh(t) = 2f(2t) - 1$.

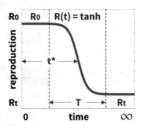

$$\begin{aligned}
\dot{S} &= -R(t)\,\gamma\,S\,I \\
\dot{E} &= +R(t)\,\gamma\,S\,I - \alpha\,E \\
\dot{I} &= \qquad\quad\; +\alpha\,E - \gamma\,I \\
\dot{R} &= \qquad\qquad\qquad\;\; +\gamma\,I\,.
\end{aligned} \qquad (7.17)$$

With equation (7.16), the dynamic reproduction number takes the following hyperbolic tangent type form,

$$R(t) = R_0 - \frac{1}{2}[1 + \tanh([t - t^*]/T)][R_0 - R_t]\,. \qquad (7.18)$$

Figure 7.11 confirms that this ansatz ensures a smooth transition from the initial
basic reproduction number $R_0 = \beta_0/\gamma$ at the beginning of the outbreak to the
reduced reproduction number $R_t = \beta_t/\gamma$ in response to public health interventions,
where t^* and T are the adaptation and transition times. Rather than restricting time-
dependence to only the contact rate $\beta(t)$, some approaches propose a dynamic ansatz
for all transition rates, $\alpha(t)$, $\beta(t)$, and $\gamma(t)$, for example, by using a cubic polynomial
ansatz [28].

Figures 7.12 to 7.15 illustrate the outbreak dynamics of the dynamic SEIR model
with a time-varying dynamic contact rate $\beta(t)$ and a time-varying dynamic reproduction number $R(t)$. The gray curves highlight the hyperbolic tangent type nature
of the dynamic reproduction number $R(t)$, the dark red, red, orange, and blue curves
illustrate the dynamics of the susceptible, exposed, infectious, and recovered populations, S, E, I, and R. Unless stated otherwise, we use a latent period of $A = 2.5$ days,
an infectious period of $C = 6.5$ days, a basic reproduction number of $R_0 = 4.5$, a
reproduction number under public health interventions of $R_t = 0.75$, and adaptation
and transition times of $t^* = 20$ days and $T = 15$ days. In all simulations, according
to equation (7.18), the dynamic reproduction number $R(t)$ transitions gradually from
the initial basic reproduction number R_0 at the beginning of the outbreak to the
reduced reproduction number R_t associated with public health interventions. This
implies that, according to equation (7.16), the contact rate $\beta(t)$ transitions gradually
from the initial contact rate $\beta_0 = R_0\,\gamma = 0.69/\text{days}$ at the beginning of the outbreak
to the reduced contact rate $\beta_t = R_t\,\gamma = 1.44/\text{days}$ associated with public health
interventions. The adaptation time t^* marks the midpoint of the transition and the
transition time T is its duration. The outbreak is more pronounced and results in
larger exposed, infectious, and recovered populations E, I, and R for larger basic

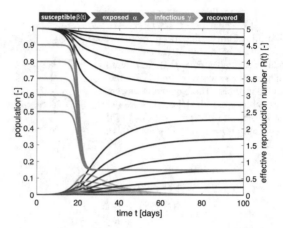

Fig. 7.12 Dynamic SEIR model. Sensitivity with respect to the basic reproduction number R_0. Increasing the basic reproduction number R_0 increases the intensity of the outbreak, and with it the number of cases I and the recovered population R. Latent period $A = 2.5$ days, infectious period $C = 6.5$ days, dynamic reproduction number $R(t) = R_0 - \frac{1}{2}[1 + \tanh([t - t^*]/T)][R_0 - R_t]$ with reduced reproduction number $R_t = 0.75$, adaptation time $t^* = 20$ days, transition time $T = 15$ days, and basic reproduction number $R_0 = [2.5, 3.0, 3.5, 4.0, 4.5, 5.0]$ [17].

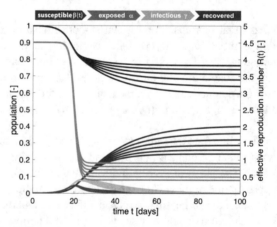

Fig. 7.13 Dynamic SEIR model. Sensitivity with respect to the reduced reproduction number R_t. Increasing the reduced reproduction number R_t decreases the effect of interventions, and increases the number of cases I and the recovered population R. Latent period $A = 2.5$ days, infectious period $C = 6.5$ days, dynamic reproduction number $R(t) = R_0 - \frac{1}{2}[1 + \tanh([t - t^*]/T)][R_0 - R_t]$ with basic reproduction number $R_0 = 4.5$, adaptation time $t^* = 20$ days, transition time $T = 15$ days, and reduced reproduction number $R_t = [0.4, 0.5, 0.6, 0.7, 0.8, 0.9]$ [17].

reproduction numbers R_0 as we see in Figure 7.12, for larger reduced reproduction numbers R_t as we see in Figure 7.13, and for larger adaptation times t^* as we see in Figure 7.14, and for larger transition times T as we can see in Figure 7.15.

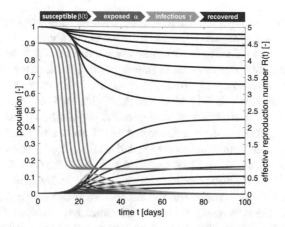

Fig. 7.14 Dynamic SEIR model. Sensitivity with respect to the adaptation time t^*. Increasing the adaptation time t^* to interventions delays the reduction to a lower reproduction number R_t, which increases the number of cases I and the recovered population R. Latent period $A = 2.5$ days, infectious period $C = 6.5$ days, dynamic reproduction number $R(t) = R_0 - \frac{1}{2}[1 + \tanh([t - t^*]/T)][R_0 - R_t]$ with basic reproduction number $R_0 = 4.5$, reduced reproduction number $R_t = 0.75$, transition time $T = 15$ days, and adaptation time $t^* = [10, 12, 14, 16, 18, 20, 22]$ days [17].

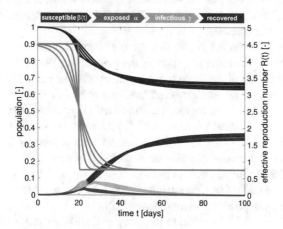

Fig. 7.15 Dynamic SEIR model. Sensitivity with respect to the transition time T. Increasing the transition time T slows down the effect of interventions, and slightly increases the number of cases I and the recovered population R. Latent period $A = 2.5$ days, infectious period $C = 6.5$ days, dynamic reproduction number $R(t) = R_0 - \frac{1}{2}[1 + \tanh([t - t^*]/T)][R_0 - R_t]$ with basic reproduction number $R_0 = 4.5$, reduced reproduction number $R_t = 0.75$, adaptation time $t^* = 20$ days, and transition time $T = [1, 15, 30, 45, 60]$ days [17].

7.8 Example: Early COVID-19 outbreak dynamics in China

China was the first country to report COVID-19 cases. The country managed the early outbreak with strict measures and rapidly reduced the number of cases by mid spring 2020. Very early in the pandemic, the Chinese case data captured all three phases–the increase, peak, and decrease of the infectious population–and was one of the richest datasets to identify model parameters and learn about the outbreak. The reported data describe the temporal evolution of confirmed, recovered, active, and death cases starting January 22, 2020 [12]. By April 4, there were 81,639 confirmed cases, 76,755 recovered, 1,558 active, and 3,326 deaths. From these data, we can map out the temporal evolution of the infectious population $I(t)$ as the difference between the confirmed cases minus the recovered and deaths, and the recovered population $R(t)$ as the sum of the recovered and deaths, for all Chinese provinces. To simulate the outbreak with our SEIR model, in addition to the exposed and infectious periods A and C and the basic reproduction number R_0, we introduce two additional parameters, the community spread ρ and the affected population η [22].

We motivate the *community spread* by the sensitivity analysis in Section 7.5. From Figure 7.8, we know that the outbreak dynamics of the SEIR model depend critically on the initial conditions, specifically on the size of the exposed population E_0 at the day on which the first infectious case is reported, $I_0 \geq 0$. However, since the exposed population is asymptomatic, its initial value E_0 is unknown. To quantify the initial exposed population, we introduce the parameter $\rho = E_0/I_0$, the initial community spread. It defines the fraction of exposed versus infectious individuals at the first day of reported cases and is a measure of hidden community spreading at the beginning of the outbreak.

We motivate the *affected population* by the tight outbreak control that closed infectious regions to the remaining part of the population. We map the SEIR model with a total population of one onto the absolute number of cases for each reporting region by introducing a normalization parameter $\eta = N^*/N$, the affected population. It defines the fraction of the region-specific potentially susceptible population N^* relative to the total regional population N [25], and is a measure for containment of the outbreak. This results in an SEIR model with a total of five parameters: the exposed period $A = 1/\alpha$, the infectious period $C = 1/\gamma$, the basic reproduction number $R = C/B$, the initial latent population $\rho = E_0/I_0$, and the affected population $\eta = N^*/N$. We identify these parameters using a Levenberg-Marquardt algorithm and ignore data from potential second waves [20].

Figures 7.16 and 7.17 summarize the early outbreak dynamics of the COVID-19 pandemic in the Chinese provinces of Hubei and Hunan. The dots indicate the reported infectious and recovered populations, the lines highlight the simulated susceptible, exposed, infectious, and recovered populations. Hubei and Hunan have populations of 58.5 and 67.4 million, of which 1.5 million and 15.6 thousand people became infected throughout the first two months of the pandemic, with maximum infectious populations of 52,554 in Hubei and 671 in Hunan. The latent and infectious periods in both provinces takes values of $A = 2.3$ days and $A = 2.8$ days, which seem to be reasonable values for COVID-19. The infectious periods of $C = 22.7$ days and $C = 15.9$ days seem to be higher than commonly reported values on the order of $C = 6.5$ days, which suggests that the reported recovery lacks behind the clinical recovery of COVID-19 patients. The basic reproduction numbers of $R_0 = 11.5$ and $R_0 = 14.6$ are on the high end when compared to Table 1.3, which could

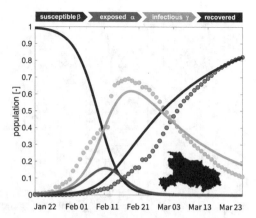

Fig. 7.16 Early COVID-19 outbreak in Hubei. Reported infectious and recovered populations and simulated susceptible, exposed, infectious, and recovered populations. Simulations are based on a province-specific parameter identification with a latent and infectious periods $A = 2.3$ days and $C = 22.7$ days, a basic reproduction number $R_0 = C/B = 11.5$, an initial community spread $\rho = E_0/I_0 = 0.0$, and an affected population $\eta = N^*/N = 0.0013$.

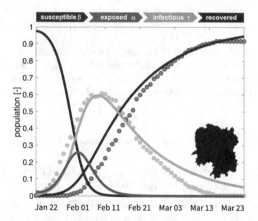

Fig. 7.17 Early COVID-19 outbreak in Hunan. Reported infectious and recovered populations and simulated susceptible, exposed, infectious, and recovered populations. Simulations are based on a province-specific parameter identification with a latent and infectious periods $A = 2.8$ days and $C = 15.9$ days, a basic reproduction number $R_0 = C/B = 14.6$, an initial community spread $\rho = E_0/I_0 = 4.9$, and an affected population $\eta = N^*/N = 0.000016$.

be a combined result of both, the long infectious periods and the early stage of the global outbreak during which the real dimension of COVID-19 was unknown and community mitigation strategies were still in their infancy. The initial latent population is zero in Hubei, $\rho = 0.0$, where the outbreak started, and larger than zero in Hunan, $\rho = 4.9$, which implies that some form of community transmission happened before the first cases were reported. The fraction of the affected population was $\eta = 0.0013$ for Hubei and $\eta = 0.000016$ for Hunan, suggesting that containment

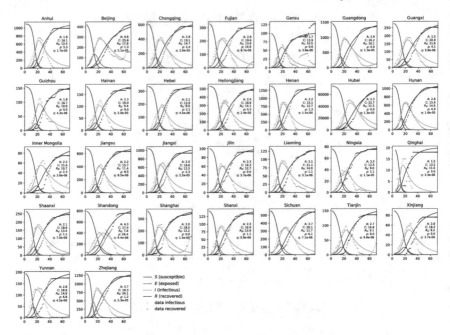

Fig. 7.18 Early COVID-19 outbreak in China. Reported infectious and recovered populations and simulated susceptible, exposed, infectious, and recovered populations. Simulations are based on a province-specific parameter identification of the latent period A, infectious period C, basic reproduction number R_0, community spread ρ, and affected population η for all 30 provinces [20].

of the outbreak was better managed in Hunan where the outbreak hit slightly later than in Hubei [20].

Figure 7.18 summarizes the early outbreak dynamics of the COVID-19 pandemic across all 30 Chinese provinces. The dots indicate the reported infectious and recovered populations, the lines highlight the simulated susceptible, exposed, infectious, and recovered populations. Each graph reports the five province-specific parameters, the latent and infectious periods A and C, the basic reproduction number R_0, the community spread ρ, and the affected population η [20]. Notably, the community spread ρ is smallest and the affected population η is largest in the province of Hubei, the first province that reported cases of COVID-19, which we have analyzed in detail in Figure 7.16. Small values of ρ indicate a close monitoring of the COVID-19 outbreak, with very few undetected cases at the reporting of the first infectious case. The largest value of $\rho = 26.4$ suggests that, at the onset of the outbreak, a relatively large number of cases in the province of Shandong was undetected. The affected population η is a province-specific measure for the containment of the outbreak. Naturally, this number is largest in the province of Hubei, with $\eta = 1.3 \cdot 10^{-3}$, and, because of a rapid transition to strict containment, much smaller in all other provinces, for example in Hunan, which we have analyzed in detail in Figure 7.17.

Table 7.1 Early COVID-19 outbreak in China. Latent period A, contact period B, infectious period C, basic reproduction number $R_0 = C/B$, community spread $\rho = E_0/I_0$, and affected population $\eta = N^*/N$; means ± standard deviations across all 30 provinces in Figure 7.18.

parameter	mean ± std	interpretation
A [days]	2.56 ± 0.72	latent period
B [days]	1.47 ± 0.32	contact period
C [days]	17.82 ± 2.95	infectious period
R_0 [-]	12.58 ± 3.17	basic reproduction number
ρ [-]	3.19 ± 5.44	community spread
η [-]	$5.19 \cdot 10^{-5} \pm 2.23 \cdot 10^{-4}$	affected population

Table 7.1 summarizes the parameters for the early COVID-19 outbreak in China. Averaged over all Chinese provinces, the analysis identified a latent period of $A = 2.56 \pm 0.72$ days, a contact period of $B = 1.47 \pm 0.32$ days, an infectious period of $C = 17.82 \pm 2.95$ days, a basic reproduction number of $R_0 = C/B = 12.58 \pm 3.17$, a community spread of $\rho = E_0/I_0 = 3.19 \pm 5.44$, and an affected population of $\eta = N^*/N = 5.19 \cdot 10^{-5} \pm 2.23 \cdot 10^{-4}$ [20]. Interestingly, these parameter values were identified for one of the first complete data sets of a COVID-19 outbreak, from initial seeding and growth, via a peak of the infectious population, to a gradual decay. The latent period of $A = 2.56 \pm 0.72$ days lies within the range of values that have later been confirmed by other studies. The infectious period of $C = 17.82 \pm 2.95$ days is about two to three times higher than the clinically recorded and now widely accepted infectious period, which suggests that there is a time lack in reporting of recovery in the available data.

Our identified basic reproduction number of $R_0 = C/B = 12.58 \pm 3.17$ is on the order of the basic reproduction number for the measles and pertussis of 12-18, but larger than those for smallpox and rubella of 6-7. From the examples in Chapters 3 and 4, we know that the SEIR model generally predicts larger basic reproduction numbers than the SIR model. Nonetheless, during the early stages of an outbreak, a reproduction number of $R_0 = 12$ is alarmingly high. Fortunately, several months into the pandemic, the reported reproduction number had dropped to values closer to one [17]. It is important to keep in mind that the later reproduction numbers naturally include community mitigation strategies and political interventions. They can vastly underestimate the true basic reproduction number R_0 and with it the herd immunity threshold, $H = 1 - 1/R_0$, and the number of people that need to acquire immunity through either infection or vaccination. To address this limitation, the dynamic SEIR model in Section 7.7 introduces a time-varying dynamic reproduction number $R(t)$. In the next section, we adopt the dynamic SEIR model to simulate the effects of COVID-19 outbreak control in Germany.

Table 7.2 Early COVID-19 outbreak in Germany. Newly reported COVID-19 cases in Germany from Wednesday, February 26, the first day of reported cases, to Tuesday, June 9, 2020, with school closure on March 13, tight border control on March 17, and lockdown-like measures on March 22, 2020; case data ΔN_I are reported as seven-day moving averages.

	Wed	Thu	Fri	Sat	Sun	Mon	Tue
week 1	2	4	9	9	16	21	27
week 2	33	71	85	103	130	152	194
week 3	244	314	429	543	682	864	1114
week 4	1481	1796	2310	2538	2723	3112	3375
week 5	3581	4088	4432	5047	5366	5404	5546
week 6	5808	5837	5755	5485	5384	5213	5122
week 7	5045	4778	4430	4194	3962	3814	3506
week 8	3066	2780	2747	2610	2555	2428	2320
week 9	2271	2204	1944	1827	1718	1670	1637
week 10	1556	1412	1296	1208	1128	1056	1014
week 11	946	917	930	908	888	918	881
week 12	848	792	730	703	681	674	665
week 13	633	578	574	534	525	500	495
week 14	480	491	472	472	453	425	400
week 15	362	353	342	343	339	349	346

7.9 Example: Early COVID-19 outbreak control in Germany

Germany was one was one of the first countries in Europe to see cases of COVID-19 [16]. Having learnt from China, the country rapidly implemented several outbreak control strategies and observed a gradual reduction of new infections. We use the example of Germany to compare two different approaches to simulate outbreak control: the *potentially susceptible population* approach and the *dynamic reproduction number* approach.

Table 7.2 summarizes the seven-day moving average of the newly reported COVID-19 cases in Germany [9], from Wednesday, February 26, 2020, the first day of reported cases, to Tuesday, June, 9, 2020. This example builds on a problem in Chapter 4 with case data in Table 4.1, but now displayed for a window of the first 15 weeks of the outbreak. Similar to Chapter 4, we first calculate the total infectious population for each day t_n within each interval,

$$N_I(t_n) = \sum_{i=n+1-C}^{n} \Delta N_I(t_i),$$

by summing the newly reported cases ΔN_I for the $C = 7$ previous days.

Potentially susceptible population approach. In the analysis of the German case data in Figure 4.2 of Chapter 4, we have estimated the weekly averages of the basic reproduction number in the first five weeks of the outbreak from February 29 to April 4, 2020 to $R_0 = [7.97, 4.75, 4.62, 2.62, 1.63]$. We approximate the early basic reproduction number as constant, $R_0 = 4$, within the entire interval, run a simulation, and record the maximum infectious population numerically as $I_{max} = 0.2742$. Germany has a total population of $N = 82.91$ million people, but, clearly,

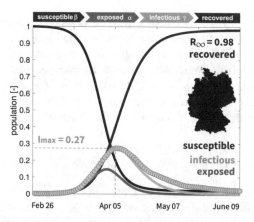

Fig. 7.19 Early COVID-19 outbreak control in Germany with potentially susceptible population N^*. Reported infectious population and simulated susceptible, exposed, infectious, and recovered populations. Simulations are based on latent and infectious periods of $A = 3$ days and $C = 7$ days, a basic reproduction number of $R_0 = 4$, an initial exposed population of $E_0 = 0.0007$, and a potentially susceptible population of $N^* = 143,031$.

because of the gradual lockdowns, only a fraction N^* of the entire population was potentially susceptible to the virus. We map the SEIR model with a total population of one onto the potentially susceptible population of N^*, which we calculate from mapping the numerically estimated maximum infectious population $I_{max} = 0.2742$ into the maximum number of infectious people, $\max\{N_I\} = 39,219$,

$$I_{max} = \frac{\max\{N_I\}}{N^*} \quad \text{thus} \quad N^* = \frac{\max\{N_I\}}{I_{max}} = \frac{39,219}{0.2742} = 143,031 .$$

To characterize the level of containment and compare it to other countries, we can calculate the *affected population* η, the ratio between the potentially susceptible population N^* and the total population N,

$$\eta = \frac{N^*}{N} \quad \text{thus} \quad \eta = \frac{143,031}{82,910,000} = 0.0017 = 0.17\% .$$

An affected population of $\eta = N * /N = 0.17\%$ confirms that, during the first wave of the outbreak, political measures were quite effective in containing the spread of COVID-19 and reducing the number of new infections.

Figure 7.19 displays a good agreement between the curves of the SEIR model and the dots of the recorded case data. The infectious population increases gradually during the first three weeks of the outbreak, then grows rapidly for another two weeks, reaches a peak, and steadily declines. On March 13, an almost nationwide school closure went into effect, on March 17, tight boarder controls were implemented, and on March 22, most of Germany was in a lockdown-like state. This reduced the basic reproduction number from values well above four, $R_0 = 7.97, 4.75, 4.62$, to $R_0 = 2.62, 1.63$. Figure 7.19 nicely highlights the success of these measures.

The classical SEIR model, combined with the potentially susceptible population approach, displays a good agreement with the recorded case data, especially before reaching the peak of infections. By design, the maximum infectious population $I_{max} = 0.27$ of the SEIR simulation matches the maximum infectious population of the reported data $\max\{N_I\}/N^* = 0.27$. Both peaks occur on April 5. For the model to peak on this day, we assumed that 100 individuals were exposed at the beginning of the outbreak, $E_0 = 100/N^*$. The SEIR model predicts a final recovered population,

$$R_\infty = 1 + W(-S_0\,R_0\,\exp(-R_0[\,1 - R_0\,]))/R_0 = 0.98\,,$$

both analytically and numerically. For a potential susceptible population of $N^* = 143{,}031$, this would imply that a total of $N_{R_\infty} = R_\infty \cdot N^* = 140{,}200$ individuals would have recovered at endemic equilibrium. Summing the reported case numbers ΔN_I in Table 7.2 yields $N_R = 185{,}226$ recovered cases, resulting in a model error of 22.8%. Figure 7.19 suggests that the model overestimates the effects of containment, especially in the post-peak regime where the reported cases clearly exceed the simulated infectious population.

This example shows that the potentially susceptible population approach is a reasonable ad hoc strategy to map our zero-to-one model populations into a population of closest contacts during short and strict lockdown periods. For longer and softer lockdowns, during which essential workers, health care professionals, and other members of the community will necessarily be in contact with a wider group of individuals, the assumption of a fixed potentially susceptible population N^* can significantly overestimate the success of mitigation strategies, especially long term.

Dynamic reproduction number approach. To address the limitations of the ad hoc potentially susceptible population approach, we will now illustrate a more mechanistic approach, the dynamic SEIR model from Section (7.7), with a time-varying dynamic basic reproduction number $R(t)$. Figure 4.2 of Chapter 4 illustrates the time evolution of the logarithmic infectious population N_I, from which we can estimate the growth rates,

$$G = \frac{\ln(N_I(t_{n+1})) - \ln(N_I(t_n))}{t_{n+1} - t_n}\,,$$

between any two days within the interval to $G = [\,0.58, 0.49, 0.41, 0.39, 0.35, 0.33, 0.31, \ldots, -0.039\,]$. From these growth rates, we can estimate the daily basic reproduction numbers from equation (4.13),

$$R_0 = 1 + G\,[\,A + C\,] + G^2\,AC\,,$$

assuming latent and infectious periods of $A = 3$ days and $C = 7$ days, $R_0 = [\,13.99, 10.96, 8.59, 7.97, 7.12, 6.51, 6.11, \ldots, 0.64\,]$. Figure 7.20 illustrates the estimated daily reproduction number R_0 and its hyperbolic tangent approximation [17],

$$R(t) = R_0 - \frac{1}{2}[1 + \tanh([t - t^*]/T)][R_0 - R_t]\,,$$

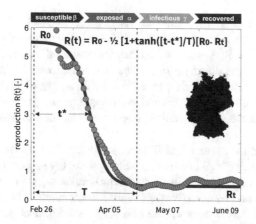

Fig. 7.20 Early COVID-19 outbreak control in Germany with dynamic reproduction number
$R(t)$. Estimated daily reproduction number, $R_0 = 1 + G[A + C] + G^2 AC$, and hyperbolic tangent type
approximation for the dynamic reproduction number, $R(t) = R_0 - \frac{1}{2}[1 + \tanh([t - t^*]/T)][R_0 - R_t]$,
with basic reproduction number $R_0 = 5.5$, reduced reproduction number $R_t = 0.5$, adaptation time
$t^* = 25$ days, and transition time $T = 50$ days.

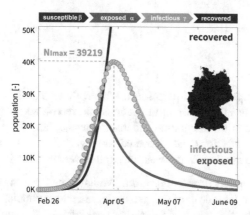

Fig. 7.21 Early COVID-19 outbreak control in Germany with dynamic reproduction number
$R(t)$. Reported infectious population and simulated exposed, infectious, and recovered populations.
Simulations are based on latent and infectious periods of $A = 3$ days and $C = 7$ days, and a dynamic
reproduction number, $R(t) = R_0 - \frac{1}{2}[1 + \tanh([t - t^*]/T)][R_0 - R_t]$, with basic reproduction number
$R_0 = 5.5$, reduced reproduction number $R_t = 0.5$, adaptation time $t^* = 25$ days, and transition time
$T = 50$ days.

for a basic reproduction number of $R_0 = 5.5$, a reduced reproduction number of
$R_t = 0.5$, an adaptation time of $t^* = 25$ days, and a transition time of $T = 50$ days.
Clearly, with the exception of the very first few days, the hyperbolic tangent ansatz
provides a reasonably good approximation of the discrete daily reproduction numbers
with a smooth transition from the early basic reproduction number of $R_0 = 5.5$ to
the reduced reproduction number of $R_t = 0.5$.

Figure 7.21 summarizes the same COVID-19 outbreak data for Germany as Figure 7.19, but now simulated using the dynamic SEIR model with the time-varying basic reproduction number $R(t)$. Since we no longer constrain the potentially susceptible population N^*, we can directly plot the absolute infectious population N_I derived from the newly reported cases ΔN_I in Table 4.1 with a peak of 39,219 infections on April 5, 2020. Figure 7.21 confirms that the reduced reproduction number R_t allows us to flatten the curve without artificially constraining the susceptible populations [9].

Taken together, the dynamics SEIR model in Figure 7.21 provides a more mechanistic approach than the ad hoc scaling with a potentially susceptible population in Figure 7.19. It allows us to directly capture behavioral changes that result in reduced contact rates $\beta(t)$ and reduced dynamic reproduction numbers $R(t)$ [19]. However, the dynamic SEIR model comes at the price of three additional parameters that are not easy to fit by hand. At the same time, these parameters, the reduced reproduction number and the adaptation and transition times provide valuable insight into the effectiveness of behavioral changes and political interventions [12]. In Chapter 12 we will show how we can systematically learn these parameters from reported case data using Bayesian methods.

Problems

7.1 Explicit vs. implicit time integration. We have seen that the explicit time integration overestimates the maximum infectious and recovered populations I_{max} and R_∞ and that implicit time integration underestimates I_{max} and R_∞. If we fit our SEIR model to reported infectious and recovered data, what does this imply for the fitted basic reproduction number R_0 if we use an explicit versus implicit time integration?

7.2 Explicit vs. implicit time integration. We have seen that the explicit time integration predicts a later peak of the maximum infectious populations I_{max} and the implicit time integration predicts and earlier peak of I_{max}. If fit our SEIR model to reported infectious data, what does this imply for the fitted infectious period C if we use an explicit versus implicit time integration?

7.3 Dynamic SEIR model. The COVID-19 outbreak in Germany displayed a hyperbolic-tangent type reproduction number, $R(t) = R_0 - \frac{1}{2}[1+\tanh([t-t^*]/T)][R_0 - R_t]$, with a basic reproduction number of $R_0 = 5.5$, a reduced reproduction number of $R_t = 0.5$, an adaptation time of $t^* = 25$ days, and a transition time of $T = 50$ days. On what day did the dynamic reproduction number drop below $R(t) = 2.5$? Compare your answer against Figure 7.20.

7.4 Dynamic SEIR model. Assume the COVID-19 outbreak in the Netherlands displayed a hyperbolic-tangent type basic reproduction number, $R(t) = R_0 - \frac{1}{2}[1 + \tanh([t - t^*]/T)][R_0 - R_t]$, with a basic reproduction number of $R_0 = 4.5$, a reduced reproduction number of $R_t = 0.5$, an adaptation time of $t^* = 24$ days, and an infectious period of $C = 7$ days. What was the contact rate after $t = 12$ days?

7.5 Dynamic SEIR model. Assume the COVID-19 outbreak in Austria displayed a hyperbolic-tangent type reproduction number, $R(t) = R_0 - \frac{1}{2}[1 + \tanh([t - t^*]/T)][R_0 - R_t]$, with initial and reduced contact periods of $B_0 = 1.0769$ days and $B_t = 14$ days, and infectious period of $C_0 = 7$ days, an adaptation time of $t^* = 20$ days, and a transition time of $T = 48$ days. What was the dynamic reproduction $R(t)$ number after $t = 20$ days?

7.6 Model your own country, state, or city. Draw the COVID-19 case data from the first wave of the outbreak of your own country, state, or city. From the daily new cases ΔN_I, calculate the infectious population N_I assuming latent and infectious periods of $A = 3$ days and $C = 7$ days. Calculate the daily growth rate, G, and estimate the daily basic reproduction number, R_0. Plot the daily basic reproduction number over time. Compare your plot against Figure 7.20. Try to fit a hyperbolic tangent ansatz, $R(t) = R_0 - \frac{1}{2}[1 + \tanh([t - t^*]/T)][R_0 - R_t]$. What are your basic reproduction number, reduced reproduction number, adaptation and transition times? Does the hyperbolic tangent ansatz provide a good fit?

7.7 Time step size. We have seen that the result of the numerical time integration depends on the time step size Δt. Implement the explicit SEIR model. Simulate an outbreak with latent and infectious periods of $A = 3$ days and $C = 7$ days, an initial infectious population of $I_0 = 0.001$ and varying reproduction numbers $R_0 = 2, 3, 4, 5, 10, 15$. Gradually increase the time step size $\Delta t = 0.01, 0.1, 1, 2, 5, 10$ days. Make informed recommendations for a reasonable time step size Δt.

7.8 Early exponential growth. Simulate the time evolution of the susceptible, infectious, and recovered populations for an SEIR model with latent and infectious periods of $A = 3$ days and $C = 7$ days, a basic reproduction number of $R_0 = 3$, and an initial infectious population of $I_0 = 0.001$. Plot the populations in time. Add the infectious population of the early exponential growth model, $I(t) = I_0 \exp(Gt)$ with $G = \frac{1}{2}\sqrt{(\alpha - \gamma)^2 + 4\alpha\beta} - \frac{1}{2}[\alpha + \gamma]$. Compare your results against Figure 7.22. When do the infectious populations start to deviate? When does the error between the SEIR model and the exponential growth model exceed 0.1% and 1%? What is the size of the susceptible population at the time of these two errors? Make recommendations when we can confidently use the exponential growth model as an approximation.

7.9 Maximum infectiousness. A survey at Stanford University reported that 7.5% of undergraduates were exposed to an influenza with exposed and infectious periods of $A = 2$ days and $C = 6$ days at the beginning of the year, and 52.5% were susceptible at the end. Determine the basic reproduction number R_0 and the contact period B analytically using the SEIR equations from Chapter 4. Simulate the outbreak of the influenza numerically using either an explicit or implicit SEIR model. Map the evolution of the infectious population in time $I(t)$. On which day $t(I_{max})$ does the infectious population reach its maximum I_{max}? What is a reasonable time step size to reproduce the analytical maximum infectious population I_{max}?

7.10 Maximum infectiousness. A French boarding school reported an influenza that spread through 1224 of its 1632 students. Simulate and map the evolution of the

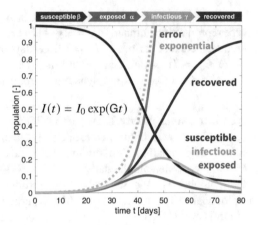

Fig. 7.22 Classical SEIR model vs exponential growth model. Solid dark red, red, orange, and blue curves represent the numerical solutions of the susceptible, exposed, infectious, and recovered populations of the SEIR model; the dotted orange line represents the analytical solution of early exponential growth, $I(t) = I_0 \exp(Gt)$; dark grey line highlights the error between both models. Infectious period $C = 7$ days, basic reproduction number $R_0 = 3$, initial infectious population $I_0 = 0.001$.

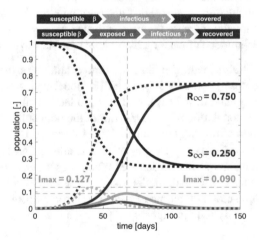

Fig. 7.23 Classical SIR and SEIR models. Infectious population during an influenza. The influenza spread through 1224 of 1632 students assuming one student was initially with exposed or infectious with latent and infectious periods of $A = 2$ days and $C = 5$ days.

infectious population $I(t)$ in time if, on the first day of class, a single student was exposed and the latent and infectious periods were $A = 2$ days and $C = 5$ days. On which day does the outbreak peak? Compare your results against Figure 7.23.

7.11 Maximum infectiousness. A French boarding school reported an influenza that spread through 1224 of its 1632 students. Simulate and map the evolution of the infectious population $I(t)$ in time using first the SEIR model from the previous

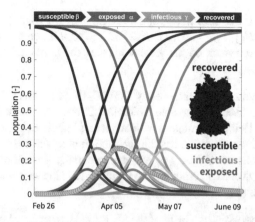

Fig. 7.24 Early COVID-19 outbreak in Germany with potentially susceptible population N^*. Reported infectious population and simulated susceptible, exposed, infectious, and recovered populations. Latent and infectious periods $A = 3$ days and $C = 7$ days, basic reproduction number $R_0 = 4$, and initial exposed population $E_0 = 1/N^*, 10/N^*, 100/N^*, 1000/N^*$ with $N^* = 143,031$, from light to dark.

problem and then the SIR model from Figure 6.11 in Chapter 6. Compare the maximum infectious populations I_{max} and the times $t(I_{max})$ of the SEIR and SIR models. Compare your results against Figure 7.23. Comment on how model selection could influence disease management and planning.

7.12 Maximum infectiousness. Two visitors introduce COVID-19 to Stanford campus with a population of 5000 students. Each infectious individual meets on average 0.4286 students per day and is infectious for seven days. The university plans to isolate all infectious individuals after a latent period of three days. Simulate the outbreak using the SEIR model. Estimate the maximum number of required isolation units and the day of the peak of the outbreak. Compare your results against the estimates for the SIR model in Chapter 3.1, explain the differences, and discuss the impact of model selection on outbreak control.

7.13 Affected population and outbreak control. For the early COVID-19 outbreak in Germany with the seven-day moving average case data in Table 7.2, redo the Germany example from Section 7.9, but now with a basic reproduction number of $R_0 = 5$. Estimate the maximum infectious population I_{max}, the potentially susceptible population N^*, and the affected population η, but keep $A = 3$ days, $C = 7$ days, and $E_0 = 100/N^*$ fixed. Discuss how increasing the basic reproduction number affects the potentially susceptible and affected populations.

7.14 Initial conditions and community spreading. For the early COVID-19 outbreak in Germany with the seven-day moving average case data in Table 7.2, implement the SEIR model and recreate Figure 7.19 for a simulation with latent and infectious periods of $A = 3$ days and $C = 7$ days and a basic reproduction number of

$R_0 = 4$. Vary the initial infectious population $E_0 = 1/N^*, 10/N^*, 100/N^*, 1000/N^*$ with $N^* = \max\{N_I\}/I_{max} = 39,219/0.2742 = 143,031$, compare your graphs to Figure 7.24, and comment on the results. Discuss the implications of initial undetected community spreading.

7.15 Outbreak dynamics. For the early COVID-19 outbreak in Germany with the seven-day moving average case data in Table 7.2, implement the SEIR model and recreate Figure 7.19 for a simulation with latent and infectious periods of $A = 3$ days and $C = 7$ days, but now with a basic reproduction number of $R_0 = 5$. Estimate the maximum infectious population I_{max} and the potentially susceptible population N^*. How does the graph change, compared to Figure 7.19? How can you adjust the initial exposed population E_0 to improve the fit?

7.16 Parameter sensitivity. The fit of the SEIR model to reported case data depends on five parameters, the latent and infectious periods A and C, the basic reproduction number R_0, the community spread ρ, and the potentially susceptible population N^*. Discuss how increasing R_0, ρ, and N^* affects the fit of the SEIR model to reported case data.

7.17 Outbreak dynamics. For the early COVID-19 outbreak in Austria with the seven-day moving average case data in Table 6.1 in Chapter 6, implement the SEIR model and create a Figure similar to Figure 6.8 for a simulation with latent and infectious periods of $A = 3$ days and $C = 7$ days and a basic reproduction number of $R_0 = 4$. Estimate the initial exposed population E_0 and the potentially susceptible population N^*. Try to fit the model such that the peak I_{max} and the timing of the peak $t(I_{max})$ match the reported infectious population N_I/N^*. Compare your results against the SIR model in Figure 6.8 and comment on the post-peak fit. Do you think the SIR model provides a better fit?

7.18 Model your own country, state, or city. Draw the COVID-19 case data from the first wave of the outbreak of your own country, state, or city. From the daily new cases ΔN_I, calculate the infectious population N_I assuming latent and infectious periods of $A = 3$ days and $C = 7$ days. Implement the SEIR model and fit the basic reproduction number R_0, the initial exposed population E_0, and the potentially susceptible population N^*. Try to fit the model such that the peak I_{max} and the timing of the peak $t(I_{max})$ match the reported infectious population N_I/N^*. Comment on the post-peak fit of the model. Do you think a dynamic SEIR model would provide a better fit?

References

1. Anderson RM, May RM (1982) Directly transmitted infectious diseases: control by vaccination. Science 215:1053-1060.
2. Anderson RM, May RM (1991) Infectious Diseases of Humans. Oxford University Press, Oxford.

3. Brauer F, Castillo-Chavez C (2001) Mathematical Models in Population Biology and Epidemiology. Springer-Verlag New York.
4. Brauer F, van den Dreissche P, Wu J (2008) Mathematical Epidemiology. Springer-Verlag Berlin Heidelberg.
5. Brauer F, Castillo-Chavez C, Feng Z (2019) Mathematical Models in Epidemiology. Springer-Verlag New York.
6. Deuflhard P (2011) Newton Methods for Nonlinear Problems. Springer-Verlag Berlin Heidelberg.
7. Dietz K (1993) The estimation of the basic reproduction number for infectious diseases. Statistical Methods in Medical Research 2:23-41.
8. European Centre for Disease Prevention and Control. Situation update worldwide. `https://www.ecdc.europa.eu/en/geographical-distribution-2019-ncov-cases` accessed: June 1, 2021.
9. Fang Y, Nie Y, Penny M (2020) Transmission dynamics of the COVID-19 outbreak and effectiveness of government interventions: a data-driven analysis. Journal of Medical Virology, 6:645-659.
10. Fauci AS, Lane HC, Redfield RR (2020) Covid-19–Navigating the uncharted. New England Journal of Medicine 382:1268-1269.
11. Goodwine B (2011) Engineering Differential Equations. Springer-Verlag New York.
12. Griffiths D, Higham DJ (2010) Numerical Methods for Ordinary Differential Equations. Springer-Verlag London.
13. Hethcote HW (1989) Three basic epidemiological models. Biomathematics: Applied Mathematical Ecology, 18:119-144.
14. Hethcote HW (2000) The mathematics of infectious diseases. SIAM Review 42:599-653.
15. Hundsdorfer W, Verwer J (2003) Numerical Solution of Time-Dependent Advection-Diffusion-Reaction Equations. Springer-Verlag Berlin Heidelberg.
16. Ioannidis JPA, Cripps S, Tanner MA (2021) Forecasting for COVID-19 has failed. International Journal of Forecasting, in press.
17. Johns Hopkins University (2021) Coronavirus COVID-19 Global Cases by the Center for Systems Science and Engineering. `https://coronavirus.jhu.edu/map.html`, `https://github.com/CSSEGISandData/covid-19` assessed: June 1, 2021.
18. Kuhl E (2020) Data-driven modeling of COVID-19 – Lessons learned. Extreme Mechanics Letters 40:100921.
19. Linka K, Peirlinck M, Sahli Costabal F, Kuhl E (2020) Outbreak dynamics of COVID-19 in Europe and the effect of travel restrictions. Computer Methods in Biomechanics and Biomedical Engineering 23: 710-717.
20. Linka K, Peirlinck M, Kuhl E (2020) The reproduction number of COVID-19 and its correlation with public heath interventions. Computational Mechanics 66:1035-1050.
21. Linka K, Goriely A, Kuhl E (2021) Global and local mobility as a barometer for COVID-19 dynamics. Biomechanics and Modeling in Mechanobiology 20:651–669.
22. Liu J, Shang, X (2020) Computational Epidemiology. Springer International Publishing.
23. Maier BF, Brockmann D (2020) Effective containment explains sub-exponential growth in confirmed cases of recent COVID-19 outbreak in mainland China. Science 368:742-746.
24. Moin P (2000) Fundamentals of Engineering Numerical Analysis. Cambridge University Press.
25. National Bureau of Statistics of China. Annual Population by Province. `http://data.stats.gov.cn`. accessed: March 20, 2020.
26. Peirlinck M, Linka K, Sahli Costabal F, Kuhl E (2020) Outbreak dynamics of COVID-19 in China and the United States. Biomechanics and Modeling in Mechanobiology 19:2179-2193.
27. Peirlinck M, Linka K, Sahli Costabal F, Bendavid E, Bhattacharya J, Ioannidis J, Kuhl E (2020) Visualizing the invisible: The effect of asymptomatic transmission on the outbreak dynamics of COVID-19. Computer Methods in Applied Mechanics and Engineering 372:113410.
28. Wang Z, Zhang X, Teichert GH, Carrasco-Teja M, Garikipati K (2020) System inference for the spatio-temporal evolution of infectious diseases: Michigan in the time of COVID-19. Computational Mechanics 66:1153-1176.

Chapter 8
The computational SEIIR model

Abstract The SEIIR model is a compartment model with five populations, the susceptible, exposed, symptomatic and asymptomatic infectious, and recovered groups S, E, I_s, I_a and R. It characterizes infectious diseases with a significant group of individuals that remain asymptomatic upon infection, but can still infect others. Since the SEIIR model has no analytical solution for the time course of its populations, we discretize it in time using finite differences and apply explicit time integration schemes to solve it. We distinguish two cases, the special case where the disease dynamics of the symptomatic and asymptomatic groups are similar, and the general case where the disease dynamics are different. To illustrate the features of the SEIIR model, we simulate the early COVID-19 outbreak in the Netherlands, one of the first countries that systematically estimated asymptomatic transmission using seroprevalence studies. The learning objectives of this chapter on computational SEIIR modeling are to

- understand the concepts of asymptomatic transmission and undercount
- discretize the classical SEIIR model in time and solve it
- illustrate model sensitivity with respect to the undercount and infectious periods
- simulate the early outbreak dynamics of COVID-19 using seroprevalence studies and reported case data
- discuss limitations of SEIIR modeling

By the end of the chapter, you will be able to computationally analyze, simulate, and predict the outbreak dynamics of infectious diseases with asymptomatic transmission like COVID-19 using seroprevalence studies and reported case data.

8.1 Introduction of the SEIIR model

Epidemiologists traditionally distinguish between infection and disease because the factors that govern their occurrence may be different, and infection without disease is common with many viruses [9]. Here we associate the classical notion of *in-*

© The Author(s), under exclusive license to Springer Nature Switzerland AG 2021   147
E. Kuhl, *Computational Epidemiology*, https://doi.org/10.1007/978-3-030-82890-5_8

fection, the multiplication of an agent within the host, with both symptomatic and asymptomatic cases and the notion of *disease*, the host response to infection that his is severe enough to evoke a recognizable pattern of clinical symptoms, with symptomatic cases only. A model that accounts for both symptomatic and asymptomatic transmission is the SEIIR model [26]. Figure 8.1 shows that it consists of five populations, the susceptible group S, the exposed group E, the symptomatic infectious group I_s, the asymptomatic infectious group I_a, and the recovered group R [3]. In the

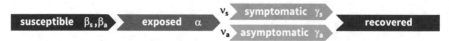

Fig. 8.1 Classical SEIIR model. The classical SEIIR model contains five compartments for the susceptible, exposed, symptomatic infectious, asymptomatic infectious, and recovered populations S, E, I_s, I_s, and R. The transition rates between the compartments, the contact, latent, and infectious rates, β, α, and γ, are inverses of the contact, latent, and infectious periods, $B = 1/\beta$, $A = 1/\alpha$, and $C = 1/\gamma$. However, only a fraction, $v_s = 1 - v_a$, becomes symptomatic upon exposure, and a fraction, v_a, remains asymptomatic. The symptomatic and asymptomatic groups have the same latent rate, $A = 1/\alpha$, but individual contact periods, $B_s = 1/\beta_s$ and $B_a = 1/\beta_a$ and individual infectious periods, $C_s = 1/\gamma_s$ and $C_a = 1/\gamma_a$.

SEIIR model, the transition between the susceptible, exposed, symptomatic, asymptomatic, and recovered groups is governed by five ordinary differential equations [3],

$$
\begin{aligned}
\dot{S} &= -S\,[\,\beta_s\,I_s + \beta_a\,I_a\,] \\
\dot{E} &= +S\,[\,\beta_s\,I_s + \beta_a\,I_a\,] - \alpha\,E \\
\dot{I}_s &= \qquad\qquad\qquad\quad + v_s\,\alpha\,E - \gamma_s\,I_s \\
\dot{I}_a &= \qquad\qquad\qquad\quad + v_a\,\alpha\,E - \gamma_a\,I_a \\
\dot{R} &= \qquad\qquad\qquad\qquad\qquad + \gamma_s\,I_s + \gamma_a\,I_a\,.
\end{aligned}
\tag{8.1}
$$

In this simple format, the SEIIR model neglects all vital dynamics, it does not account for births or natural deaths, which implies that $\dot{S} + \dot{E} + \dot{I}_s + \dot{I}_a + \dot{R} \doteq 0$ and $S + E + I_s + I_a + R = \text{const.} = 1$ [4]. Upon exposure, only a fraction, $v_s = 1 - v_a$, becomes symptomatic and a fraction, v_a, remains asymptomatic [1]. Their ratio between the asymptomatic and symptomatic fractions defines the *undercount* U,

$$
\mathsf{U} = \frac{v_a}{v_s} = \frac{1 - v_s}{v_s} \quad \text{with} \quad v_s + v_a = 1,
\tag{8.2}
$$

The undercount characterizes the number of asymptomatic individuals per single symptomatic individual. We assume that the symptomatic and asymptomatic groups, I_s and I_a, can both generate new infections. We postulate that they have the same latent period, $A = 1/\alpha$, but can have individual contact periods, $B_s = 1/\beta_s$ and $B_a = 1/\beta_a$ to mimic their different contact behavior, and individual infectious periods, $C_s = 1/\gamma_s$ and $C_a = 1/\gamma_a$, to mimic their different infectiousness [26]. From the infectious fractions, v_s and v_a, we can derive the overall contact and infectious rates, β and γ, as functions of their individual symptomatic and asymptomatic counterparts, β_s, β_a,

γ_s, and γ_a,

$$\beta = v_s \beta_s + v_a \beta_a \quad \text{and} \quad \gamma = v_s \gamma_s + v_a \gamma_a . \tag{8.3}$$

Similarly, we can express the overall contact and infectious periods, B and C, in terms of their symptomatic and asymptomatic counterparts, B_s, B_a, C_s, and C_a,

$$B = \frac{B_a B_s}{v_s B_a + v_a B_s} \quad \text{and} \quad C = \frac{C_a C_s}{v_s C_a + v_a C_s} . \tag{8.4}$$

Naturally, the different dynamics for the symptomatic and asymptomatic populations also affect the basic reproduction number R_0, the number of new infections caused by a single one individual in an otherwise uninfected, susceptible population [6],

$$R_0 = \frac{C}{B} = \frac{C_a C_s}{B_a B_s} \frac{v_s B_a + v_a B_s}{v_s C_a + v_a C_s} = \frac{v_s \beta_s + v_a \beta_a}{v_s \gamma_s + v_a \gamma_a} = \frac{\beta}{\gamma} . \tag{8.5}$$

This definition of the basic reproduction number has significant implications: For a large asymptomatic group, $v_a \to 1$, the basic reproduction number approaches the ratio between the infectious and contact periods of the asymptomatic population, $R_0 \to C_a/B_a$, which could be significantly larger than the basic reproduction number of the symptomatic population, $R_0 = C_s/B_s$, which we generally see reported in the literature [17].

Example. Asymptomatic transmission during the 1957-1958 pandemic.
The 1957-1958 pandemic had a latent period of $A = 1.9$ days and an infectious period of $C = 4.1$ days. About one third of all infections remained asymptomatic, $v_s = 2/3$, and the average symptomatic attack ratio was $A_s = v_s[1 - S_\infty] = 0.326$. The infectious period was similar for symptomatic and asymptomatic individuals, $C = C_s = C_a$, but symptomatic individuals were twice as likely to have contact as asymptomatic individuals, $\beta_s = 2\beta_a$. Assume a population of $N = 2,000$ of which twelve were initially infectious, eight of them symptomatic and four asymptomatic. Estimate the final sizes S_∞ and R_∞, the basic reproduction number R_0, and the contact times B_s and B_a.

Solution. We use the definition of the symptomatic attack ratio, $A_s = v_s[1 - S_\infty]$, to calculate the final sizes,

$$S_\infty = 1 - \frac{A_s}{v_s} = 1 - \frac{3 \cdot 0.326}{2} = 0.511 \quad \text{and} \quad R_\infty = \frac{A_s}{v_s} = \frac{3 \cdot 0.326}{2} = 0.489 .$$

and the number of susceptible and recovered individuals,

$$N_S = S_\infty \cdot N = 0.511 \cdot 2,000 = 1,022 \quad \text{and} \quad N_R = R_\infty \cdot N = 0.489 \cdot 2,000 = 978$$

at the end of the outbreak, see Figure 8.2. From the final size relation, $\log(S_0/S_\infty) = R_0[1 - S_\infty]$, we estimate the basic reproduction number,

$$R_0 = \log(S_0/S_\infty)/[1 - S_\infty] = 1.361 ,$$

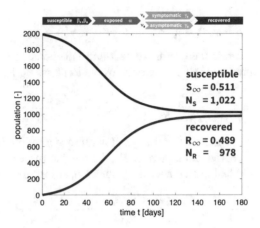

Fig. 8.2 Example. Asymptomatic transmission during the 1957-1958 pandemic. Outbreak dynamics of the 1957-1958 pandemic in a population of 2,000. For a symptomatic fraction, ν_s, and a given symptomatic attack ratio, $A_s = \nu_s[\,1 - S_\infty\,] = 0.326$, we can calculate the final sizes, $S_\infty = 0.511$ and $R_\infty = 0.489$. The outbreak decays after half a year.

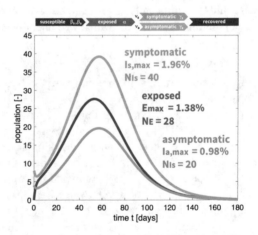

Fig. 8.3 Example. Asymptomatic transmission during the 1957-1958 pandemic. Outbreak dynamics of the 1957-1958 pandemic in a population of 2,000. Using the final size relation $\log(S_0/S_\infty) = R_0[\,1 - S_\infty\,]$, we can estimate the basic reproduction number $R_0 = 1.361$, and the symptomatic and asymptomatic contact periods $B_s = 2.511$ days and $B_a = 5.022$ days. The maximum exposed, symptomatic, and asymptomatic populations are $E_{max} = 1.38\%$, $I_{s,max} = 1.96\%$, and $I_{a,max} = 0.98\%$ all occurring within the first two months of the outbreak.

to evaluate the symptomatic contact rate β_s, using the general equation for the basic reproduction number (8.5), where we replace $\beta_a = 0.5\,\beta_s$, $\gamma_s = 1/C_s$, and $\gamma_a = 1/C_a$,

$$R_0 = \frac{\nu_s\,\beta_s + \nu_a\,\beta_a}{\nu_s/C_s + \nu_a/C_a} \qquad \text{thus} \qquad \beta_s = \frac{6\,R_0}{5\,C} = 0.398\ .$$

This results in contact periods of

$$B_s = 2.511 \text{days} \quad \text{and} \quad B_a = 5.022 \text{days},$$

for the symptomatic and asymptomatic groups. From the simulation, we conclude that the maximum exposed, symptomatic, and asymptomatic populations were $E_{max} = 1.38\%$, $I_{s,max} = 1.96\%$, and $I_{a,max} = 0.98\%$ corresponding to 28, 40, and 20 individuals, with all peaks occurring within the first 60 days of the outbreak, see Figure 8.3.

8.2 Discrete SEIIR model.

To solve the dynamics of the SEIIR model in time, we apply a finite difference approximation and replace the lefthand sides of the system of equations (8.1) by the discrete differences, $(\dot{\circ}) = [(\circ)_{n+1} - (\circ)_n]/\Delta t$, in terms of the unknown new populations, $(\circ)_{n+1}$, the known previous populations, $(\circ)_n$, and the discrete time step size, $\Delta t = t_{n+1} - t_n$, for all five populations $(\circ) = S, E, I_s, I_a, R$,

$$\dot{S} = \frac{S_{n+1} - S_n}{\Delta t} \quad \dot{E} = \frac{E_{n+1} - E_n}{\Delta t} \quad \dot{I}_s = \frac{I_{s,n+1} - I_{s,n}}{\Delta t} \quad \dot{I}_a = \frac{I_{a,n+1} - I_{a,n}}{\Delta t} \quad \dot{R} = \frac{R_{n+1} - R_n}{\Delta t}.$$

(8.6)

In the spirit of an explicit time integration, we evaluate the righthand sides of the system of equations (8.1) at the known previous time point t_n [22],

$$\dot{S} = -[\beta_s I_{s,n} + \beta_a I_{a,n}] S_n I_n \qquad \dot{E} = +[\beta_s I_{s,n} + \beta_a I_{a,n}] S_n I_n - \alpha E_n$$
$$\dot{I}_s = +\nu_s \alpha E_n - \gamma_s I_{s,n} \qquad \dot{I}_a = +\nu_a \alpha E_n - \gamma_a I_{1,n} \qquad \dot{R} = +[\gamma_s I_{s,n} + \gamma_a I_{a,n}].$$

(8.7)

This explicit time integration results in the following discrete system of equations for the SEIR model [26],

$$
\begin{aligned}
S_{n+1} &= [1 - [\beta_s I_{s,n} + \beta_a I_{a,n}] \Delta t] S_n \\
E_{n+1} &= +[\beta_s I_{s,n} + \beta_a I_{a,n}] \Delta t] S_n + [1 - \alpha \Delta t] E_n \\
I_{s,n+1} &= \qquad\qquad + \nu_s \alpha \Delta t \, E_n + [1 - \gamma_s \Delta t] I_{s,n} \\
I_{a,n+1} &= \qquad\qquad + \nu_a \alpha \Delta t \, E_n + [1 - \gamma_a \Delta t] I_{a,n} \\
R_{n+1} &= \qquad\qquad\qquad + [\gamma_s I_{s,n} + \gamma_a I_{a,n}] \Delta t + R_n.
\end{aligned}
$$

(8.8)

To explore the effect of asymptomatic transmission on the overall outbreak dynamics, we adopt the SEIIR model and perform a sensitivity analysis with respect to the two most relevant parameters associated with asymptomatic transmission, the symptomatic fraction ν_s and the asymptomatic infectious period C_a. To highlight the effect of both parameters, we visualize the exposed, symptomatic and asymptomatic infectious, and recovered populations of our SEIIR model,

$$\dot{S} = -S\,\beta(t)\,[\,I_s + I_a\,]$$
$$\dot{E} = +S\,\beta(t)\,[\,I_s + I_a\,] - \quad \alpha\,E$$
$$\dot{I}_s = \qquad\qquad\qquad + v_s\,\alpha\,E - \gamma_s\,I_s \qquad\qquad (8.9)$$
$$\dot{I}_a = \qquad\qquad\qquad + v_a\,\alpha\,E - \gamma_a\,I_a$$
$$\dot{R} = \qquad\qquad\qquad\qquad\qquad + \gamma_s\,I_s + \gamma_a\,I_a\;.$$

We assume that the differences between symptomatic and asymptomatic transmission manifest themselves exclusively in the infectious periods $C_s = 1/\gamma_s$ and $C_a = 1/\gamma_a$ while the symptomatic and asymptomatic contact rates are equal and can change dynamically in time, $\beta(t) = \beta_s(t) = \beta_a(t)$. We adopt a non-constant, monotonically decreasing, dynamic contact rate $\beta(t)$ of hyperbolic tangent type [17],

$$\beta(t) = \beta_0 - \tfrac{1}{2}[1 + \tanh([t - t^*]/T)][\beta_0 - \beta_t], \qquad (8.10)$$

where β_0 is the basic contact rate, β_t is the reduced contact rate, t^* is the adaptation time, and T is the adaptation speed. Figure 8.4 illustrates the five compartments, the transition rates, and the symptomatic and asymptomatic fractions of this dynamic SEIIR model. To mimic realistic outbreak dynamics, we adopt the basic contact

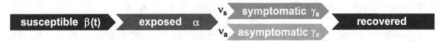

Fig. 8.4 Dynamic SEIIR model. The dynamic SEIIR model contains five compartments for the susceptible, exposed, symptomatic infectious, asymptomatic infectious, and recovered populations. The transition rates between the compartments, $\beta(t)$, α, and γ, are inverses of the dynamic contact period, $B(t) = 1/\beta(t)$, the latent period, $A = 1/\alpha$, and the infectious period, $C = 1/\gamma$. However, only a fraction, $v_s = 1 - v_a$, becomes symptomatic upon exposure, and a fraction, v_a, remains asymptomatic. The symptomatic and asymptomatic groups have the same latent period, $A = 1/\alpha$, and the same dynamic contact period, $B(t) = 1/\beta(t)$, but individual infectious periods, $C_s = 1/\gamma_s$ and $C_a = 1/\gamma_a$.

rate $\beta_0 = 0.65$ /days, the reduced contact rate $\beta_t = 0.10$ /days, the adaptation time $t^* = 18.61$ days, and the adaptation speed $T = 10.82$ /days from the mean values during the first wave of the COVID-19 pandemic across the 27 countries of the European Union from Table 12.2 [17]. We select a latent period of $A = 1/\alpha$ $= 2.5$ days and a symptomatic infectious period of $C_s = 1/\gamma_a = 6.5$ days. We conceptually distinguish two scenarios: the special case for which the symptomatic and asymptomatic populations display identical disease dynamics, $C_a = C_s$, in Section 8.3 and the general case for which these disease dynamics are different, $C_a \neq C_s$, in Section 8.4.

8.3 Similar symptomatic and asymptomatic disease dynamics

For the special case with similar disease dynamics of the symptomatic and asymptomatic populations, $\beta_s = \beta_a = \beta$ and $\gamma_s = \gamma_a = \gamma$, we can translate the SEIIR model (8.1) into the classical SEIR model (4.17) with four compartments, the susceptible, exposed, infectious, and recovered populations [20],

$$
\begin{aligned}
\dot{S} &= -\beta\, SI \\
\dot{E} &= +\beta\, SI - \alpha\, E \\
\dot{I} &= + \alpha\, E - \gamma\, I \\
\dot{R} &= + \gamma\, I\,.
\end{aligned}
\tag{8.11}
$$

For this special case, we can back-calculate the symptomatic and asymptomatic infectious groups from equation (8.11.3),

$$
I_s = v_s\, I \quad \text{and} \quad I_a = v_a\, I = [\,1 - v_s\,]\, I \quad \text{with} \quad I = I_s + I_a\,,
\tag{8.12}
$$

and the symptomatic and asymptomatic recovered groups from equation (8.11.4),

$$
R_s = v_s\, R \quad \text{and} \quad R_a = v_a\, R = [\,1 - v_s\,]\, R \quad \text{with} \quad R = R_s + R_a\,.
\tag{8.13}
$$

This relationship is useful to estimate the final sizes of the symptomatic and asymptomatic recovered populations, $R_{s\infty} = v_s\, R_\infty$ and $R_{a\infty} = v_a\, R_\infty$ with $R_\infty = 1 + W(-S_0\, R_0 \exp(-R_0[1 - R_0]))/R_0$, from the final size relation (4.31) of the general SEIR model in Chapter 4.17.

Figures 8.5 and 8.6 highlight the effects of asymptomatic transmission for varying symptomatic and asymptomatic fractions v_s and v_a at similar infectious periods, $C_a = C_s$ [26]. The darkest curves correspond to the special case of only symptomatic transmission, $v_s = 1.0$, in Figure 8.5 and only asymptomatic transmission, $v_a = 1.0$, in Figure 8.6; vice versa, the lightest curves are $v_s = 0.0$ and $v_a = 0.0$. The blue curves illustrates the symptomatic part of the recovered population, $R_s = v_s\, R$, for varying v_s and v_a, which is the fraction of the recovered population that is typically reported in the public data bases [9, 12]. The steepest blue curve is the total recovered population, $R = R_s + R_a = [\,v_s + v_a\,]\, R$, which, for this special case with $C_a = C_s$, is independent of the symptomatic and asymptomatic fractions v_s and v_a. For example, for a symptomatic fraction of $v_s = 0.2$, the ratio between both curves is $R/R_s = 5$, which implies that five times more individuals have actually already gone through the disease than we would expect from the reported case data in public databases.

Importantly, since the reported case data only reflect the symptomatic infectious and recovered groups I_s and R_s, the true infectious and recovered populations $I = I_s/v_s$ and $R = R_s/v_s$ are unknown and could be about an order of magnitude larger than the predictions of the classical SEIR model [2]. From an individual's perspective, a smaller symptomatic group v_s, or equivalently, a larger asymptomatic group $v_a = [1 - v_s]$, could have a personal effect on the likelihood of being unknowingly exposed to the virus, especially for high-risk populations: A larger asymptomatic fraction v_a would translate into an increased risk of community transmission and

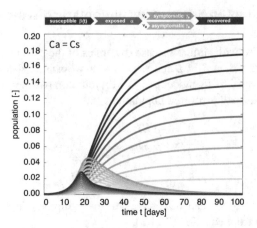

Fig. 8.5 Dynamic SEIIR model. Sensitivity with respect to the symptomatic fraction ν_s. Increasing the symptomatic fraction ν_s increases the symptomatic part of the recovered population, $R_s = \nu_s R$, shown in the blue curves. Latent period $A = 2.5$ days, infectious periods $C_s = 6.5$ days and $C_a = 6.5$ days, dynamic contact rate $\beta(t) = \beta_0 - \frac{1}{2}[1 + \tanh([t - t^*]/T)][\beta_0 - \beta_t]$ with $\beta_0 = 0.65$ /days, $\beta_t = 0.10$ /days, $t^* = 18.61$ days, and $T = 10.82$ days, and symptomatic fraction $\nu_s = [0.0, 0.1, 0.2, 0.3, ..., 1.0]$, from light to dark [26].

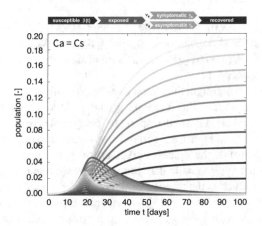

Fig. 8.6 Dynamic SEIIR model. Sensitivity with respect to the asymptomatic fraction ν_a. Increasing the asymptomatic fraction ν_a decreases the symptomatic part of the recovered population, $R_s = \nu_s R$, shown in the blue curves. Latent period $A = 2.5$ days, symptomatic and asymptomatic infectious periods $C_s = 6.5$ days and $C_a = 6.5$ days, dynamic contact rate $\beta(t) = \beta_0 - \frac{1}{2}[1 + \tanh([t - t^*]/T)][\beta_0 - \beta_t]$ with $\beta_0 = 0.65$ /days, $\beta_t = 0.10$ /days, $t^* = 18.61$ days, and $T = 10.82$ days, and symptomatic fraction $\nu_a = [0.0, 0.1, 0.2, 0.3, ..., 1.0]$, from light to dark [26].

would complicate outbreak control. From a health care perspective, however, the special case with comparable transition dynamics would not pose a threat to the health care system since the overall outbreak dynamics would remain unchanged, independent of the fraction ν_a of the asymptomatic population: A larger asymptomatic fraction would simply imply that a larger fraction of the population has already been exposed to the virus–without experiencing significant symptoms–and that the true hospitalization and mortality rates would be much lower than the reported rates [12]. Interestingly, it would also imply that herd immunity could be much closer than we would estimate from symptomatic case data alone [15].

8.4 Different symptomatic and asymptomatic disease dynamics

For the general case with different disease dynamics of the symptomatic and asymptomatic populations, we have to keep track of all five populations of the SEIIR model (8.1). For different symptomatic and asymptomatic infectious periods, $C_s \neq C_a$, we can no longer infer the dynamics of the symptomatic and asymptomatic groups from the dynamics of the overall infectious population, $I_s \neq v_s I$ and $I_a \neq v_a I$, and their maxima no longer coincide in time.

Figures 8.7 and 8.8 highlight the effects of asymptomatic transmission for varying symptomatic fractions v_s at different infectious periods, $C_a = 0.5 C_s$ and $C_a = 2 C_s$ [26]. The dark red, orange, green, and blue curves correspond to the special case of the regular SEIR model (8.11) without asymptomatic transmission, $v_s = 1.0$ and $v_a = 0.0$. The lighter curves show the effect of decreasing the symptomatic fraction towards $v_s = 0.0$ and $v_a = 1.0$. Interestingly, the effect of asymptomatic transmission on the overall outbreak dynamics varies not only with the symptomatic fraction, but is also highly sensitive to the ratio of the symptomatic and asymptomatic infectious periods. For smaller asymptomatic infectious periods, $C_a = 0.5 C_s$, the outbreak dynamics are governed by the orange symptomatic population I_s and the blue recovered population R increases with increasing symptomatic transmission v_s. For larger asymptomatic infectious periods, $C_a = 2 C_s$, the outbreak dynamics are governed by the green asymptomatic population I_a and the blue recovered population R decreases with increasing symptomatic transmission v_s.

Figures 8.9 and 8.10 highlight the effects of asymptomatic transmission for varying infectious periods C_a at different symptomatic fractions, $v_s = 0.2$ and $v_s = 0.8$ [26]. The lightest red, orange, green, and blue curves correspond to a smaller asymptomatic infectious period of $C_a = 0.5 C_s$, the darkest curves to a larger period $C_a = 2.0 C_s$. The middle curves correspond to the special case for which the asymptomatic and symptomatic populations display the same dynamics, $C_a = C_s$, such that the overall recovered population becomes independent of the symptomatic fraction. This implies that the middle blue recovered populations of Figures 8.9 and 8.10 are identical and also identical to the dark blue curves of Figures 8.7 and 8.8. For the smaller symptomatic fraction of $v_s = 0.2$ in Figure 8.9, the outbreak dynamics are governed by the green asymptomatic population I_a and the blue recovered population R increases markedly with an increasing asymptomatic infectious period C_a. For the larger symptomatic fraction of $v_s = 0.8$ in Figure 8.10, the outbreak dynamics are governed by the orange symptomatic population I_s and the blue recovered population R is relatively insensitive to the asymptomatic infectious period C_a.

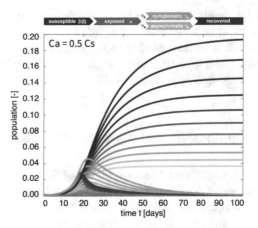

Fig. 8.7 Dynamic SEIIR model. Sensitivity with respect to the symptomatic fraction ν_s. For a smaller asymptomatic infectious period, $C_a = 0.5\,C_s$, increasing the symptomatic fraction ν_s increases the recovered symptomatic population. Latent period $A = 2.5$ days, infectious periods $C_s = 6.5$ days and $C_a = 3.25$ days, dynamic contact rate $\beta(t) = \beta_0 - \frac{1}{2}[1+\tanh([t-t^*]/T)][\beta_0-\beta_t]$ with $\beta_0 = 0.65$ /days, $\beta_t = 0.10$ /days, $t^* = 18.61$ days, and $T = 10.82$ days, and symptomatic fraction $\nu_s = [0.0, 0.1, 0.2, ..., 1.0]$, from light to dark [26].

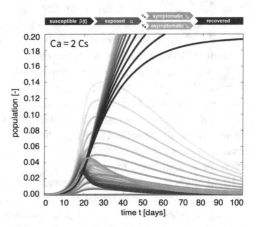

Fig. 8.8 Dynamic SEIIR model. Sensitivity with respect to the asymptomatic fraction ν_a. For a larger asymptomatic infectious period, $C_a = 2\,C_s$, increasing the symptomatic fraction ν_s decreases the recovered population. Latent period $A = 2.5$ days, infectious periods $C_s = 6.5$ and $C_a = 13$ days, dynamic contact rate $\beta(t) = \beta_0 - \frac{1}{2}[1+\tanh([t-t^*]/T)][\beta_0-\beta_t]$ with $\beta_0 = 0.65$ /days, $\beta_t = 0.10$ /days, $t^* = 18.61$ days, and $T = 10.82$ days, and symptomatic fraction $\nu_s = [0.0, 0.1, 0.2, ..., 1.0]$, from light to dark [26].

Taken together, this sensitivity analysis shows that the outbreak dynamics of an infectious disease are highly sensitive to asymptomatic transmission [10]. This sensitivity becomes more pronounced the more the asymptomatic infectious period differs from its symptomatic counterpart [15]. This difference could be a natural result of asymptomatic individuals isolating themselves less since they are unaware

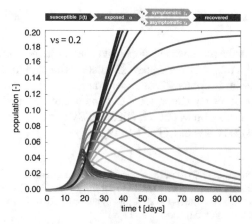

Fig. 8.9 Dynamic SEIIR model. Sensitivity with respect to the asymptomatic infectious period C_a. For a smaller symptomatic fraction of $\nu_s = 0.2$, increasing the asymptomatic infectious period C_a markedly increases the recovered population R. Latent period $A = 2.5$ days, symptomatic infectious period $C_s = 6.5$ days, dynamic contact rate $\beta(t) = \beta_0 - \frac{1}{2}[1 + \tanh([t - t^*]/T)][\beta_0 - \beta_t]$ with $\beta_0 = 0.65$ /days, $\beta_t = 0.10$ /days, $t^* = 18.61$ days, and $T = 10.82$ days, asymptomatic fraction $\nu_s = 0.2$, and asymptomatic infectious period $C_a = [3.25, ..., 6.50, ..., 13.0]$, from light to dark [26].

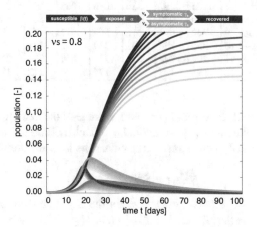

Fig. 8.10 Dynamic SEIIR model. Sensitivity with respect to the asymptomatic infectious period C_a. For a larger symptomatic fraction of $\nu_s = 0.8$, increasing the asymptomatic infectious period C_a moderately increases the recovered population R. Latent period $A = 2.5$ days, symptomatic infectious period $C_s = 6.5$ days, dynamic contact rate $\beta(t) = \beta_0 - \frac{1}{2}[1 + \tanh([t - t^*]/T)][\beta_0 - \beta_t]$ with $\beta_0 = 0.65$ /days, $\beta_t = 0.10$ /days, $t^* = 18.61$ days, and $T = 10.82$ days, asymptomatic fraction $\nu_a = 0.8$, and asymptomatic infectious period $C_a = [3.25, ..., 6.50, ..., 13.0]$, from light to dark [26].

that they can infect others [19]. Knowing the precise dynamics of asymptomatic transmission is therefore critical for a successful management of a pandemic [12].

8.5 Example: Early COVID-19 outbreak in the Netherlands

The Netherlands are the most densely populated country in the European Union with 17.2 million inhabitants, 421 per square kilometer. The country recorded the first case of COVID-19 on February 27, and then implemented physical distancing and lockdown-type measures on March 15, 2020, and became part of the Europe-wide travel regulations starting on March 17, 2020 [5]. By mid May, the Netherlands had seen more than 44,000 cases of COVID-19 [9].

Table 8.1 Asymptomatic COVID-19 transmission in the Netherlands. Newly reported COVID-19 cases in the Netherlands from Sunday, March 1 to Saturday, May 23, 2020; with social distancing and lockdown interventions on March 15, 2020 and tight border control on March 17, 2020; case data ΔN_I are reported as seven-day moving averages.

	Sun	Mon	Tue	Wed	Thu	Fri	Sat
week 1	2	2	4	5	12	18	26
week 2	36	43	52	66	76	97	110
week 3	124	156	189	221	264	313	381
week 4	439	476	551	623	710	802	875
week 5	952	1000	1005	1029	1038	1017	981
week 6	998	1007	998	991	1009	1054	1112
week 7	1105	1107	1120	1086	1065	1050	1025
week 8	1010	979	959	956	931	869	800
week 9	742	691	612	565	513	465	435
week 10	390	360	382	359	352	328	307
week 11	294	288	271	270	244	227	213
week 12	195	193	181	177	174	172	171

Table 8.1 summarizes the seven-day moving average of the newly reported COVID-19 cases in the Netherlands [9], from Sunday, March 1, 2020 to Saturday, May, 23, 2020. From these daily new cases, we calculate the total infectious population N_I for every day t_n,

$$N_I(t_n) = \sum_{i=n+1-C}^{n} \Delta N_I(t_i),$$

by summing the newly reported cases ΔN_I for the $C = 7$ previous days. We assume latent and infectious periods of $A = 3$ days and $C = 7$ days, a basic reproduction number of $R_0 = 2.5$, and similar transition dynamics for the symptomatic and asymptomatic groups, $C_s = C_a = C$ and $B_s = B_a = C/R_0$. During the first two weeks of April, a seroprevalence study based on 7,361 blood plasma samples [20], resulted in a symptomatic fraction of $v_s = 17.31\%$ with a 95% confidence interval of 15.44%-19.95% [20]. We approximate the undercount as five, and calculate the symptomatic and asymptomatic fractions of v_s and v_a,

$$U = \frac{v_a}{v_s} = 5 \quad \text{thus} \quad v_s = \frac{1}{6} \quad \text{and} \quad v_a = \frac{5}{6}.$$

Figure 8.11 shows the reported infectious population and the simulated susceptible, exposed, symptomatic and asymptomatic infectious, and recovered populations of

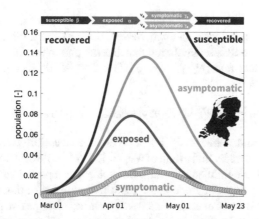

Fig. 8.11 Asymptomatic COVID-19 transmission in the Netherlands. Reported infectious population and simulated susceptible, exposed, symptomatic and asymptomatic infectious, and recovered populations. Simulations are based on latent and infectious periods of $A = 3$ days and $C = 7$ days, a basic reproduction number of $R_0 = 2.5$, an undercount of five, $U = v_a : v_s = 5$, similar transition dynamics for the symptomatic and asymptomatic groups, $C_s = C_a = C$ and $B_s = B_a = C/R_0$, and a potentially susceptible population of $N^* = 320,000$.

the static SEIIR model (8.1). From the simulation, we record the maximum total, symptomatic, and asymptomatic infectious populations,

$$I = 0.1643 \quad \text{thus} \quad I_{s,\max} = \tfrac{1}{6} I = 0.0274 \quad \text{and} \quad I_{a,\max} = \tfrac{5}{6} I = 0.1369 \, ,$$

all on day 43, April 12, 2020. The Netherlands have a total population of $N = 17.23$ million people, but, clearly, because of the gradual lockdowns, only a fraction N^* of the entire population was potentially susceptible to the virus [16]. We map the SEIIR model with a total population of one onto the potentially susceptible population of N^*, which we calculate from mapping the numerically estimated maximum symptomatic infectious population $I_{s,\max} = 0.0274$ into the maximum number of infectious people, $\max\{N_I\} = 7,649$,

$$I_{s,\max} = \frac{\max\{N_I\}}{N^*} \quad \text{thus} \quad N^* = \frac{\max\{N_I\}}{I_{s,\max}} = \frac{7,649}{0.0274} = 279,160 \, .$$

Since the reported case data have a flat maximum with no unique peak, we found a better vertical fit for a slightly larger potentially susceptible population of $N^* = 320,000$ with $I_{s,\max} = 0.0239$, along with initial conditions of $E_0 = 0.004$, $I_{s0} = 0.004$, and $I_{a0} = 0.004$ for the horizontal fit. To characterize the level of containment and compare it to other countries, we calculate the *affected population* η, the ratio between the potentially susceptible population N^* and the total population N,

$$\eta = \frac{N^*}{N} \quad \text{thus} \quad \eta = \frac{320,000}{17,230,000} = 0.0186 = 1.86\% \, .$$

Figure 8.11 displays a good agreement between the curves of the SEIIR model and the dots of the recorded case data. By design, the maximum infectious population $I_{max} = 0.0239$ of the SEIIR simulation matches the maximum infectious population of the reported data $\max\{N_I\}/N^* = 0.0239$. The SEIIR model predicts a final recovered population,

$$R_\infty = 1 + W(-S_0 \, R_0 \, \exp(-R_0[\, 1 - R_0 \,]))/R_0 = 0.89 \,,$$

both analytically and numerically. For a potential susceptible population of $N^* = 320{,}000$, this would imply that a total of $N_{R_\infty} = R_\infty \cdot N^* = 284{,}800$ individuals would have recovered at endemic equilibrium. Summing the reported case numbers ΔN_I in Table 8.1 yields 44,497 symptomatic recovered cases, resulting in 222,485 total recovered cases and a model error of 28.0%. We can see that the five populations have not yet fully converged within the 84-day time window of Figure 8.11 and that, for a larger time window, the number of reported cases would naturally increase and reduce this error. Figure 8.11 also illustrates the effect of asymptomatic transmission, which, for similar transition dynamics, simply increases the total infectious and recovered populations,

$$I = I_s/\nu_s = 6\,I_s \quad \text{and} \quad R = R_s/\nu_s = 6\,R_s \,,$$

in our case by a factor six. This implies that six times more individuals would actually have been affected by the disease than we would assume from the reported case data in Table 8.11.

Problems

8.1 Asymptomatic transmission. In a Virtual Press Conference on June 8, 2020, Dr. Maria van Kerkhove, the technical lead of COVID-19 response of the WHO, said about the COVID-19 pandemic [21]: "We have a number of reports from countries who are doing very detailed contact tracing. They're following asymptomatic cases, they're following contacts and they're not finding secondary transmission onward. It's very rare." How did our view on the asymptomatic transmission of COVID-19 change? Discuss how the presence of asymptomatic transmission affects outbreak control strategies.

8.2 Undercount and symptomatic fraction. In the literature, try to find studies that report the undercount or the symptomatic fraction of COVID-19. What are the largest and smallest values you can find for the undercount U? What are the corresponding values for the symptomatic fraction ν_s? What does this imply for health management and outbreak control?

8.3 Testing vs asymptomatic transmission. The testing frequency affects the fraction of unreported asyptomatic cases. Discuss how increasing testing changes the undercount and the symptomatic fraction ν_s. What does this imply for communities that perform regular surveillance testing, for example, college campuses?

8.4 Classical SIIR model. Develop the equations for the SIIR model in Figure 8.12 that uses the classical SIR model from Chapter 3 but further divides the infectious group into a symptomatic and an asymptomatic group based on the fractions v_s and v_a. Write down the system of four ordinary differential equations that define the susceptible, symptomatic and asymptomatic infectious, and recovered groups S, I_s, I_r, and R.

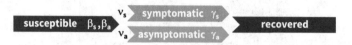

Fig. 8.12 Classical SIIR model. The classical SIIR model contains four compartments for the susceptible, symptomatic infectious, asymptomatic infectious, and recovered populations, S, I_s, I_a, and R. The transition rates between the compartments, the contact and infectious rates, β and γ, are inverses of the contact and infectious periods, $B = 1/\beta$ and $C = 1/\gamma$. However, only a fraction, $v_s = 1 - v_a$, becomes symptomatic upon exposure, and a fraction, v_a, remains asymptomatic. The symptomatic and asymptomatic groups can have individual contact periods, $B_s = 1/\beta_s$ and $B_a = 1/\beta_a$, and individual infectious periods, $C_s = 1/\gamma_s$ and $C_a = 1/\gamma_a$.

8.5 Classical SIIR model. For the SIIR model in Figure 8.12 that you have developed in the previous problem, derive the overall contact and infectious rates β and γ as functions of their symptomatic and asymptomatic counterparts β_s, β_a, γ_s and γ_a. Derive the overall basic reproduction number R_0. Discuss the implications of small and large asymptomatic groups, $v_a \rightarrow 0$ and $v_a \rightarrow 1$.

8.6 Linearization. Evaluate the righthand sides of the SEIIR model (8.1) at the unknown current time point t_{n+1}. Rephrase the implicit SEIIR model in residual form and linearize the residual **R** with respect to the unknown populations $\mathbf{P}_{n+1} = [\,S_{n+1}, E_{n+1}, I_{s,n+1}, I_{a,n+1}, R_{n+1}\,]$ to derive the tangent matrix **K** of the Newton method.

8.7 Final sizes. For a COVID-19 outbreak on Stanford campus, the average symptomatic attack ratio was $A_s = v_s[\,1 - S_\infty\,] = 0.25$, but two third of all infections remained asymptomatic. Assume latent and infectious periods of $A = 3.0$ days and $C = 7.0$ days, with similar infectious periods for symptomatic and asymptomatic individuals. The campus population was $N = 5,000$ of which twelve were initially infectious. Estimate the final sizes S_∞ and R_∞ and the absolute numbers of susceptible and recovered individuals at the end of the academic year.

8.8 Basic reproduction number and contact times. For a COVID-19 outbreak on Stanford campus, the average symptomatic attack ratio was $A_s = v_s[\,1 - S_\infty\,] = 0.25$, but two third of all infections remained asymptomatic. Assume latent and infectious periods of $A = 3.0$ days and $C = 7.0$ days, with similar infectious periods for symptomatic and asymptomatic individuals. Asymptomatic individuals were twice as likely to have contact as symptomatic individuals, $\beta_a = 2\beta_s$. The campus population was $N = 5,000$ of which twelve were initially infectious. Estimate the basic reproduction number R_0 and the symptomatic and asymptomatic contact times B_s and B_a.

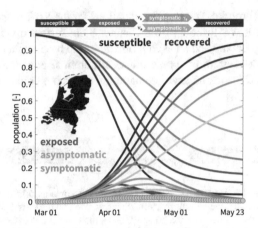

Fig. 8.13 Asymptomatic COVID-19 transmission in the Netherlands. Reported infectious population and simulated susceptible, exposed, symptomatic and asymptomatic infectious, and recovered populations. Simulations are based on latent and contact periods of $A = 3$ days and $B = 2.8$ days, nd undercount of five, $U = v_a : v_s = 5$, similar contact dynamics, $B_s = B_a = B$, but different infectious dynamics, $C_s = 7$ days and $C_a = 5, 6, 7, 10$ days, from light to dark.

8.9 Similar transition dynamics. Implement the SEIIR model with different transition dynamics (8.8). For the early COVID-19 outbreak in the Netherlands from Table 8.1 with an undercount of $v_a : v_s = 5$, recreate Figure 8.11 assuming latent and infectious periods of $A = 3$ days and $C = 7$ days, a basic reproduction number of $R_0 = 2.5$, similar transition dynamics for the symptomatic and asymptomatic groups, $C_s = C_a = C$ and $B_s = B_a = C/R_0$, and initial conditions of $E_0 = I_{s0} = I_{a0} = 0.0024$.

8.10 Basic reproduction number and final size. The early COVID-19 outbreak in the Netherlands had an undercount of $v_a : v_s = 5$, and latent and contact periods of $A = 3$ days and $B = 2.8$ days. Assume similar contact dynamics, $B_s = B_a = B$, but different infectious dynamics, $C_s = 7$ days and $C_a = 5, 6, 7, 10$ days. Calculate the basic reproduction number R_0 and the final size of the total recovered population R_∞ for all four cases.

8.11 Different transition dynamics. Implement the SEIIR model with different transition dynamics (8.8). For the early COVID-19 outbreak in the Netherlands from Table 8.1 with an undercount of $v_a : v_s = 5$, plot the evolution of the susceptible, exposed, symptomatic, asymptomatic, and recovered groups assuming latent and contact periods of $A = 3$ days and $B = 2.8$ days, and initial conditions of $E_0 = I_{s0} = I_{a0} = 0.0024$. Assume similar contact dynamics, $B_s = B_a = B$, but different infectious dynamics, $C_s = 7$ days and $C_a = 4, 5, 6, 7, 10$ days. Compare the final size of the total recovered population R_∞ against your results of the previous problem. Compare your results to Figure 8.13. How does the peak of infection change, both in magnitude and time, with increasing asymptomatic infectious periods C_a.

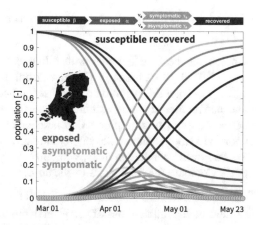

Fig. 8.14 Asymptomatic COVID-19 transmission in the Netherlands. Reported infectious population and simulated susceptible, exposed, symptomatic and asymptomatic infectious, and recovered populations. Simulations are based on latent and contact periods of $A = 3$ days and $B = 2.8$ days, an undercount of five, $U = \nu_a : \nu_s = 5$, similar infectious dynamics, $C_s = C_a = C$, but different contact dynamics, $B_s = 2.80$ days and $B_a = 2.10, 2.45, 2.80, 3.15, 3.50$ days, from light to dark.

8.12 Basic reproduction number and final size. The early COVID-19 outbreak in the Netherlands had an undercount of $\nu_a : \nu_s = 5$, and latent and contact periods of $A = 3$ days and $B = 2.8$ days. Assume similar contact dynamics, $B_s = B_a = B$, but different infectious dynamics, $C_s = 7$ days and $C_a = 5, 6, 7, 10$ days. Calculate the basic reproduction number R_0 and the final size of the total recovered population R_∞ for all four cases.

8.13 Different transition dynamics. Implement the SEIIR model with different transition dynamics (8.8). For the early COVID-19 outbreak in the Netherlands from Table 8.1 with an undercount of $\nu_a : \nu_s = 5$, plot the evolution of the susceptible, exposed, symptomatic, asymptomatic, and recovered groups assuming latent and infectious periods of $A = 3$ days and $C = 7$ days, and initial conditions of $E_0 = I_{s0} = I_{a0} = 0.0024$. Assume similar infectious dynamics, $C_s = C_a = C$, but different contact dynamics, $B_s = 2.80$ days and $B_a = 2.10, 2.45, 2.80, 3.15, 3.50$ days. Compare the final size of the total recovered population R_∞ against your results of the previous problem. Compare your results to Figure 8.14. How does the peak of infection change, both in magnitude and time, with increasing asymptomatic contact periods B_a.

8.14 Undercount. Implement the SEIIR model with different transition dynamics (8.8). For the early COVID-19 outbreak in Germany from Table 7.2 with an undercount of $U = \nu_a : \nu_s = 5$, plot the evolution of the susceptible, exposed, symptomatic, asymptomatic, and recovered groups assuming latent and infectious periods of $A = 3$ days and $C = 7$ days, a basic reproduction number of $R_0 = 4$, similar transition dynamics for the symptomatic and asymptomatic groups, $C_s = C_a = C$ and $B_s = B_a = C/R_0$, and initial conditions of $E_0 = 0.0007$. Vary the undercount to

$v_a : v_s = 2, 4, 6, 8, 10$. Compare the maximum total infectious population I_{max}. Could you have predicted this?

8.15 Undercount. Implement the SEIIR model with different transition dynamics (8.8). For the early COVID-19 outbreak in Germany from Table 7.2 with an undercount of $U = v_a : v_s = 5$, plot the evolution of the susceptible, exposed, symptomatic, asymptomatic, and recovered groups assuming latent and infectious periods of $A = 3$ days and $C = 7$ days, a basic reproduction number of $R_0 = 4$, similar transition dynamics for the symptomatic and asymptomatic groups, $C_s = C_a = C$ and $B_s = B_a = C/R_0$, and initial conditions of $E_0 = 0.0007$. Vary the undercount to $v_a : v_s = 2, 4, 6, 8, 10$. Compare the finite sizes of the susceptible and recovered groups S_∞ and R_∞.

8.16 Herd immunity. Discuss the implications of a large undercount on herd immunity.

References

1. Arino J, Brauer F, van den Driessche P, Watmough J, Wu J (2006) Simple models for containment of a pandemic. Journal of the Royal Society Interface 3:453-457.
2. Bendavid E, Mulaney B, Sood N, Shah S, Ling E, Bromley-Dulfano R, Lai C, Weissberg Z, Saavedra-Walker R, Tedrow J, Tversky D, Bogan A, Kupiec T, Eichner D, Gupta R, Ioannidis JPA, Bhattacharya J (2020). COVID-19 antibody seroprevalence in Santa Clara County, California. medRxiv doi:10.1101/2020.04.14.20062463.
3. Brauer F (2006) Some simple epidemic models. Mathematical Biosciences and Engineering 3:1-15.
4. Brauer F, van den Dreissche P, Wu J (2008) Mathematical Epidemiology. Springer-Verlag Berlin Heidelberg.
5. Brauer F, Castillo-Chavez C, Feng Z (2019) Mathematical Models in Epidemiology. Springer-Verlag New York.
6. Delamater PL, Street EJ, Leslie TF, Yang YT, Jacobsen KH (2019) Complexity of the basic reproduction number (R_0). Emerging Infectious Diseases 25:1-4.
7. European Centre for Disease Prevention and Control. Situation update worldwide. https://www.ecdc.europa.eu/en/geographical-distribution-2019-ncov-cases accessed: June 1, 2021.
8. European Commission. COVID-19: Temporary restriction on non-essential travel to the EU. Communication from the Commission to the European Parliament, the European Council and the Council. Brussels, March 16, 2020.
9. Evans AS (1976) Viral Infections of Humans. Epidemiology and Control. Plenum Medical Book Company, New York and London.
10. Ioannidis JPA (2021) Reconciling estimates of global spread and infection fatality rates of COVID- 19: An overview of systematic evaluations. European Journal of Clinical Investigation 51:e13554.
11. Ioannidis JPA, Cripps S, Tanner MA (2021) Forecasting for COVID-19 has failed. International Journal of Forecasting, in press.
12. Johns Hopkins University (2021) Coronavirus COVID-19 Global Cases by the Center for Systems Science and Engineering. https://coronavirus.jhu.edu/map.html, https://github.com/CSSEGISandData/covid-19 assessed: June 1, 2021.
13. Kuhl E (2020) Data-driven modeling of COVID-19 – Lessons learned. Extreme Mechanics Letters 40:100921.

14. Linka K, Peirlinck M, Sahli Costabal F, Kuhl E (2020) Outbreak dynamics of COVID-19 in Europe and the effect of travel restrictions. Computer Methods in Biomechanics and Biomedical Engineering 23: 710-717.
15. Linka K, Peirlinck M, Kuhl E (2020) The reproduction number of COVID-19 and its correlation with public heath interventions. Computational Mechanics 66:1035-1050.
16. Moin P (2000) Fundamentals of Engineering Numerical Analysis. Cambridge University Press.
17. Peirlinck M, Linka K, Sahli Costabal F, Kuhl E (2020) Outbreak dynamics of COVID-19 in China and the United States. Biomechanics and Modeling in Mechanobiology 19:2179-2193.
18. Peirlinck M, Linka K, Sahli Costabal F, Bendavid E, Bhattacharya J, Ioannidis J, Kuhl E (2020) Visualizing the invisible: The effect of asymptomatic transmission on the outbreak dynamics of COVID-19. Computer Methods in Applied Mechanics and Engineering 372:113410.
19. Prem K, Liu Y, Russell TW, Kucharski AJ, Eggo RM, Davies N (2020) The effect of control strategies to reduce social mixing on outcomes of the COVID-19 epidemic in Wuhan, China: a modelling study. Lancet Public Health 5:E261-270.
20. Slot E, Hogema BM, Reusken CBEM, Reimerink JH, Molier M, Karregat HM, Ijist J, Novotny VMJ, van Lier RAW, Zaaijer HL (2020) Low SARS-CoV-2 seroprevalence in blood donors in the early COVID-19 epidemic in the Netherlands. Nature Communications 11:5744.
21. World Health Organization. COVID-19 Virtual Press Conference, June 8, 2020. https://www.who.int/docs/default-source/coronaviruse/transcripts/who-audio-emergencies-coronavirus-press-conference-08jun2020.pdf?sfvrsn=f6fd460a$_0$; accessed: June 1, 2021.

Part III
Network epidemiology

Chapter 9
Introduction to network epidemiology

Abstract Most compartment models in epidemiology characterize the space-time evolution of their populations through a set of coupled nonlinear partial differential equations. These equations generally have no analytical solution and we have to solve them numerically. Here we introduce the basic concepts of numerical methods for partial differential equations and illustrate network diffusion and finite element methods to solve them. To demonstrate the features of these two approaches, we derive and compare explicit and implicit network diffusion and finite element methods for the SIS model. The learning objectives of this chapter on network epidemiology are to

- discretize nonlinear partial differential equations in space and time
- solve the discrete equations using the network diffusion method
- calculate adjacency and degree matrices and the graph Laplacian
- solve the discrete equations using the finite element method
- calculate time, flux, and source vectors and tangent matrices
- understand the difference between network diffusion and finite element methods
- discuss limitations of numerical methods for network epidemiology

By the end of the chapter, you will be able to discretize, linearize, and solve nonlinear partial differential equations and analyze, simulate, and predict the spatio-temporal pattern of simple infectious diseases.

9.1 Numerical methods for partial differential equations

Partial differential equations are ubiquitous in the quantitative sciences. For more than 200 years, *partial differential equations* have provided insight into the scientific understanding of heat, diffusion, sound, vibration, elasticity, electrostatics, electrodynamics, and fluid dynamics [4]. In contrast to *ordinary differential equations* [1], partial differential equations contain differentials with respect to several independent variables [26]. For most physics-based problems, these variables are the time t and the position $\mathbf{x}$. Most partial differential equations in physics and engineering are *second-order equations* of the general form,

$$\xi\ddot{c} + \eta\dot{c} = \text{div}(\boldsymbol{D} \cdot \nabla c) + f(c), \tag{9.1}$$

where $c = c(\mathbf{x}, t)$ is the state of the system, and $\dot{c} = \text{d}(\circ)/\text{d}t$ and $\ddot{c} = \text{d}^2(\circ)/\text{d}t^2$ are the first and second derivatives in time. The term $\text{div}(\nabla c)$ denotes the second derivative in space associated with a *flux* or diffusion, $\boldsymbol{D}$ is the *diffusion tensor*, and $f(c)$ is the *source* or reaction term. The notion second order implies that the highest order of derivatives of c, either with respect to time or space, is two.

Gradient, divergence, and Laplacian. Spatial derivatives are central to second order partial differential equations in physics and engineering. They naturally induce the notions of gradient and divergence.

Gradient. The gradient of a scalar-valued field $f : \mathcal{B} \to \mathbb{R}$ is a vector-valued field $\nabla f : \mathcal{B} \to \mathbb{R}^3$ with components

$$\nabla f(\mathbf{x}) = \frac{\partial f(\mathbf{x})}{\partial x_i} = f_{,i} = [\, f_{,1}, f_{,2}, f_{,3}\,]^{\text{t}}\,.$$

The gradient of a vector-valued field $f : \mathcal{B} \to \mathbb{R}^3$ is a second order tensor field $\nabla \mathbf{f} : \mathcal{B} \to \mathbb{R}^{3\times3}$ with components

$$\nabla \mathbf{f}(\mathbf{x}) = \frac{\partial f(x_i)}{\partial x_j} = f_{i,j} = \begin{bmatrix} f_{1,1} & f_{1,2} & f_{1,3} \\ f_{2,1} & f_{2,2} & f_{2,3} \\ f_{3,1} & f_{3,2} & f_{3,3} \end{bmatrix}\,.$$

In general, the gradient operator increases the order, from scalar to vector, from vector to second order tensor. In physical terms, the gradient of a scalar field is a vector field that contains information about the direction and magnitude of the steepest slope at any given point. In computational epidemiology, we take the gradient of the population fields, S, E, I, or R to estimate in which direction and with which intensity an infection might spread.

Divergence. The divergence of a vector-valued field $\mathbf{f} : \mathcal{B} \to \mathbb{R}^3$ is a scalar-valued field $\text{div}(\mathbf{f}) : \mathcal{B} \to \mathbb{R}$, sometimes also denoted as $\nabla \cdot \mathbf{f}$, with

$$\text{div}(\mathbf{f}(\mathbf{x})) = \nabla \mathbf{f}(\mathbf{x}) : \boldsymbol{I} = \frac{\partial f_i(\mathbf{x})}{\partial x_i} = f_{i,i} = f_{1,1} + f_{2,2} + f_{3,3}\,.$$

The divergence of a tensor-valued field $\boldsymbol{F} : \mathcal{B} \to \mathbb{R}^{3\times3}$ is a vector-valued field $\text{div}(\boldsymbol{F}) : \mathcal{B} \to \mathbb{R}^3$, sometimes also denoted as $\nabla \cdot \boldsymbol{F}$, with components

$$\text{div}(\boldsymbol{F}(\mathbf{x})) = \nabla \boldsymbol{F}(\mathbf{x}) : \boldsymbol{I} = \frac{\partial F_{ij}(\mathbf{x})}{\partial x_j} = F_{ij,j} = \begin{bmatrix} F_{11,1} + F_{12,2} + F_{13,3} \\ F_{21,1} + F_{22,2} + F_{23,3} \\ F_{31,1} + F_{32,2} + F_{33,3} \end{bmatrix}\,.$$

Fig. 9.1 Pierre Simon Laplace was born on March 23, 1749 in Normandy, France. He is one of the greatest scientists of all time. Amongst his many contributions to engineering, mathematics, statistics, physics, and astronomy, one of his major accomplishments was the his five-volume monograph *Mécanique Céleste* in which he reinterpreted the geometric study of classical mechanics using calculus. He formulated Laplace's equation and the Laplacian differential operator was named after him. Pierre Simon Laplace was named marquis in 1817 and died on March 5, 1827 in Paris, France.

In general, the divergence operator decreases the order, from vector to scalar, from second order tensor to vector. In physical terms, the divergence of a vector field is a scalar field that contains information about the outflux at any given point. In computational epidemiology, we take the divergence of the population fluxes, ∇S, ∇E, ∇I, or ∇R to estimate whether more infectious individuals travel to or from a specific location.

Laplacian. The Laplacian of a scalar-valued field $f : \mathcal{B} \to \mathbb{R}$ is a scalar-valued field $\Delta f : \mathcal{B} \to \mathbb{R}$, sometimes also denoted as $\nabla^2 f$, with

$$\Delta(f) = \text{div}(\nabla f(\mathbf{x})) = f_{,ii} = f_{,11} + f_{,22} + f_{,33} \, .$$

The Laplacian of a vector-valued field $\mathbf{f} : \mathcal{B} \to \mathbb{R}^3$ is a vector-valued field $\Delta \mathbf{F} : \mathcal{B} \to \mathbb{R}^3$, sometimes also denoted as $\nabla^2 \mathbf{f}$, with components

$$\Delta(\mathbf{f}(\mathbf{x})) = \text{div}(\nabla \mathbf{f}(\mathbf{x})) = f_{i,jj} = \begin{bmatrix} f_{1,11} + f_{1,22} + f_{1,33} \\ f_{2,11} + f_{2,22} + f_{2,33} \\ f_{3,11} + f_{3,22} + f_{3,33} \end{bmatrix} \, .$$

In general, the Laplacian operator maintains the order, from scalar to scalar, from vector to vector. In physical terms, the Laplacian represents the flux density of the gradient flow of a function and tells us how much the steepness of a field is changing. In computational epidemiology, we apply the Laplacian operator to population fields, S, E, I, or R to estimate the net rate at which an infection travels toward or away from a specific location.

Second order partial differential equations are widely used in physics and engineering [4]. They fall into three categories: elliptic, parabolic, and hyperbolic. Knowing the type of a partial differential equation is important because it provides insight into the smoothness of the solution, the propagation of information, and the effects of initial and boundary conditions.

Elliptic partial differential equations have no characteristic curves. Their solutions are smooth and do not have discontinuities. Perturbations at any point instantaneously affect the solution in the entire domain. With $\xi = 0$ and $\eta = 0$, they take the general form

$$\mathrm{div}(\boldsymbol{D} \cdot \nabla c) = f(c).$$

For the special case of $\boldsymbol{D} = \kappa \boldsymbol{I}$, the divergence term simplifies to $\mathrm{div}(\boldsymbol{D} \cdot \nabla c) = \kappa \, \mathrm{div}(\nabla c) = \kappa \, \Delta$, where κ is the diffusion coefficient and $\Delta = \mathrm{div}(\nabla)$ denotes the Laplace operator. For a heterogeneous righthand side, $f \neq 0$, this results in the Poisson equation $\Delta c = f$. For a homogeneous righthand side, $f = 0$, this results in the Laplace equation $\Delta c = 0$. Typical examples of elliptic equations are the steady state heat equation, the Navier-Stokes equations, and the equations of electrostatics and elastostatics.

Parabolic partial differential equations have a single characteristic curve. Their solutions become smooth and discontinuities disappear as time progresses. Perturbations at any point gradually smoothen out in time. With $\xi = 0$, they take the general form

$$\dot{c} = \mathrm{div}(\boldsymbol{D} \cdot \nabla c) + f(c).$$

Typical examples of parabolic equations are the heat equation, the diffusion equation, and the linear Schrödinger equation. Compartment models in computational epidemiology typically result in diffusion equations with nonlinear reaction or source terms $f(c)$ and fall into this category of parabolic equations.

Hyperbolic partial differential equations have two characteristic curves. Their solutions retain any discontinuity of the initial conditions. Perturbations do not affect the entire domain instantaneously; instead they travel at a finite propagation speed along the characteristics of the equation. They take the general form

$$\ddot{c} + \eta \dot{c} = \mathrm{div}(\boldsymbol{D} \cdot \nabla c) + f(c).$$

Typical examples of hyperbolic equations are the wave equation in acoustics and vibrations and the equations of elastodynamics.

Compartment models in computational epidemiology typically contain no second order derivatives in time. This implies that the first term in equation (9.1) vanishes, $\xi = 0$, and the remaining equation is a *parabolic partial differential equation* of the following form,

$$\dot{c} = \mathrm{div}(\boldsymbol{D} \cdot \nabla c) + f(c). \tag{9.2}$$

Only very few partial differential equations have an explicit analytical solution. In fact, compartment models typically require numerical methods to approximate their solution computationally. To do so, we discretize the set of governing equations both in space and time, convert them into a system of algebraic equations, and apply numerical methods to approximate their solution [22]. The most common numerical methods to solve partial differential equations are finite difference methods, finite volume methods, and finite element methods.

Finite difference methods discretize the domain of interest using uniform grids, convert the Laplace operator into stencils that connect neighboring nodes, add local source terms, and solve for the time evolution of the unknown at discrete, usually equidistant nodes. The major advantages of finite difference methods are their conceptual simplicity and their computational efficiency; however, finite difference methods are difficult to generalize to complex geometries and to higher order.

Finite volume methods discretize the domain of interest using volume elements, convert the divergence term into surface fluxes into and out of each volume element, add local source terms, and solve for the time evolution of the unknown in each volume element. The major advantages of finite volume methods are their conceptual simplicity, their inherent conservation properties, and their flexibility when analyzing complex domains with unstructured meshes; however, finite volume methods are difficult to generalize to higher order.

Finite element methods discretize the domain of interest using finite elements, approximate the weak form of the governing equation in an integral sense, and solve for the time evolution of the unknown at the nodes that connect the elements. The major advantages of finite element methods are their geometric flexibility when analyzing complex domains and their potential to adjust the order of accuracy by using or combining different element types; however, finite element methods are the most complex numerical method of these three.

The choice of the numerical method depends on the underlying problem and the nature of the partial differential equation. For the parabolic equations with linear flux and nonlinear source terms we consider in this book, linear finite difference, finite volume, and finite element methods result a in similar set of discrete equations. In the following sections, we illustrate an even simpler method, the network diffusion method [24], and contrast it with the most complex representative of these three methods, the finite element method [16], in combination with both explicit and implicit time integration.

9.2 Network diffusion method

Network diffusion models are popular numerical methods to solve partial differential equations on discrete directed or undirected graphs [24]. In computational epidemiology, network diffusion models are widely used because they are inherently discrete, conceptually simple, and computationally inexpensive [16]. They model the spreading of a disease across a continent, country, or state through the mobility of populations within a *weighted undirected graph* G with n_{nd} nodes and n_{el} edges [18]. The nodes represent individual regions and the edges the connections between them [16]. To illustrate the concept of network diffusion modeling, we begin with the continuous nonlinear partial differential equation for reaction diffusion problems for an unknown field $c(\mathbf{x}, t)$ with initial conditions c_0 at time t_0,

$$\dot{c} = \frac{dc}{dt} = \mathrm{div}(\boldsymbol{D} \cdot \nabla c) + f(c) \quad \text{with} \quad c(t_0) = c_0. \tag{9.3}$$

Here, $\mathrm{div}(\boldsymbol{D} \cdot \nabla c)$ and $f(c)$ are the flux or diffusion and the source or reaction terms, $\boldsymbol{D} \cdot \nabla c$ is the flux, $\boldsymbol{D}$ the diffusion tensor, and ∇c the gradient of the unknown c. To discretize equation (9.3) in space, we introduce the unknown c_I as global unknown at each $I = 1, ..., n_{nd}$ node of a weighted graph $\mathcal{G}$. We summarize the connectivity of the graph $\mathcal{G}$ in terms of the *adjacency matrix* A_{IJ}, the weighted connection between two nodes I and J, and the *degree matrix* $D_{II} = \mathrm{diag} \sum_{J=1, J \neq I}^{n_{nd}} A_{IJ}$, the weighted number of incoming and outgoing connections of node I. The difference between the degree matrix D_{IJ} and the adjacency matrix A_{IJ} defines the weighted *graph Laplacian* L_{IJ}[24],

$$L_{IJ} = D_{IJ} - A_{IJ} \quad \text{with} \quad D_{II} = \mathrm{diag} \sum_{J=1, J \neq I}^{n_{nd}} A_{IJ}. \tag{9.4}$$

The weighted graph Laplacian is a $n_{nd} \times n_{nd}$ matrix that introduces a discrete approximation of the diffusion term in equation (9.3), $\mathrm{div}(\boldsymbol{D} \cdot \nabla c) = -\kappa \sum_{J=1}^{n_{nd}} L_{IJ} c_J$, where κ is the diffusion coefficient [19]. With it, we rewrite the nonlinear field equation (9.3) as a network diffusion model, a discrete set of n_{nd} equations with n_{nd} unknowns at the I nodes of the network [6],

$$\dot{c}_I = -\kappa \sum_{J=1}^{n_{nd}} L_{IJ} c_J + f_I(c_I) \quad \forall I = 1, ..., n_{nd}. \tag{9.5}$$

Similar to the ordinary differential equations in Chapter 5, we solve this set of equations in time by partitioning the time interval $\mathcal{T}$ into n_{step} discrete time steps, $\mathcal{T} = \bigcup_{n=1}^{n_{step}} [t_n, t_{n+1}]$, where the subscripts $(\circ)_n$ and $(\circ)_{n+1}$ are associated with the beginning and the end of the current time step. We assume that we know the nodal unknowns $c_{I,n}$ at the beginning of the time step t_n and approximate the first order time derivative using finite differences,

$$\dot{c}_I \approx \frac{c_{I,n+1} - c_{I,n}}{\Delta t} \quad \text{with} \quad \Delta t = t_{n+1} - t_n, \tag{9.6}$$

where Δt denotes the time step size and n is the increment counter.

Explicit forward Euler method. In the spirit of an explicit time integration, we can evaluate the righthand side exclusively at the known time point t_n,

$$\frac{c_{I,n+1} - c_{I,n}}{\Delta t} = -\kappa \sum_{J=1}^{n_{nd}} L_{IJ} c_{J,n} + f_I(c_{I,n}), \tag{9.7}$$

which results in an *explicit* equation for the nodal unknowns $c_{I,n+1}$ at the new time point t_{n+1},

$$c_{I,n+1} = c_{I,n} - \kappa \sum_{J=1}^{n_{nd}} L_{IJ} c_{J,n} \Delta t + f_I(c_{I,n}) \Delta t. \tag{9.8}$$

In the context of network epidemiology, the nodal unknown c_I can be a single population, for example the infectious population of the SIS model, or multiple populations that interact locally at each node, for example the susceptible, exposed, infectious, and recovered populations of the SEIR model [8]. For the SIS model of

Sections 2 and 5 with a single unknown at each node, the source or reaction term at each node would become $f_I = \beta [\, 1 - c_{I,\mathrm{n}} \,]\, c_{I,\mathrm{n}} - \gamma c_{I,\mathrm{n}}$.

Implicit backward Euler method. Alternatively, we can adopt an implicit time integration and can evaluate the righthand side at the unknown time point t_{n+1},

$$\frac{c_{I,\mathrm{n}+1} - c_{I,\mathrm{n}}}{\Delta t} = -\kappa \sum_{J=1}^{n_{\mathrm{nd}}} L_{IJ}\, c_{J,\mathrm{n}+1} + f_I(c_{I,\mathrm{n}+1}), \tag{9.9}$$

which results in a set of *implicit* equations for the nodal unknowns $c_{I,\mathrm{n}+1}$ at the new time point t_{n+1}. Similar to Chapter 5 for ordinary differential equations, we can adopt a *Newton method* to solve for the incremental iterative updates of the unknowns at each node. The resulting discrete residual R_I resembles the residual for ordinary differential equations (5.8), but now evaluated at all $I = 1,..n_{\mathrm{nd}}$ nodes, connected through the network diffusion term,

$$R_I = \frac{1}{\Delta t}[c_{I,\mathrm{n}+1} - c_{I,\mathrm{n}}] + \kappa \sum_{J=1}^{n_{\mathrm{nd}}} L_{IJ}\, c_{J,\mathrm{n}+1} - f_I(c_{I,\mathrm{n}+1}). \tag{9.10}$$

From its *linearization* with respect to the nodal unknowns, we obtain the discrete tangent matrix K_{IJ} for the Newton method of the network diffusion model,

$$K_{IJ} = \frac{dR_I}{dc_J} = \frac{1}{\Delta t}I_{IJ} + \kappa L_{IJ} - \frac{df_I}{dc_I}I_{IJ}, \tag{9.11}$$

where I_{IJ} is the $n_{\mathrm{nd}} \times n_{\mathrm{nd}}$ unit matrix and df_I/dc_I is the derivative of the nodal source term with respect to the nodal unknowns. With the residual vector R_I and the tangent matrix K_{IJ}, we solve for the incremental iterative update of the nodal unknowns,

$$c_I^{k+1} = c_I^k + dc_I \quad \text{with} \quad dc_I = -\sum_{J=1}^{n_{\mathrm{nd}}} K_{JI}^{-1} \cdot R_J. \tag{9.12}$$

We iterate until the norm of the residual, $||R_I|| < \mathrm{tol}$, is smaller than a user defined tolerance, tol, and then proceed to the next time step. An incremental iterative solution within an implicit time integration scheme results in two nested loops, an outer loop for all time steps n and an inner loop one for all iterations k.

Network model of five states. Figure 9.2 illustrates the discrete graph $\mathcal{G}$ of California, Texas, New York, Georgia, and Minnesota with $n_{\mathrm{nd}} = 5$ nodes and the $n_{\mathrm{el}} = 8$ most travelled edges. We estimate the weights of the edges from the annual incoming and outgoing passenger air travel [3]. The size and color of the nodes represent the degree D_{II}, the thickness and number of the edges represents the adjacency A_{IJ} normalized to one. For this particular example, when sorting the states by population, $I = [\,\mathrm{California, Texas, New\ York, Georgia, Minnesota}\,]$, the adjacency matrix A_{IJ} and the degree matrix, $D_{II} = \mathrm{diag} \sum_{J=1}^{n_{\mathrm{nd}}} A_{IJ}$, take the following forms,

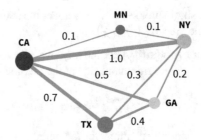

Fig. 9.2 Network diffusion model of California, Texas, New York, Georgia, and Minnesota. Discrete graph G with five nodes and the eight most traveled edges. The size and color of the nodes represent the degree D_{IJ}; the thickness and number of the edges represent the adjacency A_{IJ} estimated from weighted annual incoming and outgoing passenger air travel.

$$
A_{IJ} = \begin{bmatrix} 0.0 & 0.7 & 1.0 & 0.5 & 0.1 \\ 0.7 & 0.0 & 0.3 & 0.4 & 0.0 \\ 1.0 & 0.3 & 0.0 & 0.2 & 0.1 \\ 0.5 & 0.4 & 0.2 & 0.0 & 0.0 \\ 0.1 & 0.0 & 0.1 & 0.0 & 0.0 \end{bmatrix} \quad D_{IJ} = \begin{bmatrix} 2.3 & 0.0 & 0.0 & 0.0 & 0.0 \\ 0.0 & 1.4 & 0.0 & 0.0 & 0.0 \\ 0.0 & 0.0 & 1.6 & 0.0 & 0.0 \\ 0.0 & 0.0 & 0.0 & 1.1 & 0.0 \\ 0.0 & 0.0 & 0.0 & 0.0 & 0.2 \end{bmatrix}, \quad (9.13)
$$

The adjacency matrix A_{IJ} highlights the strongest connection between California, the first row and column, and New York, the third row and column and the weakest connection between Minnesota, the fifth row and column and California and New York. The zero entries on the diagonal indicate that the model neglects travel within states. The zero off-diagonal entries reveal that the model has no direct connections between Minnesota and Texas and between Minnesota and Georgia. The degree matrix D_{IJ} is a diagonal matrix that contains the sum of each row, the passengers traveling to and from each node. The degree is largest in California and smallest in Minnesota. From the adjacency and degree matrices A_{IJ} and D_{IJ}, we calculate the graph Laplacian $L_{IJ} = D_{IJ} - A_{IJ}$,

$$
L_{IJ} = \begin{bmatrix} +2.3 & -0.7 & -1.0 & -0.5 & -0.1 \\ -0.7 & +1.4 & -0.3 & -0.4 & 0.0 \\ -1.0 & -0.3 & +1.6 & -0.2 & -0.1 \\ -0.5 & -0.4 & -0.2 & +1.1 & 0.0 \\ -0.1 & 0.0 & -0.1 & 0.0 & +0.2 \end{bmatrix} \quad (9.14)
$$

normalize it to one, and scale it by the travel coefficient κ. This results in a mobility network, for which infectious diseases spread across the graph G proportionally to the annual air passenger travel statistics within the five states.

Network model of the United States. Figure 9.3 illustrates the discrete graph G of the entire United States with $n_{nd} = 50$ nodes and the $n_{el} = 200$ most travelled edges [3]. The size and color of the nodes represent the degree D_{II}, the thickness of the edges represents the adjacency A_{IJ}. For this passenger travel-weighted graph, the degree ranges from 100 million in California to less than 1 million in Delaware,

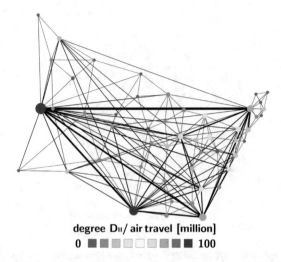

degree D_{II}/ air travel [million]

0 ■■■■□□□■■■ 100

Fig. 9.3 Network diffusion model of the United States. Discrete graph $\mathcal{G}$ of the United States with 50 nodes and the 200 most traveled edges. The size and color of the nodes represent the degree D_{II}; the thickness of the edges represents the adjacency A_{IJ} estimated from annual incoming and outgoing passenger air travel [20].

Vermont, West Virginia, and Wyoming, with a mean degree of $\bar{D}_{II} = 16$ million per node. From the adjacency and degree matrices A_{IJ} and D_{IJ}, we calculate the graph Laplacian $L_{IJ} = D_{IJ} - A_{IJ}$, normalize it to one, and scale it by the travel coefficient κ. This results in a mobility network, for which infectious diseases spread across the graph $\mathcal{G}$ proportionally to the annual air passenger travel statistics within the United States [20].

Network model of the European Union. Figure 9.4 illustrates the discrete graph $\mathcal{G}$ of the European Union with $n_{nd} = 27$ nodes and $n_{el} = 172$ edges [10]. The size and color of the nodes represent the degree D_{II}, the thickness of the edges represents the adjacency A_{IJ}. For this passenger travel-weighted graph, the degree ranges from 222 million in Germany, 221 million in Spain, 162 million in France, and 153 million in Italy to just 3 million in Estonia and Slovakia, and 2 million in Slovenia, with a mean degree of $\bar{D}_{II} = 48 \pm 64$ million per node. From the adjacency and degree matrices A_{IJ} and D_{IJ}, we calculate the graph Laplacian $L_{IJ} = D_{IJ} - A_{IJ}$, normalize it to one, and scale it by the travel coefficient κ. This results in a mobility network, for which infectious diseases spread across the graph $\mathcal{G}$ proportionally to the annual air passenger travel statistics within the European Union [16].

Example. Network diffusion analysis of the SIS model. To illustrate the network discretization of the SIS model [18], we consider the nonlinear ordinary differential for the infectious population, $\dot{I} = \beta[1-I]I - \gamma I$, of the local SIS model from Chapter 2, and introduce the infectious populations I_I as global unknown at each $I = 1, ..., n_{nd}$ node of our weighted graph $\mathcal{G}$. To connect the

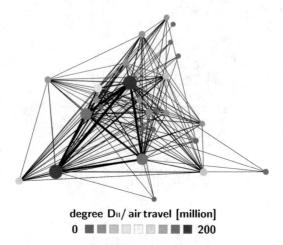

degree D₍ₗₗ₎/ air travel [million]

0 ■■■■■□□■■■ 200

Fig. 9.4 Network epidemiology model of the European Union. Discrete graph G of the Eurpean Union with 27 nodes and the 172 most travelled edges. The size and color of the nodes represent the degree D_{II}; the thickness of the edges represents the adjacency A_{IJ} estimated from annual incoming and outgoing passenger air travel [16].

Table 9.1 Classical SIS diffusion model. Pseudocode for network diffusion method. Incremental iterative Newton algorithm to calculate network diffusion using the SIS model where c and cn represent the current and previous infectious population, L and I are the nd×nd graph Laplacian and identity matrices, and R and K are the residual and iteration matrix of the Newton method, and kappa, beta, Iinf, and dt are the diffusion coefficient, contact rate, finite size, and time step size.

```
L = diag(sum(A)) - A; I = eye(nd); tol=1e-14;
for n = 1:nstep;
      res = 1;
      while res > tol
            R = (c-cn)/dt + kappa*L*c - beta*(Iinf*c-diag(c*c'));
            K = I /dt + kappa*L - beta*(Iinf*I-2*diag(c)*I);
            res = norm(R);
      c = c - K \ R;
end
cn = c;
```

individual nodes, we add the discrete diffusion term, $-\kappa\sum_{J=1}^{n_{nd}} L_{IJ}\,I_J$, where L_{IJ} is the discrete graph Laplacian, the $n_{nd} \times n_{nd}$ matrix that tells us how fast infectious individuals travel from one node to another. This results in the discrete network diffusion model, a set of n_{nd} equations with n_{nd} unknowns,

$$\dot{I}_I = -\kappa\sum_{J=1}^{n_{nd}} L_{IJ}\,I_J + \beta I_I\,[\,I_\infty - I_I\,],$$

where $I_\infty = 1 - 1/R_0 = 1 - \gamma/\beta$ denotes the finite size at endemic equilibrium. We discretize our network model in time using finite differences, $\dot{I}_I = \frac{1}{\Delta t}[I_{I,n+1} - I_{I,n}]$, an implicit Euler backward scheme, and a Newton

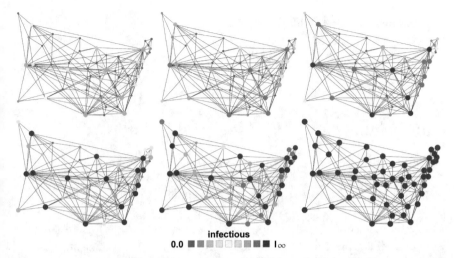

Fig. 9.5 Classical SIS diffusion model. Network diffusion analysis of disease spreading across the United States. Discrete network of the United States with $n_{nd} = 50$ nodes and the $n_{el} = 200$ most traveled edges. From the initial seeding region in New York, the disease gradually spreads across the United States by air travel weighted diffusion. Infectious period $C = 5$ days, basic reproduction number $R_0 = 2.0$, and diffusion coefficient $\kappa = 0.05$.

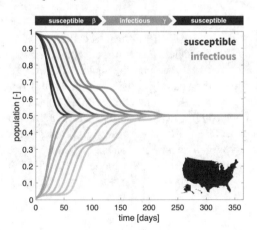

Fig. 9.6 Classical SIS diffusion model. Network diffusion method. Sensitivity with respect to the diffusion coefficient κ. Small diffusion coefficients κ generate a successive state-wise infection associated with a step-wise increase in the infectious population, see Figure 9.7. Increasing the diffusion coefficient κ smoothes the infection across the states and accelerates convergence to the endemic equilibrium, but the final sizes S_∞ and I_∞ remain unchanged, see Figure 9.8. Infectious period $C = 5$ days, initial infectious population $I_0 = 0.01$, basic reproduction number $R_0 = 2.0$, and diffusion coefficient $\kappa = 10^{-7}, 10^{-6}, 10^{-5}, 10^{-4}, 10^{-3}, 10^{-2}, 10^{-1}, 1$ from light to dark.

method to solve for the incremental iterative updates of the infectious population at each node. The resulting discrete residual R_I resembles the residual of the local SIS model in Chapter 2, but now evaluated at all $I = 1, ..n_{nd}$ nodes, connected through the diffusion term,

$$R_I = \frac{1}{\Delta t}[I_{I,n+1} - I_{I,n}] + \kappa \sum_{J=1}^{n_{nd}} L_{IJ} I_{J,n+1} - \beta I_{I,n+1}[I_\infty - I_{I,n+1}].$$

The tangent matrix for the Newton method takes the following format,

$$K_{IJ} = \frac{dR_I}{dI_J} = \frac{1}{\Delta t} I_{IJ} + \kappa L_{IJ} - \beta [I_\infty I_{IJ} - 2\operatorname{diag}(I_{n+1}) I_{IJ}],$$

where I_{IJ} is the $n_{nd} \times n_{nd}$ unit matrix and $\operatorname{diag}(I_{n+1})$ is a diagonal matrix with the infectious populations $I_{I,n+1}$ at all n_{nd} nodes. Table 9.1 summarizes the pseudocode for the network diffusion method of the SIS diffusion model.

Figure 9.5 illustrates the spreading pattern of the infectious population across the United States represented through a discrete network with $n_{nd} = 50$ nodes, one for each state, and the $n_{el} = 200$ most traveled edges, weighted by passenger air travel. From the initial seeding region in New York, the disease first spreads to the most connected states, California, Texas, Florida, Chicago, and Georgia, and from there, gradually across the entire United States.

Figure 9.6 highlights the sensitivity of the overall infectious population with respect to the diffusion coefficient κ. Small diffusion coefficients κ generate a successive state-wise infection associated with a step-wise increase in the infectious population. Increasing the diffusion coefficient κ smoothes the infection across all states and accelerates convergence to the endemic equilibrium, but the final sizes S_∞ and I_∞ remain unchanged. Figures 9.7 and 9.8 illustrate the different spreading patterns of slow and fast diffusion. Slow diffusion generates a successive state-wise infection with an almost binary minimum-maximum infectious pattern; fast diffusion generates a gradual infection with an almost homogeneous infection pattern across all states. Notably, the associated dark curves in of the largest diffusion coefficient in Figure 9.6 are almost identical to the solution for the local SIS model in Figure 2.3 of Chapter 2.

9.3 Finite element method

The finite element method is a widely used numerical method to solve partial differential equations in mathematics, biology, physics, and engineering [8]. It is particularly popular when solving initial boundary value problems in complex two or three dimensional domains [22]. From a mathematical point, finite element methods use variational calculus to approximate a solution by minimizing an error function [16]. From a practical point, finite element methods subdivide a large system into smaller simpler parts, the finite elements, and approximates the solution through an algebraic system of equations that represents the entire problem [2]. In this section,

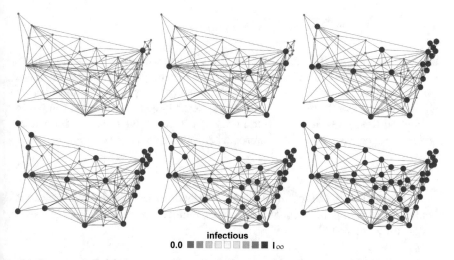

Fig. 9.7 Classical SIS diffusion model. Network diffusion analysis of slow disease spreading across the United States. A small diffusion to infection ratio $\kappa : C$ generates a successive state-wise infection associated with a step-wise increase in the infectious population. Infectious period $C = 5$ days, basic reproduction number $R_0 = 2.0$, and diffusion coefficient $\kappa = 0.0000001$.

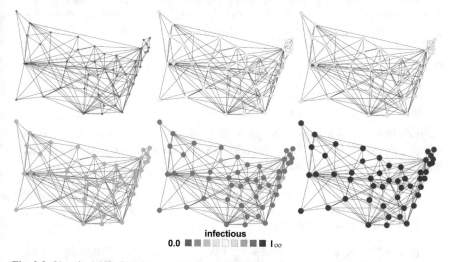

Fig. 9.8 Classical SIS diffusion model. Network diffusion analysis of fast disease spreading across the United States. A large diffusion to infection ratio $\kappa : C$ generates a homogeneous infection across all states associated with a smooth increase in the infectious population. Infectious period $C = 5$ days, basic reproduction number $R_0 = 2.0$, and diffusion coefficient $\kappa = 1.0$.

we illustrate the concept of the finite element method by discretizing and solving a nonlinear partial differential equation. We begin with the continuous nonlinear partial differential equation for reaction diffusion problems for an unknown field $c(\mathbf{x}, t)$ with initial conditions c_0 at time t_0,

$$\dot{c} = \frac{dc}{dt} = \text{div}(\boldsymbol{D} \cdot \nabla c) + f(c) \quad \text{with} \quad c(t_0) = c_0 \,. \tag{9.15}$$

Here, $\text{div}(\boldsymbol{D} \cdot \nabla c)$ and $f(c)$ are the flux or diffusion and the source or reaction terms, $\boldsymbol{D} \cdot \nabla c$ is the flux, $\boldsymbol{D}$ the diffusion tensor, and ∇c the gradient of the unknown c. In epidemiology, the source or reaction term $f(c)$ is typically nonlinear in c, and we cannot solve the resulting algebraic system of equations directly. Similar to the implicit time integration in Section 5.3 of Chapter 5 for ordinary differential equations, we convert the partial differential equation into its *residual form* R,

$$\text{R} = \dot{c} - \text{div}(\boldsymbol{D} \cdot \nabla c) - f(c) \doteq 0 \,. \tag{9.16}$$

The fundamental basis of the finite element method is the *weak form* G, which we obtain by multiplying the residual R with the test function δc and integrating it over the domain of interest $\mathcal{B}$,

$$\text{G} = \int_{\mathcal{B}} \delta c \, \dot{c} \, dV - \int_{\mathcal{B}} \delta c \, \text{div}(\boldsymbol{D} \cdot \nabla c) \, dV - \int_{\mathcal{B}} \delta c \, f(c) \, dV \doteq 0 \,. \tag{9.17}$$

At this point, the diffusion term, $\delta c \, \text{div}(\boldsymbol{D} \cdot \nabla c)$, is non-symmetric with no spatial derivative on the test function δc and two spatial derivatives on the trial function c. To convert the diffusion term into a symmetric format with gradients on both test and trial functions, we integrate it by parts, $\delta c \, \text{div}(\boldsymbol{D} \cdot \nabla c) = \nabla(\delta c \, \boldsymbol{D} \cdot \nabla c) - \nabla \delta c \cdot \boldsymbol{D} \cdot \nabla c$, and assume homogeneous Neumann boundary conditions, $\nabla c \cdot \mathbf{n} = 0$, such that $\delta c \, \text{div}(\boldsymbol{D} \cdot \nabla c) = -\nabla \delta c \cdot \boldsymbol{D} \cdot \nabla c$. This results in the symmetric discrete weak form,

$$\text{G} = \int_{\mathcal{B}} \delta c \, \dot{c} \, dV + \int_{\mathcal{B}} \nabla \delta c \cdot \boldsymbol{D} \cdot \nabla c \, dV - \int_{\mathcal{B}} \delta c \, f(c) \, dV \doteq 0 \,. \tag{9.18}$$

To discretize the weak form in in space, we partition the domain of interest $\mathcal{B}$ into n_{el} finite elements, $\mathcal{B} = \bigcup_{e=1}^{n_{\text{el}}} \mathcal{B}_e$, and adopt an element-wise interpolation of the test functions δc and trial functions c,

$$\delta c = \sum_{i=1}^{n_{\text{en}}} N_i \, \delta c_i \quad \text{and} \quad c = \sum_{j=1}^{n_{\text{en}}} N_j \, c_j \,, \tag{9.19}$$

and of their spatial and temporal gradients,

$$\nabla \delta c = \sum_{i=1}^{n_{\text{en}}} \nabla N_i \, \delta c_i \quad \text{and} \quad \nabla c = \sum_{j=1}^{n_{\text{en}}} \nabla N_j \, c_j \quad \text{and} \quad \dot{c} = \sum_{j=1}^{n_{\text{en}}} N_j \, \dot{c}_j \,, \tag{9.20}$$

where N_i and N_j are the finite element shape functions at the n_{en} element nodes i and j and $\nabla N_i = dN_i/dx$ and $\nabla N_j = dN_j/dx$ are their spatial gradients. This results in the following discrete weak form of the residual at each finite element node I,

$$\text{R}_I = \underset{e=1}{\overset{n_{\text{el}}}{\text{A}}} \int_{\mathcal{B}_e} N_i \, \dot{c} + \nabla N_i \cdot \boldsymbol{D} \cdot \nabla c - N_i \, f(c) \, dV_e \doteq 0_I \quad \forall \, I = 1, \dots, n_{\text{nd}} \,, \tag{9.21}$$

where the operator $\underset{e=1}{\overset{n_{\text{el}}}{\text{A}}}$ denotes the assembly of the element residuals at the $i = 1, \dots, n_{\text{en}}$ element nodes to system residual at the $I = 1, \dots, n_{\text{nd}}$ global nodes. To discretize the weak from in time, we divide the time interval $\mathcal{T}$ into n_{step} discrete time steps, $\mathcal{T} = \bigcup_{n=1}^{n_{step}} [\,t_n, t_{n+1}\,]$, where the subscripts $(\circ)_n$ and $(\circ)_{n+1}$ are associated with the beginning and the end of the current time step. We approximate the first order time derivative of the unknown c using finite differences,

$$\dot{c} \approx \frac{c_{n+1} - c_n}{\Delta t} \quad \text{with} \quad \Delta t := t_{n+1} - t_n \qquad (9.22)$$

where Δt is the time step size and n is the increment counter.

Explicit forward Euler method. To evaluate the residual equation (9.21), we can adopt an explicit time integration and evaluate the source and flux terms exclusively at the known time point t_n,

$$R_I = \mathop{A}_{e=1}^{n_{el}} \int_{\mathcal{B}_e} N_i \frac{c_{n+1} - c_n}{\Delta t} + \nabla N_i \cdot \boldsymbol{D} \cdot \nabla c_n - N_i \, f(c_n) \, dV_e \doteq 0_I, \qquad (9.23)$$

which results in an *explicit* equation for the nodal unknowns $c_{I,n+1}$ at the new time point t_{n+1},

$$c_{I,n+1} = c_{I,n} - \mathop{A}_{e=1}^{n_{el}} \int_{\mathcal{B}_e} \nabla N_i \cdot \boldsymbol{D} \cdot \nabla c_n \, dV_e \, \Delta t + f_I(c_{I,n}) \, \Delta t. \qquad (9.24)$$

The evaluation of the source term depends on the selection of the finite element type. We will illustrate this step for the simplest finite element type, linear truss elements, at the end of this section.

Implicit backward Euler method. Alternatively, we can adopt an implicit backward Euler scheme and evaluate the source and flux terms at the end of the current time step t_{n+1},

$$R_I = \mathop{A}_{e=1}^{n_{el}} \int_{\mathcal{B}_e} N_i \frac{c_{n+1} - c_n}{\Delta t} + \nabla N_i \cdot \boldsymbol{D} \cdot \nabla c_{n+1} - N_i \, f(c_{n+1}) \, dV_e \doteq 0_I \qquad (9.25)$$

This is an *implicit* equation for the nodal unknowns $c_{I,n+1}$, which we solve using an incremental iterative Newton method. We linearize the residual with respect to the nodal unknowns c_J, such that $R_I^{k+1} = R_I^k + \sum_{J=1}^{n_{nd}} K_{IJ} \cdot dc_J \doteq 0$, where K_{IJ} is the global tangent matrix of the Newton method,

$$K_{IJ} = \frac{dR_I}{dc_J} = \mathop{A}_{e=1}^{n_{el}} \int_{\mathcal{B}_e} N_i \frac{1}{\Delta t} N_j + \nabla N_i \cdot \boldsymbol{D} \cdot \nabla N_j - N_i \frac{df(c)}{dc} N_j \, dV_e, \qquad (9.26)$$

where the operator $\mathop{A}_{e=1}^{n_{el}}$ denotes the assembly of the element matrices at the $i, j = 1, ..., n_{en}$ element nodes to the system matrix at the $I, J, = 1, ..., n_{nd}$ global nodes and $df(c)/dc$ is the derivative of the source term f. Solving the linearized residual equation, provides the incremental iterative update of the unknowns,

$$c_J^{k+1} = c_J^k + dc_J \quad \text{with} \quad dc_J = -\sum_{I=1}^{n_{nd}} K_{JI}^{-1} \cdot R_I^k, \qquad (9.27)$$

at all n_{nd} global nodes J. We iterate until the norm of the residual, $||R_I|| < \text{tol}$, is smaller than a user defined tolerance tol and then proceed to the next time step.

Linear truss elements. The simplest discretization is a linear finite element approximation for the unknown field c, with linear trusses, triangles, or tetrahedra for one-, two-, and three-dimensional problems. Linear one-dimensional truss elements

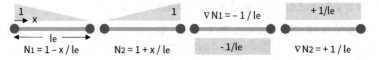

Fig. 9.9 Finite element shape functions for linear truss elements. The Lagrangian shape functions, N_1 and N_2, of linear truss elements are one at their respective node and zero at the other node, and their gradients, ∇N_1 and ∇N_2 are constant within the element.

have two nodes, $n_{en} = 2$, per element. Figure 9.9 illustrates their Lagrangian shape functions, N_1 and N_2, which are one at their respective node and zero at the other node, and their gradients, ∇N_1 and ∇N_2, which are constant within the element,

$$N_1 = 1 - \frac{x}{l_e} \quad N_2 = +\frac{x}{l_e} \quad \nabla N_1 = \frac{dN_1}{dx} = -\frac{1}{l_e} \quad \nabla N_2 = \frac{dN_2}{dx} = +\frac{1}{l_e}, \quad (9.28)$$

where l_e is the element length.

Explicit forward Euler method. For the explicit time integration, we use the shape functions (9.28) to evaluate the explicit equation (9.24) for the nodal unknowns $c_{I,n+1}$ at the new time point t_{n+1} from the known values $c_{I,n}$ at the previous time point t_n,

$$c_{I,n+1} = c_{I,n} - \sum_{J=1}^{n_{nd}} \left[\mathop{A}_{e=1}^{n_{el}} \int_0^l \nabla N_i \cdot \boldsymbol{D} \cdot \nabla N_j \, dl_e \right] c_{J,n} \, \Delta t + f_I(c_{I,n}) \, \Delta t . \quad (9.29)$$

Here $[\mathop{A}_{e=1}^{n_{el}} \int_0^l \nabla N_i \cdot \boldsymbol{D} \cdot \nabla N_j \, dl_e]$ is an $n_{nd} \times n_{nd}$ matrix that collectively contains information about the diffusion in the system. Specifically, for isotropic diffusion with $\boldsymbol{D} = \kappa \boldsymbol{I}$ such that $\boldsymbol{D} \cdot \nabla N_j = \kappa \nabla N_j$, we can integrate the $\nabla N \nabla N$-type flux matrix analytically,

$$\boldsymbol{K}^{\nabla N} = \mathop{A}_{e=1}^{n_{el}} \boldsymbol{K}_e^{\nabla N} \text{ with } \boldsymbol{K}_e^{\nabla N} = \begin{bmatrix} \int_0^l \nabla N_1 \nabla N_1 dl_e & \int_0^l \nabla N_1 \nabla N_2 dl_e \\ \int_0^l \nabla N_2 \nabla N_1 dl_e & \int_0^l \nabla N_2 \nabla N_2 dl_e \end{bmatrix} = \frac{1}{l_e} \begin{bmatrix} +1 & -1 \\ -1 & +1 \end{bmatrix}.$$
$$(9.30)$$

With this analytical expression, the update equation (9.29) takes the following simple expression,

$$c_{I,n+1} = c_{I,n} - \kappa \sum_{J=1}^{n_{nd}} K_{IJ}^{\nabla N} c_{J,n} \, \Delta t + f_I(c_{I,n}) \, \Delta t . \quad (9.31)$$

Interestingly, this explicit update equation for linear one-dimensional finite elements elements (9.31) has a similar format as the explicit update equation of the network diffusion method (9.8). We explore this analogy in detail in Section 9.4.

Implicit backward Euler method. For the implicit time integration, we use the shape functions (9.28) to evaluate the residual vector (9.25) and the tangent matrix (9.26). First, we evaluate the N-type time and source terms and the ∇N-type flux term of the element contributions to the residual vector (9.25),

$$\mathbf{R}_e^N = \begin{bmatrix} \int_0^l N_1 [\, F_1 N_1 + F_2 N_2\,]\, dl_e \\ \int_0^l N_2 [\, F_1 N_1 + F_2 N_2\,]\, dl_e \end{bmatrix} \quad \text{and} \quad \mathbf{R}_e^{\nabla N} = \begin{bmatrix} \int_0^l \nabla N_1 \cdot [\, F_1 \nabla N_1 + F_2 \nabla N_2\,]\, dl_e \\ \int_0^l \nabla N_2 \cdot [\, F_1 \nabla N_1 + F_2 \nabla N_2\,]\, dl_e \end{bmatrix},$$

(9.32)

where $F = [\, c_{n+1} - c_n\,]/\Delta t$ for the time term, $F = \mathbf{D} \cdot \nabla c_{n+1} = \kappa \nabla c_{n+1}$ for the flux term, and $F = f(c_{n+1})$ for the source term. Here we have assumed isotropic diffusion with $\mathbf{D} = \kappa \mathbf{I}$, such that $\mathbf{D} \cdot \nabla c = \kappa \nabla c$. For linear truss elements with constant F terms within the element, we can analytically integrate these two residuals $\mathbf{R}_e^N$ and $\mathbf{R}_e^{\nabla N}$. Upon integrating the first vector, we obtain the consistent element residual vector, $\mathbf{R}_e^N = l_e [\, 2/6 F_1 + 1/6 F_2; 1/6 F_1 + 2/6 F_2\,]$, which is often conveniently simplified to the lumped residual vector, $\mathbf{R}_e^N = l_e/2 [\, F_1; F_2\,]$, for which values at each node just contribute to their own node. With it, the residual vectors simplify to the following format,

$$\mathbf{R}_e^N = \frac{l_e}{2} \begin{bmatrix} F_1 \\ F_2 \end{bmatrix} \quad \text{and} \quad \mathbf{R}_e^{\nabla N} = \frac{1}{l_e} \begin{bmatrix} +F_1 - F_2 \\ -F_1 + F_2 \end{bmatrix}.$$

(9.33)

With these two explicit expressions, we can analytically integrate the residual vector (9.25) and obtain the following explicit representation,

$$\mathbf{R} = \mathop{\mathbf{A}}_{e=1}^{n_{el}} \mathbf{R}^e \quad \text{with} \quad \mathbf{R}^e = \frac{1}{\Delta t} \frac{l_e}{2} \begin{bmatrix} c_{1,n+1} - c_{1,n} \\ c_{2,n+1} - c_{2,n} \end{bmatrix} + \frac{\kappa}{l_e} \begin{bmatrix} +c_{1,n+1} - c_{2,n+1} \\ -c_{1,n+1} + c_{2,n+1} \end{bmatrix} - \frac{l_e}{2} \begin{bmatrix} f(c_{1,n+1}) \\ f(c_{2,n+1}) \end{bmatrix}.$$

(9.34)

Second, we evaluate the NN-type time and source terms and the $\nabla N \nabla N$-type flux term of the element contributions to the tangent matrix (9.26),

$$\mathbf{K}_e^N = \begin{bmatrix} \int_0^l N_1 N_1 dl_e & \int_0^l N_1 N_2 dl_e \\ \int_0^l N_2 N_1 dl_e & \int_0^l N_2 N_2 dl_e \end{bmatrix} \quad \text{and} \quad \mathbf{K}_e^{\nabla N} = \begin{bmatrix} \int_0^l \nabla N_1 \nabla N_1 dl_e & \int_0^l \nabla N_1 \nabla N_2 dl_e \\ \int_0^l \nabla N_2 \nabla N_1 dl_e & \int_0^l \nabla N_2 \nabla N_2 dl_e \end{bmatrix}$$

(9.35)

Upon integrating the first matrix, we obtain the consistent element tangent matrix $\mathbf{K}_e^N = l_e [\, 2/6, 1/6; 1/6, 2/6\,]$, which, when working with the simplified element residual vector (9.33.1), simplifies to the lumped element tangent matrix, $\mathbf{K}_e^N = l_e/2 [\, 1, 0; 0, 1\,]$, a diagonal matrix with the sum of each row forming the diagonal entry. Because of its simplicity, the lumped element vector and lumped element matrix are often used in many practical engineering applications. With it, the element tangent matrices simplify to the following format,

$$\mathbf{K}_e^N = \frac{l_e}{2} \begin{bmatrix} +1 & 0 \\ 0 & +1 \end{bmatrix} \quad \text{and} \quad \mathbf{K}_e^{\nabla N} = \frac{1}{l_e} \begin{bmatrix} +1 & -1 \\ -1 & +1 \end{bmatrix}.$$

(9.36)

With these two explicit expressions, we can analytically integrate the system matrix (9.26) and obtain the following explicit representation,

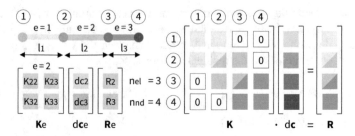

Fig. 9.10 Finite element assembly. Finite element discretization with four nodes and three linear truss elements. Colors indicate the assembly of the three element residual vectors $\mathbf{R}_e$ and element tangent matrices $\mathbf{K}_e$ to the system residual vector $\mathbf{R}$ and system tangent matrix $\mathbf{K}$ that form the global system, $\mathbf{K} \cdot d\mathbf{c} = \mathbf{R}$, for the incremental iterative update of the nodal unknowns $\mathbf{c}$.

$$\mathbf{K} = \overset{n_{el}}{\underset{e=1}{\text{A}}} \mathbf{K}^e \text{ with } \mathbf{K}^e = \frac{1}{\Delta t} \frac{l_e}{2} \begin{bmatrix} +1 & 0 \\ 0 & +1 \end{bmatrix} + \frac{\kappa}{l_e} \begin{bmatrix} +1 & -1 \\ -1 & +1 \end{bmatrix} - \frac{df(c_{n+1})}{dc_{n+1}} \frac{l_e}{2} \begin{bmatrix} +1 & 0 \\ 0 & +1 \end{bmatrix}.$$
(9.37)

Clearly, the three terms of the continuous governing equation (9.15), the time derivative, the flux or diffusion, and the source or reaction, are still present in the discrete element residual (9.34) and tangent (9.37).

Figure 9.10 illustrates the assembly $\text{A}_{e=1}^{n_{el}}$ for a discretization with $n_{nd} = 4$ nodes and $n_{el} = 3$ linear truss elements. The colors indicate the assembly of the element residual vectors $\mathbf{R}_e$ and element tangent matrices $\mathbf{K}_e$ to the system residual vector $\mathbf{R}$ and system tangent matrix $\mathbf{K}$ that form the global system, $\mathbf{K} \cdot d\mathbf{c} = \mathbf{R}$, for the incremental iterative update of the nodal unknowns $\mathbf{c}$. The tangent matrix $\mathbf{K}$ has the dimension $n_{nd} \times n_{nd}$, is symmetric, and sparsely populated. It has non-zero entries where two nodes are connected through an element and zeros at all nodes that are not directly connected to one another. For simple linear trusses, triangles, and tetrahedra, we can integrate the residual vectors and tangent matrices analytically [8]. In some cases, higher order elements become necessary to improve the accuracy of the numerical approximation [2]. However, higher order elements, quadratic or cubic, or more advanced element types, quadrilateral or hexahedral, require a numerical integration at discrete integration points, which is beyond the scope of this book [22].

Finite element model of five states. Figure 9.11 illustrates the finite element discretization of California, Texas, New York, Georgia, and Minnesota with $n_{nd} = 5$ nodes and $n_{el} = 7$ linear truss elements [21]. The discretization is based on a *Delaunay triangulation* that connects any triple of neighboring nodes such that no other node lies inside the circumcircle through the three nodes. The size and color of the nodes represent the population of each state, the thickness and number of each edge represent the distance l_{IJ} and the inverse distance $1/l_{IJ}$ between two neighboring states I and J. For this particular example, when sorting the states by population, $I =$[California, Texas, New York, Georgia, Minnesota], the assembly of the $n_{el} = 7$ element tangent matrices $\mathbf{K}_e^N$ and $\mathbf{K}_e^{\nabla N}$ according to equations (9.35) and (9.36) yields the following system tangent matrices for the time and source terms $\mathbf{K}^N$ and

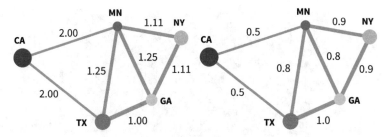

Fig. 9.11 Finite element model of California, Texas, New York, Georgia, and Minnesota. Finite element discretization with five nodes and seven linear truss elements. The size and color of the nodes represent the population; the thickness and number of the edges represent the element length, left, and inverse element length, right.

for the flux term $\mathbf{K}^{\nabla N}$,

$$
K_{IJ}^N = \frac{1}{2}
\begin{bmatrix}
4.00 & 0.00 & 0.00 & 0.00 & 0.00 \\
0.00 & 4.25 & 0.00 & 0.00 & 0.00 \\
0.00 & 0.00 & 2.22 & 0.00 & 0.00 \\
0.00 & 0.00 & 0.00 & 3.36 & 0.00 \\
0.00 & 0.00 & 0.00 & 0.00 & 5.61
\end{bmatrix}
\quad
K_{IJ}^{\nabla N} =
\begin{bmatrix}
+1.0 & -0.5 & 0.0 & 0.0 & -0.5 \\
-0.5 & +2.3 & 0.0 & -1.0 & -0.8 \\
0.0 & 0.0 & +1.8 & -0.9 & -0.9 \\
0.0 & -1.0 & -0.9 & +2.7 & -0.8 \\
-0.5 & -0.8 & -0.9 & -0.8 & +3.0
\end{bmatrix}.
$$
(9.38)

The time and source matrix $\mathbf{K}^N$ is a diagonal matrix for which every entry is half of sum of the lengths l_{IJ} of all elements connected at this node. The matrix entries are largest for the most connected node Minnesota and smallest for the least connected node New York. The flux matrix $\mathbf{K}^{\nabla N}$ is a symmetric matrix with a total of $(2\,n_{el})$ off-diagonal entries. Each element contributes to its two nodes and generates a matrix entry of the inverse element length $1/l_{IJ}$. These entries are added on the diagonal and subtracted on the off-diagonal. Diagonal entries are largest for the most connected node Minnesota and smallest for the least connected node California. Off-diagonal entries are largest for the shortest elements between Minnesota and Texas and Minnesota and Georgia and smallest between the longest elements between California and Texas and between California and Minnesota. Notably, the flux matrix $\mathbf{K}^{\nabla N}$ takes a similar structure as the graph Laplacian matrix $\mathbf{L}$ with negative off-diagonal entries and positive sums of each row on the diagonal. This example suggests that, for a similar discretization, the flux matrix of linear finite truss elements is exactly identical to the graph Laplacian matrix from Section 9.2.

Finite element model of the United States. Figure 9.12 illustrates the finite element model of the entire United States with $n_{nd} = 50$ nodes, $n_{el} = 129$ linear truss elements $n_{el} = 80$ linear triangular elements. The size and color of the nodes represent the population of each state. The discretization of the 50 states was generated using a *Delaunay triangulation*. The Delaunay triangulation connects any triple of neighboring nodes such that no other node lies inside the circumcircle through the three nodes. The circle and the blue triangle highlight the Delaunay circumcircle for the triangle of Washington, Oregon, and Idaho.

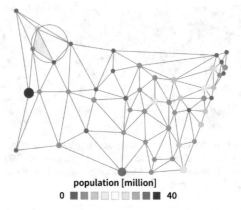

Fig. 9.12 Finite element model of the United States. Finite element discretization with 50 nodes and 129 linear trusses or 80 linear triangles. The size and color of the nodes represent the population of each state. The discretization was created by a Delaunay triangulation that connects any triple of neighboring nodes such that no other node lies inside the circumcircle through the three nodes. The circle and triangle highlight the Delaunay circumcircle for the triangle of Washington, Oregon, and Idaho.

Example. Finite element analysis of the SIS model. To illustrate the finite element discretization of the SIS model, we consider the nonlinear ordinary differential for the infectious population, $\dot{I} = \beta\,[\,1 - I\,]\,I - \gamma\,I$, of the local SIS model from Chapter 2, and add the diffusion term, $\mathrm{div}(\boldsymbol{D} \cdot \nabla I)$. This results in the nonlinear partial differential equation of reaction-diffusion type,

$$\dot{I} = \mathrm{div}(\boldsymbol{D} \cdot \nabla I) + \beta I\,[\,I_\infty - I\,],$$

where $I_\infty = 1 - 1/\mathsf{R}_0 = 1 - \gamma/\beta$ denotes the finite size at endemic equilibrium. We convert this equation into its *residual form* R,

$$\mathsf{R} = \dot{I} - \mathrm{div}(\boldsymbol{D} \cdot \nabla I) - \beta I\,[\,I_\infty - I\,] \doteq 0,$$

derive its ıweak form by multiplying it with the test function δI, and integrate it over the domain of interest $\mathcal{B}$. To convert the divergence term into a symmetric format with gradients on both test and trial functions, δI and I, we integrate it by parts, $\delta I\,\mathrm{div}(\boldsymbol{D} \cdot \nabla I) = \nabla(\delta I\,\boldsymbol{D} \cdot \nabla I) - \nabla\delta I \cdot \boldsymbol{D} \cdot \nabla I$, and assume homogeneous Neumann boundary conditions, $\nabla I \cdot \mathbf{n} = 0$. This results in the symmetric discrete *weak form* G,

$$\mathsf{G} = \int_{\mathcal{B}} \delta I\,\dot{I}\,\mathrm{d}V + \int_{\mathcal{B}} \nabla\delta I \cdot \boldsymbol{D} \cdot \nabla I\,\mathrm{d}V - \int_{\mathcal{B}} \delta I\,\beta I\,[\,I_\infty - I\,]\,\mathrm{d}V \doteq 0.$$

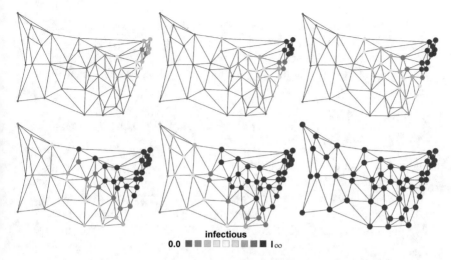

Fig. 9.13 **Classical SIS diffusion model. Finite element analysis of disease spreading across the United States.** Discrete finite element mesh of the United States with n_{nd} = 50 nodes and n_{el} = 129 linear truss elements. From the initial seeding region in New York, the disease gradually spreads across the United States, from east to west, by inverse-distance weighted diffusion.

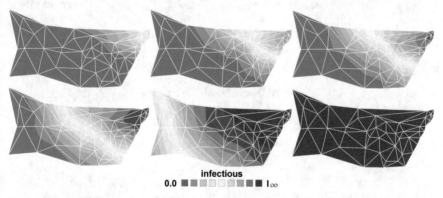

Fig. 9.14 **Classical SIS diffusion model. Finite element analysis of disease spreading across the United States.** Discrete finite element mesh of the United States with n_{nd} = 50 nodes and n_{el} = 80 linear triangular elements. From the initial seeding region in New York, the disease gradually spreads across the United States, from east to west, by inverse-distance weighted diffusion.

To discretize the weak form in space, we partition the domain of interest $\mathcal{B}$ into n_{el} finite elements, $\mathcal{B} = \bigcup_{e=1}^{n_{el}} \mathcal{B}_e$ and adopt an element-wise interpolation of the test functions, $\delta I = \sum_{i=1}^{n_{en}} N_i \, \delta I_i$, and trial functions, $I = \sum_{j=1}^{n_{en}} N_j \, I_j$, where N_i and N_j are the finite element shape functions. To discretize the weak from in time, we divide the time interval $\mathcal{T}$ into n_{step} discrete time steps, $\mathcal{T} = \bigcup_{n=1}^{n_{step}} [\, t_n, t_{n+1} \,]$, where the subscripts $(\circ)_n$ and $(\circ) = (\circ)_{n+1}$ are associated with the beginning and the end of the current time step. We discretize the time derivative of the infectious population using finite differences, $\dot{I} =$

$[I_{n+1} - I_n]/\Delta t$, where $\Delta t := t_{n+1} - t_n$ denotes the time step size. We adopt an implicit backward Euler scheme and evaluate the infectious population at the end of the current time step $t = t_{n+1}$. This results in the following discrete weak form of the residual for the infectious population at each finite element node I,

$$R_I = \overset{n_{el}}{\underset{e=1}{A}} \int_{\mathcal{B}_e} N_i \frac{1}{\Delta t} [I_{n+1} - I_n] + \nabla N_i \cdot \boldsymbol{D} \cdot \nabla I_{n+1} - N_i \beta I_{n+1} [I_\infty - I_{n+1}] \, dV_e \doteq 0_I,$$

where the operator $\overset{n_{el}}{\underset{e=1}{A}}$ denotes the assembly of the element residuals at the n_{en} element nodes i to the n_{nd} global nodes I. To solve the discrete residual, we adopt an incremental iterative Newton method and linearize the residual with respect to the unknown nodal infectious populations I_J, such that $R_I^{k+1} = R_I^k + \sum_{J=1}^{n_{nd}} K_{IJ} \cdot dI_J \doteq 0$, where K_{IJ} is the global tangent of the Newton method,

$$K_{IJ} = \frac{dR_I}{dc_J} = \overset{n_{el}}{\underset{e=1}{A}} \int_{\mathcal{B}_e} N_i \frac{1}{\Delta t} N_j + \nabla N_i \cdot \boldsymbol{D} \cdot \nabla N_j - N_i \beta [I_\infty - 2 I_{n+1}] N_j \, dV_e.$$

We use linear truss elements to evaluate the element residual vector and tangent matrix. With the source term of the SIS model, $f(c) = \beta I [I_\infty - I]$, the discrete version of the residual vector (9.34) becomes

$$\mathbf{R} = \overset{n_{el}}{\underset{e=1}{A}} \mathbf{R}_e \quad \text{with} \quad \mathbf{R}_e = \frac{1}{\Delta t} \frac{l_e}{2} \begin{bmatrix} I_1 - I_{1,n} \\ I_2 - I_{2,n} \end{bmatrix} + \frac{\kappa}{l_e} \begin{bmatrix} +I_1 - I_2 \\ -I_1 + I_2 \end{bmatrix} - \beta \frac{l_e}{2} \begin{bmatrix} I_1 [I_\infty - I_1] \\ I_2 [I_\infty - I_2] \end{bmatrix},$$

and with the linearization of the source term, $df/dc = \beta [I_\infty - 2I]$, the tangent matrix (9.37) becomes

$$\mathbf{K} = \overset{n_{el}}{\underset{e=1}{A}} \mathbf{K}_e \quad \text{with} \quad \mathbf{K}_e = \frac{1}{\Delta t} \frac{l_e}{2} \begin{bmatrix} 1 & 0 \\ 0 & 1 \end{bmatrix} + \frac{\kappa}{l_e} \begin{bmatrix} +1 & -1 \\ -1 & +1 \end{bmatrix} - \beta \frac{l_e}{2} \begin{bmatrix} [I_\infty - 2I_1] & 0 \\ 0 & [I_\infty - 2I_2] \end{bmatrix},$$

where we have omitted the subscript $(\circ)_{n+1}$ on the unknowns I_1 and I_2 for the sake of compactness. Solving the linearized residual equation, $dI_J = -\sum_{I=1}^{n_{nd}} K_{JI}^{-1} \cdot R_I^k$, provides the incremental iterative update of the infectious population, $I_J^{k+1} = I_J^k + dI_J$, at all n_{nd} global nodes J. We iterate until the norm of the residual, $||R_I|| < \text{tol}$, is smaller than a user defined tolerance tol and then proceed to the next time step.

Figure 9.13 illustrates the spreading pattern of the infectious population across the United States represented through a linear finite element model with $n_{nd} = 50$ nodes, one for each state, and $n_{el} = 129$ linear truss elements generated by a Delaunay triangulation that connects the closest neighboring nodes. The simulation uses the same parameters as the network model in Figure 9.5, an infectious period of $C = 5$ days, a basic reproduction number of $R_0 = 2.0$, and a diffusion coefficient of $\kappa = 0.05$. Since the connectivity is different from the network model, the spreading pattern is more gradual and resembles a traveling wave. From the initial seeding region in New York, the disease gradually spreads across the United States, from east to west, by distance weighted diffusion.

Figure 9.14 shows the spreading pattern of the infectious population, now using linear triangular elements instead of linear truss elements. Similar to Figures 9.5 and 9.13, the model uses $n_{nd} = 50$ nodes, but now $n_{el} = 80$ linear triangles. The discretization was generated by the same Delaunay triangulation as Figure 9.13 and the model uses the same disease parameterization. In the two-dimensional triangular analysis, the continuous spreading pattern of the disease is more transparent than in the network and truss analyses: From the initial seeding region in New York, the disease spreads gradually as a traveling wave with a radially expanding contour pattern that is characteristic of distance weighted diffusion. In classical diffusion models, nodes in the proximity of the seeding region are infected first and farthest nodes are infected last.

9.4 Comparison of network diffusion and finite element methods

We have seen that both the network diffusion method in Section 9.2 and the finite element method in Section 9.3 successfully predict patterns of disease spreading [6]. For fast spreading with large diffusion coefficients κ, spreading is close to instantaneous, the dynamics of the system are dominated by the ordinary differential equations of the epidemiology model, and both methods converge towards the homogeneous disease models in Chapter 5. For slow spreading with small diffusion coefficients κ, spreading is more gradual, the dynamics of the system are dominated by the flux term in the partial differential equation, and both methods converge to a simple diffusion process. The resulting diffusion pattern is determined by the connectivity of the nodes.

Comparison of global features. Figures 9.15 and 9.16 illustrate the sensitivity of both network diffusion and finite element methods with respect to the infectious period C and the basic reproduction number R_0. As such, both figures are the global partial differential equation analogues to Figures 2.3 and 2.4 of the local ordinary differential equation SIS model in Chapter 2. Their underlying trends are similar to the local SIS model: Increasing the infectious period C delays convergence to the endemic equilibrium, but the final sizes S_∞ and I_∞ remain unchanged. Increasing the basic reproduction number R_0 accelerates convergence to the endemic equilibrium, decreases S_∞, and increases I_∞. To match the activation times, we selected an initial infectious population of $I_0 = 0.005$ of the population of New York for the network model and $I_0 = 0.015$ for the finite element model. Choosing similar initial infectious populations of $I_0 = 0.01$ for both models would accelerate vs slow down the initial onset of the outbreak and shift all curves of the network model to the right and all curves of the the finite element model to the left, while the slopes and final sizes S_∞ and I_∞ would remain unchanged. We conclude that both network diffusion

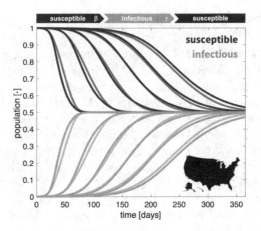

Fig. 9.15 Classical SIS diffusion model. Network diffusion method vs finite element method. Sensitivity with respect to the infectious period C**.** Increasing the infectious period C delays convergence to the endemic equilibrium, but the final sizes S_∞ and I_∞ remain unchanged. Basic reproduction number $R_0 = 2.0$, diffusion coefficient $\kappa = 0.05$, and infectious period $C = 5, 10, 15, 20, 25, 30$ days. Network diffusion simulation in dark and finite element simulation in light colors.

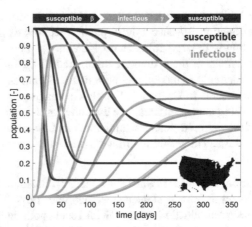

Fig. 9.16 Classical SIS diffusion model. Network diffusion method vs finite element method. Sensitivity with respect to the basic reproduction number R_0**.** Increasing the basic reproduction number R_0 accelerates convergence to the endemic equilibrium, decreases S_∞, and increases I_∞. Infectious period $C = 20$ days, diffusion coefficient $\kappa = 0.05$, and basic reproduction number $R_0 = 1.7, 2.0, 2.4, 3.0, 5.0, 10.0$. Network diffusion simulation in dark and finite element simulation in light colors.

and finite element models can produce similar global results when integrated across the entire domain, in this case the United States, when diffusion is relatively fast. In today's world, where infectious individuals travel from coast to coast by plane rather than by horse carriage, the time scales associated with global diffusion can easily outpace the time scales associated with the local disease dynamics [12]. In the next Chapter, we will see that restricting travel is a powerful tool in disease management to control the spreading of a disease.

Comparison of local features. While the global features of the network diffusion and finite element methods can be quite similar, there can be important local differences that become clear when comparing the disease spreading patterns in Figures 9.5 and 9.13. For both methods, the degrees of freedom, in this case the infectious population, live on the nodes, which is different, for example, for finite volume methods. The side-by-side comparison of the network diffusion and finite element models of five states in Figures 9.2 and 9.11 shows that network methods can connect any pair of nodes whereas finite element methods typically only connect the closest neighbors. This becomes most apparent when comparing the network diffusion and finite element models of the entire United States in Figures 9.3 and 9.12. However, with the same connectivity and similar diffusion characteristics, network methods and linear one-dimensional finite element methods can actually be parameterized to be entirely identical to the extent that all residual and matrix entries are exactly the same. We have shown that the explicit update equation for the nodal unknowns of the network model (9.8) is identical to the equation for the one-dimensional linear finite element model (9.31). We can also see this qualitatively when comparing the discrete graph Laplacian matrix of the network model (9.14) to the discrete Laplacian matrix of the finite element model (9.38).

Comparison of computational features. From the derivations in Sections 9.2 and 9.3 it is clear that finite element methods can be a lot more cumbersome to achieve the same results as network diffusion methods [6]. In addition, finite element methods are limited to symmetric diffusion, whereas network diffusion methods can easily incorporate nonsymmetric diffusion by using directed graphs. This raises the question why we would even consider finite element methods to solve boundary value problems. Finite element methods are widely used to solve engineering problems associated with balance equations for which the rate of change of a scalar- or vector-valued unknown is balanced with source and flux terms [2]. While finite element methods are naturally more complex and computationally more expensive, they are flexible in dimension and order [8]. Our example in Figure 9.14 has shown that it is possible to use, or even combine, different element types, trusses, triangles, tetrahedra, quadrilaterals, or hexahedra [21]. Here we only used linear elements but it is straightforward to generalize the concept to higher order approximations that are quadratic, cubic, or quartic. While the finite element method can more easily be extended to multiple dimensions or higher order, the network diffusion method can more easily incorporate mechanistic diffusion characteristics in terms of directionality, anisotropy, and heterogeneity, which is critical to accurately model the spreading of infectious diseases.

Taken together, unlike the selection of the time integration scheme in Chapter 5, the selection of the space discretization scheme can actually affect the physics of the solution. While several scientists have applied finite element methods to model the spreading of infectious diseases, the major criticism–aside from high computational cost–is the realism of radial diffusion patterns around the initial seeding region. Clearly, the purely isotropic diffusion in Figure 9.13 does not produce very realistic spreading patterns. In reality, diffusion is likely not only anisotropic but also highly heterogeneous. Successful finite element models for disease spreading critically depend on reasonable and meaningful diffusion parameters that can be difficult to calibrate or learn from disease data. In network diffusion models, diffusion is inherently correlated to mobility, through car traffic, public transport, or air travel between different cities, states, or countries [17]. Mobility data and travel statistics exist and provide valuable insight into the spreading characteristics of infectious diseases [19]. In the following sections, we show how we can use network diffusion modeling, combined with passenger air travel, to understand outbreak patterns and explain the effect of travel restrictions.

Problems

9.1 Element tangent matrix of the finite element method. Show that the $K_{e,12}^N = \int_0^l N_1 N_2 dl$ term of the element tangent matrix $\mathbf{K}_e^N$ for the time derivative and source terms of linear truss elements with linear shape functions, $N_1 = 1 - x/l_e$ and $N_2 = x/l_e$, is $[l_e/6]$.

9.2 Element tangent matrix of the finite element method. Show that the $K_{e,11}^{\nabla N} = \int_0^l \nabla N_1 \nabla N_2 dl$ term of the element tangent matrix $\mathbf{K}_e^{\nabla N}$ for the flux term of linear truss elements with linear shape functions, $N_1 = 1 - x/l_e$, is $[1/l_e]$.

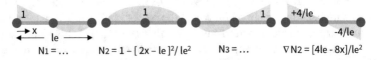

$$N_1 = \ldots \qquad N_2 = 1 - [\,2x - l_e\,]^2/\,l_e^2 \qquad N_3 = \ldots \qquad \nabla N_2 = [4l_e - 8x]/l_e^2$$

Fig. 9.17 Finite element shape functions for quadratic truss elements. The Lagrangian shape functions, N_1, N_2, and N_2, of quadratic truss elements are one at their respective node and zero at both other nodes, and their gradients, ∇N_1, ∇N_2, and ∇N_3 are linear within the element.

9.3 Quadratic truss elements. Lagrangian shape functions are one on their respective node and zero on all other nodes. Show that the quadratic function $N_2 = 1 - [2x - l_e]^2/l_e^2$ with $0 \le x \le l_e$ is the Lagrangian shape function of node 2 for quadratic truss elements as illustrated in Figure 9.17.

9.4 Quadratic truss elements. Lagrangian shape functions are one on their respective node and zero on all other nodes. Derive the quadratic Lagrangian shape functions N_1 and N_3 as a function of the coordinate x with $0 \le x \le l_e$ for quadratic truss elements as illustrated in Figure 9.17.

9.5 Finite element flux tangent matrix. Show that the term $K_{e,22}^{\nabla N} = \int_0^l \nabla N_2 \nabla N_2 dl$ of the element flux tangent matrix $\mathbf{K}_e^{\nabla N}$ of quadratic truss elements with quadratic shape functions, $N_2 = 1 - [2x - l_e]^2 / l_e^2$, as illustrated in Figure 9.17 is $[16/(3l_e)]$.

9.6 Fisher Kolmogorov model. The Fisher Kolmogorov model is a popular non-linear partial differential equation to model traveling waves in ecology, physiology, combustion, crystallization, plasma physics, phase transition and biology. It defines spatio-temporal patterns of the concentration c of a quantity of interest that varies from zero to one, $0 \leq c \leq 1$, as

$$\dot{c} = \text{div}(\mathbf{D} \cdot \nabla c) + \alpha \, c \, [\, 1 - c \,],$$

where $\mathbf{D}$ is the diffusion tensor and α is a growth parameter. Identify is the source or reaction term $f(c)$ of the Fisher Kolmogorov model. Compare the source terms of the Fisher Kolmogorov model and the SIS model. What is the value that the concentration c of the Fisher Kolmogorov model converges towards for $t \to \infty$, at endemic equilibrium?

9.7 Fisher Kolmogorov model. Recent studies have adopted the Fisher Kolmogorov model to simulate the spreading of misfolded proteins across the brain's connectome network in Alzheimer's disease. Discretize the Fisher Kolmogorov model, $\dot{c} = \text{div}(\mathbf{D} \cdot \nabla c) + \alpha \, c \, [\, 1 - c \,]$, in space and time using an explicit network diffusion model. Derive the explicit equation for the nodal concentrations $c_{I,n+1}$ at the new time point t_{n+1} for given nodal concentrations $c_{I,n}$ at the previous time point t_n and a given graph Laplacian L_{IJ}, diffusion coefficient κ, and growth parameter α.

9.8 Fisher Kolmogorov model. Discretize the Fisher Kolmogorov model, $\dot{c} = \text{div}(\mathbf{D} \cdot \nabla c) + \alpha \, c \, [\, 1 - c \,]$, in space and time using an implicit network diffusion model. Derive the discrete residual R_I and the discrete tangent matrix K_{IJ} for the Network method for given nodal concentrations $c_{I,n}$ at the previous time point t_n and a given graph Laplacian L_{IJ}, diffusion coefficient κ, and growth parameter α.

9.9 FitzHugh-Nagumo model. The Fitz-Hugh Nagumo model is a widely used nonlinear partial differential equation to model traveling waves in cardiac electro-physiology. It defines spatio-temporal patterns of the electric activation potential of the heart ϕ, as

$$\dot{\phi} = \text{div}(\mathbf{D} \cdot \nabla \phi) + \gamma \, [\, \phi \, [\, 1 - \phi \,][\, \phi - \alpha \,] - r \,],$$

where $\mathbf{D}$ is the diffusion tensor, γ is a scaling factor, α is the oscillation threshold, and r is the recovery variable. Identify is the source or reaction term $f(\phi)$ of the FitzHugh-Nagumo model that is required to evaluate the discrete residual R_I. Determine the derivative of the source term $df(\phi)/d\phi$ that is required to evaluate the discrete tangent matrix K_{IJ}.

9.10 FitzHugh-Nagumo model. Researchers use the FitzHugh-Nagumo model to simulate the action potential propagation across the heart's activation network of

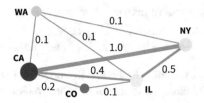

Fig. 9.18 Network diffusion model of California, New York, Illinois, Washington, and Colorado. Discrete graph G with five nodes and the eight most traveled edges. The size and color of the nodes represent the degree D_{II}; the thickness and number of the edges represent the adjacency A_{IJ} estimated from weighted annual incoming and outgoing passenger air travel.

Purkinje cells. Discretize the FitzHugh-Nagumo model, $\dot{\phi} = \mathrm{div}(\boldsymbol{D} \cdot \nabla \phi) + \gamma \,[\, \phi \,[\, 1 - \phi \,][\, \phi - \alpha \,] - r \,]$, in space and time using an explicit network diffusion model. Derive the explicit equation for the nodal action potential $\phi_{I,n+1}$ at the new time point t_{n+1} for given action potentials $\phi_{I,n}$ at the previous time point t_n and a given graph Laplacian L_{IJ}, diffusion coefficient κ, and scaling factor γ, oscillation threshold α, and recovery variable r.

9.11 Adjacency matrix, degree matrix, and graph Laplacian. For the discrete graph G of California, New York, Illinois, Washington, and Colorado in Figure 9.18 with $n_{nd} = 5$ nodes and $n_{el} = 8$ edges, how many rows and columns do the adjacency matrix A_{IJ}, the degree matrix D_{IJ}, and the graph Laplacian L_{IJ} have? How many entries in the adjacency matrix A_{IJ} are non-zero? How many entries in the degree matrix D_{IJ} are non-zero? How many entries in the graph Laplacian matrix L_{IJ} are non-zero?

9.12 Adjacency matrix, degree matrix, and graph Laplacian. For the discrete graph G of California, New York, Illinois, Washington, and Colorado in Figure 9.18 with $n_{nd} = 5$ nodes and $n_{el} = 8$ edges, calculate the adjacency matrix A_{IJ}, the degree matrix D_{IJ}, and the graph Laplacian L_{IJ} when sorting the states by population, $I = [\,$California, New York, Illinois, Washington, Colorado$\,]$. Which states have the largest and smallest degrees and what are their degrees?

9.13 Adjacency matrix, degree matrix, and graph Laplacian. For the discrete graph G of California, Florida, New York, Washington, and Arizona in Figure 9.19 with $n_{nd} = 5$ nodes and $n_{el} = 8$ edges, calculate the adjacency matrix A_{IJ}, the degree matrix D_{IJ}, and the graph Laplacian L_{IJ} when sorting the states by population, $I = [\,$California, Florida, New York, Washington, Arizona $\,]$. Which states have the largest and smallest degrees and what are their degrees?

9.14 Network diffusion vs. finite element method. For the discretization of California, Texas, New York, Georgia, and Minnesota in Figure 9.11, right, with $n_{nd} = 5$ nodes and $n_{el} = 7$ edges, calculate the inverse-distance weighted adjacency matrix $A_{IJ} = 1/l_{IJ}$, the degree matrix D_{IJ}, and the graph Laplacian L_{IJ} when sorting the

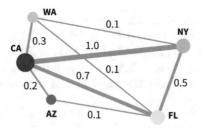

Fig. 9.19 Network diffusion model of California, Florida, New York, Washington, and Arizona.
Discrete graph $\mathcal{G}$ with five nodes and the eight most traveled edges. The size and color of the nodes
represent the degree D_{II}; the thickness and number of the edges represent the adjacency A_{IJ}
estimated from weighted annual incoming and outgoing passenger air travel.

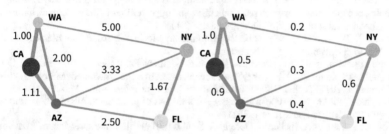

Fig. 9.20 Finite element model of California, Florida, New York, Washington, and Arizona.
Finite element discretization with five nodes and seven linear truss elements. The size and color of
the nodes represent the population; the thickness and number of the edges represent the element
length, left, and inverse element length, right.

states by population, $I = [\,\text{California, Texas, New York, Georgia, and Minnesota}\,]$.
Compare your graph Laplacian matrix L_{IJ} to the finite element flux tangent matrix
$K_{IJ}^{\nabla N}$ in equation (9.38). Comment on your observation!

9.15 Finite element mass and source tangent matrix. For the finite element discretization of California, Florida, New York, Washington, and Arizona in Figure 9.20
with $n_{\mathrm{nd}} = 5$ nodes and $n_{\mathrm{el}} = 7$ linear truss elements, calculate the mass and source
tangent matrix K_{IJ}^{N} when sorting the states by population, $I = [\,\text{California, Florida,}$
New York, Washington, Arizona$\,]$. Which states have the largest and smallest matrix
entries and what are their values?

9.16 Finite element flux tangent matrix. For the finite element discretization of
California, Florida, New York, Washington, and Arizona in Figure 9.20 with $n_{\mathrm{nd}} = 5$
nodes and $n_{\mathrm{el}} = 7$ linear truss elements, calculate the flux tangent matrix $K_{IJ}^{\nabla N}$ when
sorting the states by population, $I = [\,\text{California, Florida, New York, Washington,}$
Arizona$\,]$. Which states have the largest and smallest diagonal entries and what are
their values?

9.17 Network diffusion vs. finite element method. For the discretization of California, Florida, New York, Washington, and Arizona in Figure 9.20, right, with

$n_{nd} = 5$ nodes and $n_{el} = 7$ edges, calculate the inverse-distance weighted adjacency matrix $A_{IJ} = 1/l_{IJ}$, the degree matrix D_{IJ}, and the graph Laplacian L_{IJ} when sorting the states by population, $I = [$ California, Florida, New York, Washington, Arizona $]$. Compare your graph Laplacian matrix L_{IJ} to the finite element flux tangent matrix $K_{IJ}^{\nabla N}$ from the previous example. Comment on your observation!

References

1. Alber M, Buganza Tepole A, Cannon W, De S, Dura-Bernal S, Garikipati K, Karniadakis G, Lytton WW, Perdikaris P, Petzold L, Kuhl E (2019) Integrating machine learning and multiscale modeling: Perspectives, challenges, and opportunities in the biological, biomedical, and behavioral sciences. npj Digital Medicine 2:115.
2. Bonet J, Wood RD (1997) Nonlinear Continuum Mechanics for Finite Element Analysis Cambridge University Press, Cambridge.
3. Bureau of Transportation Statistics. Non-Stop Segment Passengers Transported by Origin-State/DestState for 2019. https://www.transtats.bts.gov. assessed: March 30, 2020.
4. Epstein M (2017) Partial Differential Equations. Springer International Publishing.
5. Eurostat. Your key to European statistics. Air transport of passengers. https://ec.europa.eu/eurostat accessed: June 1, 2021.
6. Fornari S, Schafer A, Jucker M, Goriely A, Kuhl E (2019) Prion-like spreading of Alzheimer's disease within the brain's connectome. Journal of the Royal Society Interface 16:20190356.
7. Hethcote HW (2000) The mathematics of infectious diseases. SIAM Review 42:599-653.
8. Hughes TJR (1987) The Finite Element Method: Linear Static and Dynamic Finite Element Analysis. Prentice Hall.
9. Kuhl E (2020) Data-driven modeling of COVID-19 – Lessons learned. Extreme Mechanics Letters 40:100921.
10. Linka K, Peirlinck M, Sahli Costabal F, Kuhl E (2020) Outbreak dynamics of COVID-19 in Europe and the effect of travel restrictions. Computer Methods in Biomechanics and Biomedical Engineering 23: 710-717.
11. Linka K, Peirlinck M, Kuhl E (2020) The reproduction number of COVID-19 and its correlation with public heath interventions. Computational Mechanics 66:1035-1050.
12. Linka K, Goriely A, Kuhl E (2021) Global and local mobility as a barometer for COVID-19 dynamics. Biomechanics and Modeling in Mechanobiology 20:651–669.
13. Moin P (2000) Fundamentals of Engineering Numerical Analysis. Cambridge University Press.
14. Newman MEJ (2002) Spread of epidemic disease on networks. Physical Review E 66:016128.
15. Newman M (2010) Networks: An Introduction. Oxford University Press, New York.
16. Oden JT (1971) Finite Elements of Nonlinear Continua. McGraw-Hill.
17. Pastor-Satorras R, Vespignani A (2001) Epidemic spreading in scale-free networks. Physical Review Letters 86:3200-3203.
18. Pastor-Satorras R, Castallano C, Van Mieghem P, Vespignani A (2015) Epidemic processes in complex networks. Reviews of Modern Physics 87:925-979.
19. Peirlinck M, Linka K, Sahli Costabal F, Kuhl E (2020) Outbreak dynamics of COVID-19 in China and the United States. Biomechanics and Modeling in Mechanobiology 19:2179-2193.
20. Peng GCY, Alber M, Buganza Tepole A, Cannon W, De S, Dura-Bernal S, Garikipati K, Karniadakis G, Lytton WW, Perdikaris P, Petzold L, Kuhl E (2021) Multiscale modeling meets machine learning: What can we learn? Archive of Computational Methods in Engineering 28:1017-1037.
21. Viguerie A, Veneziani A, Lorenzo G, Baroli D, Aretz-Nellesen N, Patton A, Yankeelov TE, Reali A, Hughes TJR, Auricchio F (2020) Diffusion-reaction compartmental models formulated in a continuum mechanics framework: application to COVID-19, mathematical anaysis, and numerical study. Computational Mechanics 66:1131-1152.
22. Wriggers P (2008) Nonlinear Finite Element Methods. Springer-Verlag Berlin Heidelberg.

Chapter 10
The network SEIR model

Abstract The network SEIR model is a locally resolved compartment model with four populations, the susceptible, exposed, infectious, and recovered groups S, E, I, and R, at each node of the network. It characterizes the spatio-temporal spreading of infectious diseases along the edges of the network proportional to human mobility. Since the network SEIR model has no analytical solution, we discretize it in space using a weighted Laplacian graph and apply explicit and implicit time integration schemes to solve it. To illustrate the features of the network SEIR model, we simulate the early COVID-19 outbreak in the United States and the European Union using reported case data and air travel statistics. The learning objectives of this chapter on SEIR network modeling are to

- understand the concept of network compartment modeling in epidemiology
- discretize the classical SEIR model using the network diffusion method
- solve the network SEIR equations using the explicit forward Euler method
- derive the residual network SEIR equations using the implicit Euler method
- linearize and solve the residual network SEIR equations using the Newton method
- estimate and interpret the effect of travel restrictions
- discuss limitations of network SEIR modeling

By the end of the chapter, you will be able to discretize, linearize, and solve the network SEIR model and analyze, simulate, and predict the spatio-temporal pattern of infectious diseases like COVID-19 using reported air travel data.

10.1 Network diffusion method of the SEIR model

The network SEIR model characterizes the discrete spatio-temporal evolution of susceptible, exposed, infectious, and recovered individuals [20]. It is based on the set of continuous partial differential equations for the SEIR model for the unknown population fields $S(\mathbf{x}, t)$, $E(\mathbf{x}, t)$, $I(\mathbf{x}, t)$, and $R(\mathbf{x}, t)$ with initial conditions S_0, E_0, I_0, and R_0 at time t_0,

© The Author(s), under exclusive license to Springer Nature Switzerland AG 2021
E. Kuhl, *Computational Epidemiology*, https://doi.org/10.1007/978-3-030-82890-5_10

Fig. 10.1 Network SEIR model. Each node of the network SEIR model contains four compartments for the susceptible, exposed, infectious, and recovered populations, S, E, I, and R. The transition rates between the compartments of each node are the contact, latent, and infectious rates, β, α, and γ. The diffusion between the four compartments of connected nodes I and J is a product of the diffusion coefficients κ_S, κ_E, κ_I, and κ_R and the graph Laplacian L_{IJ}.

$$
\begin{aligned}
\dot{S} &= \mathrm{div}(\boldsymbol{D}_S \cdot \nabla S) - \beta\,S\,I \\
\dot{E} &= \mathrm{div}(\boldsymbol{D}_E \cdot \nabla E) + \beta\,S\,I - \alpha\,E \\
\dot{I} &= \mathrm{div}(\boldsymbol{D}_I \cdot \nabla I) \qquad\quad + \alpha\,E - \gamma\,I \\
\dot{R} &= \mathrm{div}(\boldsymbol{D}_R \cdot \nabla R) \qquad\qquad\quad + \gamma\,I\,.
\end{aligned}
\tag{10.1}
$$

For all four populations, $\{\circ\} = \{S, E, I, R\}$, $\mathrm{div}(\boldsymbol{D}_\circ \cdot \nabla(\circ))$ denotes the flux or diffusion term, $\boldsymbol{D}_\circ \cdot \nabla(\circ)$ is the flux, $\boldsymbol{D}_\circ$ the diffusion tensor , and $\nabla(\circ)$ the gradient of the unknown population $(\circ)$. Similar to the local SEIR model in Chapter 4, β, α, and γ are the contact, latent, and infectious rates. To discretize the set of equations (10.1) in space, we introduce the unknown populations, $\boldsymbol{P}_I = [\,S_I, E_I, I_I, R_I\,]$, as global unknowns at each $I = 1, ..., n_{\mathrm{nd}}$ node of a weighted graph $\mathcal{G}$. Similar to the SIS model in Section 9.2, we summarize the connectivity of the graph $\mathcal{G}$ in terms of the adjacency matrix A_{IJ}, the weighted connection between two nodes I and J, and the degree matrix $D_{II} = \mathrm{diag}\sum_{J=1,J\neq I}^{n_{\mathrm{nd}}} A_{IJ}$, the weighted number of incoming and outgoing connections of node I [6]. The difference between the degree matrix D_{IJ} and the adjacency matrix A_{IJ} defines the weighted graph Laplacian L_{IJ} [24],

$$
L_{IJ} = D_{IJ} - A_{IJ} \quad \text{with} \quad D_{II} = \mathrm{diag}\sum\nolimits_{J=1,J\neq I}^{n_{\mathrm{nd}}} A_{IJ}\,.
\tag{10.2}
$$

The weighted graph Laplacian is a $n_{\mathrm{nd}} \times n_{\mathrm{nd}}$ matrix that introduces a discrete approximation of the diffusion terms in equation (10.1), $\mathrm{div}(\boldsymbol{D}_\circ \cdot \nabla(\circ)) = -\kappa_\circ \sum_{J=1}^{n_{\mathrm{nd}}} L_{IJ}\,c_J$, where $\kappa_\circ$ are the diffusion coefficients that can, in general, be different for each population $\{\circ\} = \{S, E, I, R\}$. We can now rewrite the set of nonlinear field equations (10.1) as the network SEIR model, a discrete set of $4\,n_{\mathrm{nd}}$ equations with $4\,n_{\mathrm{nd}}$ unknowns at the $I = 1, ..., n_{\mathrm{nd}}$ nodes of the network [16],

$$
\begin{aligned}
\dot{S}_I &= -\kappa_S \sum\nolimits_{J=1}^{n_{\mathrm{nd}}} L_{IJ}\,S_J - \beta\,S_I I_I \\
\dot{E}_I &= -\kappa_E \sum\nolimits_{J=1}^{n_{\mathrm{nd}}} L_{IJ}\,E_J + \beta\,S_I I_I - \alpha\,E_I \\
\dot{I}_I &= -\kappa_I \sum\nolimits_{J=1}^{n_{\mathrm{nd}}} L_{IJ}\,I_J \qquad\quad + \alpha\,E_I - \gamma\,I_I \\
\dot{R}_I &= -\kappa_R \sum\nolimits_{J=1}^{n_{\mathrm{nd}}} L_{IJ}\,R_J \qquad\qquad\quad + \gamma\,I_I\,.
\end{aligned}
\tag{10.3}
$$

Figure 7.1 illustrates the susceptible, exposed, infectious, and recovered populations, S, E, I, and R, at two connected nodes I and J. The transition rates between the compartments at each node are the contact rate β, the latent rate α, and the infectious rate γ in units [1/days], which are the inverses of the contact period $B = 1/\beta$, the latent period $A = 1/\alpha$, and the infectious period $C = 1/\gamma$ in units [days]. The diffusion between the four compartments of connected nodes is a product of the diffusion coefficients κ_S, κ_E, κ_I, and κ_R and the graph Laplacian L_{IJ} in units [1/day]. In this simple format, the network SEIR model neglects all vital dynamics, it does not account for births or natural deaths, such that $\dot{S} + \dot{E} + \dot{I} + \dot{R} \doteq 0$ and $S + E + I + R = \text{const.} = 1$. This implies that, without loss of information, we could eliminate the last equation, $\dot{R}_I = -\kappa_R \sum_{J=1}^{n_{nd}} L_{IJ} R_J + \gamma I_I$, and simply calculate the recovered population, $R = 1 - S - E - I$, in a postprocessing step. To discretize the set of equations (10.3) in time, we partition the time interval $\mathcal{T}$ into n_{step} discrete time steps, $\mathcal{T} = \bigcup_{n=1}^{n_{step}} [t_n, t_{n+1}]$, where the subscripts $(\circ)_n$ and $(\circ)_{n+1}$ are associated with the beginning and the end of the current time step. We assume that we know the nodal unknowns $S_{I,n}$, $E_{I,n}$, $I_{I,n}$, and $R_{I,n}$ at the beginning of the time step t_n and approximate the first order time derivative using finite differences,

$$\dot{S}_I = \frac{S_{I,n+1} - S_{I,n}}{\Delta t} \quad \dot{E}_I = \frac{E_{I,n+1} - E_{I,n}}{\Delta t} \quad \dot{I}_I = \frac{I_{I,n+1} - I_{I,n}}{\Delta t} \quad \dot{R}_I = \frac{R_{I,n+1} - R_{I,n}}{\Delta t},$$
(10.4)

where Δt denotes the time step size and n is the increment counter.

Explicit forward Euler method. For the explicit time integration, we evaluate the righthand side exclusively at the known time point t_n,

$$
\begin{aligned}
{[S_{I,n+1} - S_{I,n}]}/\Delta t &= -\kappa_S \sum_{J=1}^{n_{nd}} L_{IJ} S_{J,n} - \beta S_{I,n} I_{I,n} & &= 0 \\
{[E_{I,n+1} - E_{I,n}]}/\Delta t &= -\kappa_E \sum_{J=1}^{n_{nd}} L_{IJ} E_{J,n} + \beta S_{I,n} I_{I,n} - \alpha E_{I,n} & &= 0 \\
{[I_{I,n+1} - I_{I,n}]}/\Delta t &= -\kappa_I \sum_{J=1}^{n_{nd}} L_{IJ} I_{J,n} & + \alpha E_{I,n} - \gamma I_{I,n} &= 0 \\
{[R_{I,n+1} - R_{I,n}]}/\Delta t &= -\kappa_R \sum_{J=1}^{n_{nd}} L_{IJ} R_{J,n} & + \gamma I_{I,n} &= 0.
\end{aligned}
$$
(10.5)

which results in a set of explicit equations for the nodal unknowns $S_{I,n+1}$, $E_{I,n+1}$, $I_{I,n+1}$, and $R_{I,n+1}$ at the new time point t_{n+1},

$$
\begin{aligned}
S_{I,n+1} &= S_{I,n} - \kappa_S \sum_{J=1}^{n_{nd}} L_{IJ} S_{J,n} \Delta t - \beta S_{I,n} I_{I,n} \Delta t \\
E_{I,n+1} &= E_{I,n} - \kappa_E \sum_{J=1}^{n_{nd}} L_{IJ} E_{J,n} \Delta t + \beta S_{I,n} I_{I,n} \Delta t - \alpha E_{I,n} \Delta t \\
I_{I,n+1} &= I_{I,n} - \kappa_I \sum_{J=1}^{n_{nd}} L_{IJ} I_{J,n} \Delta t \qquad\qquad\quad + \alpha E_{I,n} \Delta t - \gamma I_{I,n} \Delta t \\
R_{I,n+1} &= R_{I,n} - \kappa_R \sum_{J=1}^{n_{nd}} L_{IJ} R_{J,n} \Delta t \qquad\qquad\qquad\qquad\quad + \gamma I_{I,n} \Delta t .
\end{aligned}
$$
(10.6)

After updating the populations, we simply advance in time to the next step, and repeat solving the update equations (10.6) for n time steps.

Implicit backward Euler method. For the implicit time integration, we evaluate the righthand side at the unknown time point t_{n+1}, and rephrase equations (10.3) as discrete residuals, $\mathbf{R}_I = [R_{S,I}, R_{E,I}, R_{I,I}, R_{R,I}]$, which resemble the residuals of the local SEIR model in equation (7.10) of Chapter 4, but now evaluated at all $I = 1, .. n_{nd}$ nodes, connected through the diffusion terms,

$$R_{S,I} = [S_{I,n+1} - S_{I,n}] / \Delta t + \kappa_S \sum_{J=1}^{n_{nd}} L_{IJ} S_{J,n+1} + \beta S_{I,n+1} I_{I,n+1} \qquad = 0$$
$$R_{E,I} = [E_{I,n+1} - E_{I,n}] / \Delta t + \kappa_E \sum_{J=1}^{n_{nd}} L_{IJ} E_{J,n+1} - \beta S_{I,n+1} I_{I,n+1} + \alpha E_{I,n+1} = 0$$
$$R_{I,I} = [I_{I,n+1} - I_{I,n}] / \Delta t + \kappa_I \sum_{J=1}^{n_{nd}} L_{IJ} I_{J,n+1} - \alpha E_{I,n+1} + \gamma I_{I,n+1} \qquad = 0$$
$$R_{R,I} = [R_{I,n+1} - R_{I,n}] / \Delta t + \kappa_R \sum_{J=1}^{n_{nd}} L_{IJ} R_{J,n+1} - \gamma I_{I,n+1} \qquad = 0.$$
$$(10.7)$$

The tangent matrix, $\mathbf{K}_{IJ} = d\mathbf{R}_I / d\mathbf{P}_J$, for the Newton method is a $4 n_{nd} \times 4 n_{nd}$ matrix that contains the derivatives of the four residuals, $\mathbf{R}_I = [R_{S,I}, R_{E,I}, R_{I,I}, R_{R,I}]$, with respect to the four populations, $\mathbf{P}_J = [S_J, E_J, I_J, R_J]$ at each node,

$$\mathbf{K}_{IJ} = \begin{bmatrix} \frac{1}{\Delta t} & 0 & 0 & 0 \\ 0 & \frac{1}{\Delta t} & 0 & 0 \\ 0 & 0 & \frac{1}{\Delta t} & 0 \\ 0 & 0 & 0 & \frac{1}{\Delta t} \end{bmatrix} I_{IJ} + \begin{bmatrix} \kappa_S & 0 & 0 & 0 \\ 0 & \kappa_E & 0 & 0 \\ 0 & 0 & \kappa_I & 0 \\ 0 & 0 & 0 & \kappa_R \end{bmatrix} L_{IJ} + \begin{bmatrix} +\beta I_{n+1} & 0 & +\beta S_{n+1} & 0 \\ -\beta I_{n+1} & +\alpha & -\beta S_{n+1} & 0 \\ 0 & -\alpha & +\gamma & 0 \\ 0 & 0 & -\gamma & 0 \end{bmatrix} I_{IJ},$$
$$(10.8)$$

where I_{IJ} and L_{IJ} are the $n_{nd} \times n_{nd}$ unit matrix and graph Laplacian matrix. With the residual vector $\mathbf{R}_I$ and the tangent matrix $\mathbf{K}_{IJ}$, we solve for the incremental iterative update of the nodal unknowns,

$$\mathbf{P}_I^{k+1} = \mathbf{P}_I^k + d\mathbf{P}_I \quad \text{with} \quad d\mathbf{P}_I = -\sum_{J=1}^{n_{nd}} \mathbf{K}_{JI}^{-1} \cdot \mathbf{R}_J . \qquad (10.9)$$

We iterate until the norm of the residual, $||\mathbf{R}_I|| < $ tol, is smaller than a user defined tolerance, tol, and the proceed to the next time step. An incremental iterative solution within an implicit time integration scheme results in two nested loops, an outer loop for all time steps n and an inner loop for all iterations k. On the one hand, within each iteration, we have to invert the $4 n_{nd} \times 4 n_{nd}$ tangent matrix $\mathbf{K}$. Depending on the number of nodes, n_{nd}, this can make the implicit scheme relatively expensive. On the other hand, we can use the number of iterations k, to adaptively adjust the time step size Δt, to make the algorithm more efficient.

10.2 Example: COVID-19 spreading across the United States

Especially during the early stages of a pandemic outbreak, passenger air travel can play a critical role in spreading a disease since traveling individuals naturally have a disproportionally high contact rate [5]. Border control is a critical measure in mitigating epidemics and prevent the spreading between cities, states, or countries [30]. In a controversial move, on February 2, 2020, the United States declared travel restrictions to and from China. Unlike China, by mid spring 2020, the United States were at the very early stage of the COVID-19 outbreak and all states were still seeing an increase of the number of daily new cases [12]. The available data describe the temporal evolution of confirmed, recovered, active, and death cases starting January 21, 2020, the first day of the outbreak in the United States. As of April 4, there were 311,357 confirmed cases, 14,825 recovered, 288,081 active, and 8,451 deaths.

Similar to Section 7.8, we map out the temporal evolution of the infectious group $I(t)$ as the difference between the confirmed cases minus the recovered and deaths in each state state [20]. When we initially performed this study, on April 4, 2020, the outbreak data were still in the very early stages and it was difficult to identify the latent and infectious periods from the available data. Instead, we used the mean latent and infectious periods $A = 2.56$ and $C = 17.82$ from the Chinese outbreak in Table 7.1. For each state, we map the SEIR model with a total population of one onto the state-specific population N. We identify three parameters for each state, the basic reproduction number $R_0 = C/B$, the ratio between the infectious and contact periods, the community spread $\rho = E_0/I_0$, the ratio between the initial exposed and infectious populations, and the outbreak delay d_0, the time between the first case nationwide and the first day of reported infections, $I_0 \geq 0$, in the state. We perform the parameter identification using the Levenberg-Marquardt algorithm. We then use these state-specific parameters for forward simulations with our network SEIR model in Figure 9.3 to explore the early spreading of COVID-19 across the United States.

Figure 10.2 shows the dynamics of the early stages of the COVID-19 outbreak in the 50 states of the United States, the District of Columbia, and the territories of Guam, Puerto Rico, and the Virgin Islands. The dots indicate the reported cases and death, the lines highlight the simulated susceptible, exposed, infectious, and recovered populations. Each graph reports three state-specific parameters, the basic reproduction number R_0, the community spread ρ, and the outbreak delay d_0. Notably, at the date of the analysis, the state of New York was seeing the most significant impact with more than 100,000 cases. Naturally, in Washington, Illinois, California, and Arizona where the first cases were reported, the community spread ρ is small. The largest community spread ρ occurred in New York, New Jersey, Michigan, and Louisiana. The largest basic reproduction numbers R_0 are identified in Idaho, Puerto Rico, Pennsylvania, and Indiana.

Table 10.1 Early COVID-19 outbreak dynamics in the United States. Contact period B, basic reproduction number $R_0 = C/B$, community spread ρ, and outbreak delay d_0, with latent period A and infectious period C from the outbreak in China in Table 7.1; means ± standard deviations across all 54 locations in Figure 10.2.

parameter	mean ± std	interpretation
A [days]	2.56 ± 0.72	latent period (China)
B [days]	3.38 ± 0.69	contact period
C [days]	17.82 ± 2.95	infectious period (China)
R_0 [-]	5.30 ± 0.95	basic reproduction number
ρ [-]	43.75 ± 126.34	community spread
d_0 [days]	41.28 ± 13.78	outbreak delay

Table 10.1 summarizes the parameters for the early stages of the COVID-19 outbreak in the United States. Averaged over all states, we found a basic reproduction number of $R_0 = C/B = 5.30 \pm 0.95$ resulting in a contact period of $B = 3.38 \pm 0.69$ days, a community spread of $\rho = E_0/I_0 = 43.75 \pm 126.34$ and an outbreak delay of

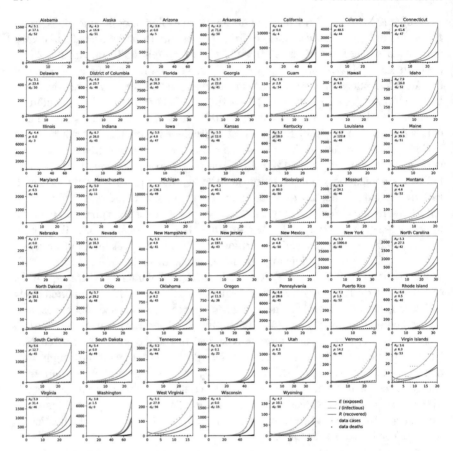

Fig. 10.2 Early COVID-19 outbreak dynamics in the United States. Reported infectious and recovered populations and simulated susceptible, exposed, infectious, and recovered populations. Simulations are based on a state-specific parameter identification of the basic reproduction number R_0 and community spread ρ for all 54 locations, for a given outbreak delay d_0 and disease specific latent and infectious periods $A = 2.56$ and $C = 17.82$ identified for the Chinese outbreak from Table 7.1 [20].

$d_0 = 41.28 \pm 13.78$ days. The basic reproduction number of $R_0 = 5.30 \pm 0.95$ suggests that COVID-19 is less infectious than the measles and pertussis with $R_0 = 12 - 18$, as infectious as rubella, smallpox, polio, and mumps with $R_0 = 4 - 7$, slightly more infectious than SARS with $R_0 = 2 - 5$, and more infectious than an influenza with $R_0 = 1.5 - 1.8$, see Table 1.3. The basic reproduction number of $R_0 = 5.30$ is significantly lower than the basic reproduction number of $R_0 = 12.58$ for China in Section 7.8. This could reflect on an increased awareness of COVID-19 transmission a few weeks into the global pandemic.

Figure 10.3 and 10.4 illustrate the basic reproduction number, outbreak delay, and community spread across all 50 United States. The basic reproduction number is largest in Idaho and Puerto Rico with $R_0 = 7.9$ and $R_0 = 7.2$ and smallest in Arizona

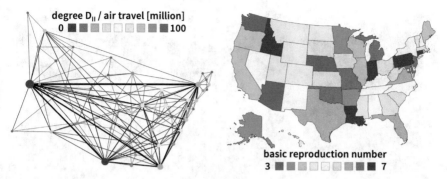

Fig. 10.3 Regional variation of basic reproduction number R_0. The mobility network of the United States with $N = 50$ nodes, color-coded by degree, and the 200 most traveled edges, with thicknesses by connectivity reveals strongest connections between California, Texas, New York, and Florida. The basic reproduction number varies between three and seven with largest values in Puerto Rico and Idaho and smallest in Arizona and Nebraska [20].

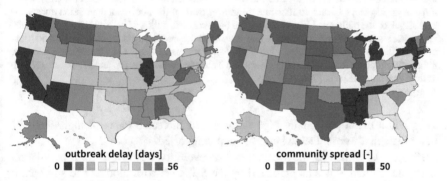

Fig. 10.4 Regional variation of outbreak delay d_0 and community spread ρ. The outbreak delay varies from 0 days in Washington, the first state affected by the outbreak, to 56 days in West Virginia, the last state affected by the outbreak. The community spread is smallest in Washington, Illinois, California, and Arizona and largest in Louisiana with 122.8, Michigan with 136.1, New Jersey with 197.1, and New York with 1,000 [20].

and Nebraska with $R_0 = 3.6$ and $R_0 = 2.5$. The outbreak delay shows that the first reported case was in the state of Washington on January 21, 2020, followed by cases in Illinois with a delay of $d_0 = 3$, California with $d_0 = 4$, and Arizona with $d_0 = 5$, shown in blue. The final states to see an outbreak were Alabama, Idaho, Montana with $d_0 = 52$ and West Virginia with $d_0 = 56$, shown in red. This illustrates that there was a significant time delay in the outbreak with many of the earlier affected states located on the West Coast. The community spread is small in the first states where the outbreak was reported, Washington, Illinois, California, and Arizona, suggesting that the reported cases were truly amongst the first cases in those states. In states where the first cases were reported later, the community spread increases. Notably, Louisiana, Michigan, New Jersey, and New York have the highest community spread with of $\eta = 122.8$, $\eta = 136.1$, $\eta = 197.1$, and $\eta = 1,000$ suggesting that these

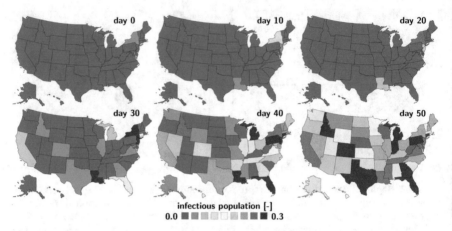

Fig. 10.5 Early COVID-19 spreading across the United States. Regional evolution of the infectious population I predicted by the SEIR network model calibrated with data from the early stages of the outbreak. Days 10 and 20 illustrate the slow growth of the infectious population during the early stages of the outbreak. New York sees the outbreak first, followed by New Jersey, Louisiana, and California. Days 30 and 40 illustrate how the outbreak spreads across the country. Day 50 illustrates that the earlier affected states, New York, New Jersey, and Louisiana already see a decrease of the infected population, while other states like Nebraska, West Virginia, and Wisconsin are still far from reaching the peak [20].

states had a disproportionally high number of undetected exposed individuals or individuals that were infected but remained unreported.

Figure 10.5 illustrates the spatio-temporal evolution of the infectious population across the United States as predicted by the SEIR network model. The simulation uses the mean parameters from Table 10.1, it begins on the day when the last state recorded an outbreak, March 17, 2020, and uses a travel coefficient of $\kappa = 0.43$. Days 10 and 20 illustrate the slow growth of the infectious population during the early stages of the outbreak. The state of New York sees the outbreak first, followed by New Jersey and Louisiana. Days 30 and 40 show how the outbreak spreads across the country. Day 50 suggests that the earlier affected states, New York, New Jersey, and Louisiana already see a decrease of the infected population, while Nebraska, West Virginia, and Wisconsin are still far from reaching the peak. Notably, these maps account for both, the outbreak delay and the travel of individuals between the different states represented through the air traffic mobility network. During the early stages of the pandemic, the outbreak pattern reflects the strong connectivity between New York, the most affected state during the early outbreak, and California, Texas, and Florida, states in which case numbers increased around day 30.

Taken together, this example shows that network epidemiology modeling is an important tool to understand the global spreading of an infectious disease [19]: During the early stages of an outbreak, it provides insights into the spreading patterns of the disease and can help policy makers to interpret the impact of travel restrictions [18] and border control [30]. In the next section, we use an SEIR network model to illustrate the effects of travel restrictions across the European Union.

10.3 Example: Travel restrictions across the European Union

On March 13, 2020, the World Health Organization declared Europe the epicenter of the COVID-19 pandemic with more reported cases and deaths than the rest of the world combined [22]. The first official case of COVID-19 in Europe was reported in France on January 24, 2020, followed by Germany and Finland only three and five days later. Within only six weeks, all 27 countries of the European Union were affected, with the last cases reported in Malta, Bulgaria, and Cyprus on March 9, 2020. At this point, there were 13,944 active cases within the European Union and the number of active cases doubled every three to four days [9]. On March 17, 2020, for the first time in its history, the European Union closed all its external borders to prevent a further spreading of the virus [5]. The decision to temporarily restrict

Fig. 10.6 Effects of travel restrictions across the European Union. By March 22, 2020, the average passenger air travel in Europe was cut in half, and by the time of this study, April 18, it was reduced by 95% in Spain, 94% in Italy, 93% in France, and 89% in Germany. Commercial airlines had to park and store thousands of grounded planes in response to the massive reduction in air travel.

all non-essential travel was by no means uncontroversial, although it was very much in line with the mitigation strategies of most of the local governments: Italy had introduced a national lockdown on March 9, Germany had implemented school and border closures starting March 13, Spain followed on March 14, and France on March 16. By 18 March, 2020, more than 250 million people in Europe were in lockdown. Strikingly, by March 22, 2020, the average passenger air travel in Europe was cut in half, and by the time of this study, April 18, it was reduced by 95% in Spain, 94% in Italy, 93% in France, and 89% in Germany [10]. Figure 10.6 illustrates the effect of the travel restrictions on commercial air traffic, where all airlines had to park and store their planes around local airports in response to the massive reduction in air travel. By mid April, 2020, the economic pressure to identify exit strategies was rising and we asked ourselves: How effective are travel restrictions in mitigating the spreading of the COVID-19 and how save is it to lift them? This motivates using a network epidemiology model to explore the effect of travel restrictions within the European Union and simulate the outbreak dynamics of COVID-19 with and without travel restrictions [16].

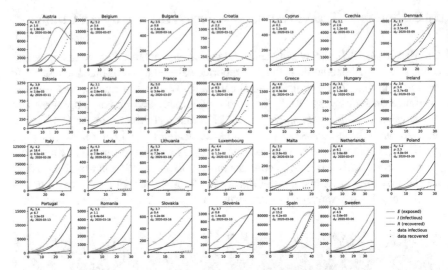

Fig. 10.7 Early COVID-19 outbreak dynamics across the European Union. Reported infectious populations and simulated susceptible, exposed, infectious, and recovered populations. Simulations are based on a country-specific parameter identification of the basic reproduction number R_0, community spread ρ, and affected population η for all 27 countries, for a given outbreak delay d_0 and disease specific latent and infectious periods $A = 2.56$ and $C = 17.82$ identified for the Chinese outbreak from Table 7.1 [16].

We draw the COVID-19 outbreak data of all 27 states of the European Union from the reported confirmed, recovered, active, and death cases from January 22, 2020 until April 13, 2020 [9]. From these data, we map out the temporal evolution of the infectious group I as the difference between the confirmed cases minus the recovered cases and deaths. At the time of the study, most of the European countries had not reached the peak of the first wave or fully recovered, and their data were incomplete to identify all three parameters of the SEIR model. We thus fix the latent and infectious periods to values, which we have previously identified for the COVID-19 outbreak in 30 Chinese provinces, $A = 2.56$ days and $C = 17.82$ days, see Table 7.1. For all 27 countries, we identify the country-specific basic reproduction number $R_0 = C/B$, the community spread $\rho = E_0/I_0$, and the affected population $\eta = N^*/N$, using a Levenberg-Marquardt algorithm. For the identified parameters, we compare two COVID-19 outbreak scenarios: outbreak dynamics with the current travel restrictions, a mobility coefficient of $\kappa = 0.00$, and a country-specific outbreak on day d_0, the day at which 0.001% of the population was reported as infected; and outbreak dynamics without travel restrictions, with a mobility coefficient of $\kappa = 0.43$, and a European outbreak on day d_0 in Italy, the first country in which 0.001% of the population was reported as infected.

Figure 10.7 illustrates the reported infectious population and the simulated exposed, infectious, and recovered populations for all 27 countries. The simulations use the basic reproduction number R_0, the initial community spread $\rho = E_0/I_0$, and the affected population $\eta = N^*/N$ identified for each country using disease specific latent and infectious periods of $A = 2.56$ days and $C = 17.82$ days. Day d_0 indicates the beginning of the outbreak at which 0.001% of the population are infected.

Table 10.2 Early COVID-19 outbreak dynamics across Europe. Latent, contact, and infectious periods A, B, and C, basic reproduction number $R_0 = C/B$, community spread ρ and affected population η with means and standard deviations for all 27 countries of the European Union.

parameter	mean $\pm$ std	interpretation
A [days]	2.56 $\pm$ 0.72	latent period
B [days]	4.07 $\pm$ 1.23	contact period
C [days]	17.82 $\pm$ 2.95	infectious period
R_0 [-]	4.62 $\pm$ 1.32	basic reproduction number
ρ [-]	3.53 $\pm$ 3.97	community spread
η [-]	0.0767 $\pm$0.2612	affected population

Table 10.2 summarizes the parameters for the early stages of the COVID-19 outbreak across the European Union. Averaged over all countries, we found a a basic reproduction number of $R_0 = C/B = 4.62 \pm 1.32$ resulting in a contact period of $B = 4.07 \pm 1.23$ days, a community spread of $\rho = E_0/I_0 = 3.53 \pm 3.97$ and an affected population of $\eta = N^*/N = 0.0767 \pm 0.2612$.

Figures 10.8 and 10.9 show the the mobility network of the European Union, the basic reproduction number $R_0 = C/B$, the community spread $\rho = E_0/I_0$, and the affected population $\eta = N^*/N$ across all 27 countries of the European Union. The basic reproduction number is largest in Austria and Germany with $R_0 = 8.7$ and $R_0 = 6.0$ and smallest in Malta and Denmark with $R_0 = 3.0$ and $R_0 = 2.7$, with a mean of $R_0 = 4.62 \pm 1.32$. The initial community spread is largest in Italy and Spain with $\rho = 18.4$ and $\rho = 15.2$ and smallest in Malta and Cyprus with $\rho = 0.2$ and $\rho = 0.1$ with a mean of $\rho = 3.53 \pm 3.97$. The affected population is largest in Ireland and Hungary with $\eta = 2.73\%$ and $\eta = 1.21\%$ and smallest in Slovakia and Bulgaria with $\eta = 0.04\%$ and $\eta = 0.02\%$, with a mean of $\eta = 0.08 \pm 0.26$.

Figure 10.10 highlights the effects of travel restrictions across Europe. The top row shows the simulated outbreak under constrained mobility with the spring 2020 travel restrictions and border control in place; the bottom row shows the outbreak under unconstrained mobility without travel restrictions. The spreading pattern in the top row follows the number of reported infections; the spreading pattern in the bottom row emerges naturally as a result of the network mobility simulation, spreading rapidly from Italy to Germany, Spain, and France. During the early stages of the pandemic, the predicted outbreak pattern in the bottom row agrees well with the outbreak pattern in the top row. During the later stages, the side-by-side comparison shows a faster spreading of the outbreak under unconstrained mobility with a massive, immediate outbreak in Central Europe and a faster spreading to the eastern and northern countries.

Freedom of movement is the cornerstone of the European Union. On March 26, 2020, the 25th anniversary of the Schengen Agreement that guarantees unrestricted movement between its member countries, all external and many internal boarders

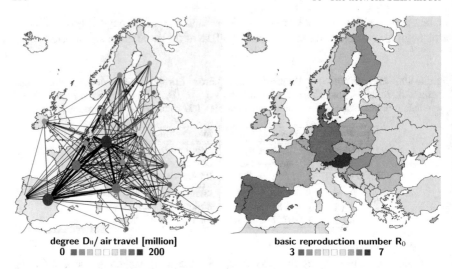

Fig. 10.8 **Outbreak dynamics of COVID-19 across Europe.** The mobility network of the European Union with $N = 27$ nodes, color-coded by degree, and the 172 most traveled edges, with thicknesses by connectivity, reveals strongest connections between Germany, Spain, Italy, and France. The basic reproduction number R_0 varies between three and seven with largest values in Austria and Germany and smallest in Malta and Denmark [16].

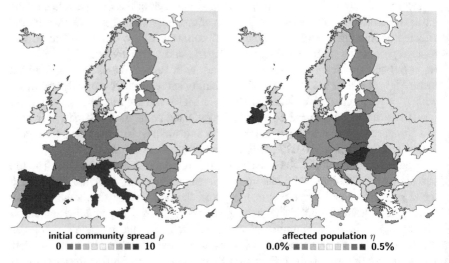

Fig. 10.9 **Outbreak dynamics of COVID-19 across Europe.** The initial community spread $\rho = E_0/I_0$ is a measure for undetected spreading at the beginning of the outbreak with largest values in Italy and Spain. The affected population $\eta = N^*/N$ is a measure of containment with smallest values in Slovakia and Bulgaria [16].

of the European Union were closed to minimize the spreading of COVID-19 [5]. These drastic measures have stimulated a wave of criticism, not only because many believe they violate international law, but also because of a strong evidence that

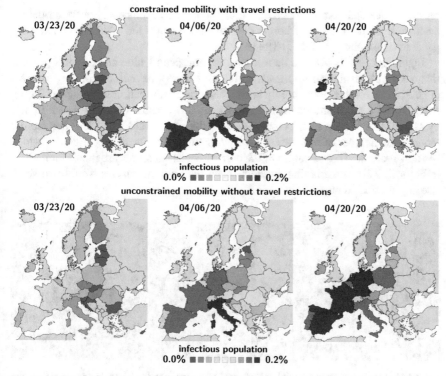

Fig. 10.10 Outbreak control of COVID-19 across Europe. The timeline of the infectious population for constrained and unconstrained mobility across all 27 countries reveals the effect of travel restrictions. Simulations are based on latent and infectious periods of $A = 2.56$ days and $C = 17.82$ days, a basic reproduction number of $R_0 = 4.38$, and mobility coefficients of $\vartheta = 0.00$ with travel restrictions and $\vartheta = 0.43$ without [16].

travel restrictions can only slow, but not stop, the spread of a pandemic [22]. A recent study based on a global metapopulation disease transmission model for the COVID-19 outbreak in China has shown that the Wuhan travel ban essentially came too late, at a point where most Chinese cities had already received many infected travelers [4]. Our study shows a similar trend for Europe, where travel restrictions were only implemented a week after every country had reported cases of COVID-19. [9]. As a natural consequence, unfortunately, no European country was protected from the outbreak.

Strikingly, our results suggest that the emerging pattern of the COVID-19 outbreak closely followed global mobility patterns of air passenger travel [19]: From its European origin in Italy, the novel coronavirus spread rapidly via the strongest network connections to Germany, Spain, and France, while slowly reaching the less connected countries, Estonia, Slovakia, and Slovenia. Although air travel is certainly not the only determinant of the outbreak dynamics, our findings indicate that mobility is a strong contributor to the global spreading of COVID-19 [4]. This is in line with recent studies that demonstrated the potential of air-travel based network models to evaluate the effectiveness of border control strategies for the 2009 H1N1 influenza pandemic [30] and for COVID-19 [18].

Figure 10.10 supports the decision of the European Union and its local governments to implement rigorous travel restrictions to delay the outbreak of the pandemic [17]. Austria, for example, rapidly introduced drastic mitigation strategies including strict border control and massive travel bans. Its national air travel was cut in half by March 20 and soon after converged to a reduction of 95% [10]. This has rapidly reduced the number of new cases to a fraction of 10% of its all time high [9], which has now motivated a gradual lift of the current constraints. Our prognosis for Austria in Figure 10.10 suggests that, without travel restrictions, Austria would still see a rise of the infected population.

Fig. 10.11 Network mobility model of Europe. Restricting mobility is especially powerful at the beginning of an outbreak, or, during the later stages of a pandemic, when new variants of a virus begin to emerge. Network epidemiology models provide valuable insight into the effects of the early spreading, constraining the spread, and releasing the constraints.

Taken together, network epidemiology modeling is a powerful tool to estimate initial spreading patterns and explore the impact of travel restrictions [19]. Compared to other discretization techniques like finite element methods [21] or methods that represent mixing through a network-based contact matrix [10], network diffusion methods are conceptually simple, straightforward to implement, and computationally inexpensive [16]. Most importantly, their diffusion matrix is inherently discrete and intimately correlated to human mobility as suggested by Figure 10.11. Here we simply illustrated the effect of air traffic mobility [1], but it is straightforward to generalize network epidemiology models to public transport and car traffic, or even local mobility within a community or across age groups [2]. Our results show that restricting mobility is especially powerful at the beginning of an outbreak, or, during the later stages of a pandemic, when new variants of a virus begin to emerge. Network epidemiology models provide valuable insight into the effects of the early spreading, constraining the spread, and releasing the constraints.

Problems

10.1 Basic reproduction number. The basic reproduction number during the early stages of the COVID-19 outbreak in the United States and Europe in Tables 10.1 and 10.2 was $R_0 = 5.30 \pm 0.95$ and $R_0 = 4.62 \pm 1.32$. In China in Table 7.1 it was $R_0 = 12.58 \pm 3.17$. Discuss at least three community mitigation strategies that could have reduced the reproduction number in the United States and Europe after learning about the early outbreak in China.

10.2 Effects of travel density. Explain, in layman's terms, why states or countries that have a high travel density–California, Texas, Germany, or France–have been affected very early by the COVID-19 pandemic. Now, explain the same phenomenon, but use technical terms, e.g., diffusion, adjacency, degree, and graph Laplacian. Which matrix best reflects the travel density of a state or country in terms of incoming and outgoing travelers?

10.3 Effects of travel restrictions. Explain, in layman's terms, why travel restrictions are most effective during the early stages of a pandemic. Now, explain the same phenomenon, but use technical terms, e.g., seeding, initial conditions, diffusion, and endemic equilibrium. Which parameters best reflect travel restrictions between different states or countries?

10.4 Size of network SEIR model. For the SEIR network model of California, Texas, New York, Georgia, and Minnesota in Figure 9.2 discretized on a graph $\mathcal{G}$ with five nodes and the eight most traveled edges, how many unknowns $\mathbf{P}_I$ are there at each node and on the entire graph? For the implicit backward Euler method, what is the size of the residual vector $\mathbf{R}_I$ at each node and for the entire graph? What is the size of the tangent matrix $\mathbf{K}_{IJ}$ at each node and for the entire graph?

10.5 Tangent matrix of network SEIR model. For the SEIR network model of California, Texas, New York, Georgia, and Minnesota in Figure 9.2 discretized on a graph $\mathcal{G}$ with five nodes and the eight most traveled edges, what is the size of the tangent matrix $\mathbf{K}_{IJ}$ for the entire graph? How many tangent matrix entries contain $1/\Delta t$ terms? How many tangent matrix entries contain α, β, or γ terms? How many tangent matrix entries contain κ terms on the diagonal? How many tangent matrix entries contain κ terms off diagonal? Justify your answers!

10.6 Network SIR model. For the SIR model in Chapter 3, discretize the set of equations on each node of the network and connect the nodes through graph Laplacian terms. Derive the equations for the time evolution of the unknowns, $\dot{S}_I$, $\dot{I}_I$, and $\dot{R}_I$. Compare your result to the network SEIR model in equation (10.3).

10.7 Explicit network SIR model. For the SIR model in Chapter 3, discretize the set of equations on each node of the network, connect the nodes through graph Laplacian terms, and discretize the equations in time using an explicit forward Euler method. Derive the update equations for the unknowns, $S_{I,n+1}$, $I_{I,n+1}$, and $R_{I,n+1}$. Compare your result to the network SEIR model in equation (10.6).

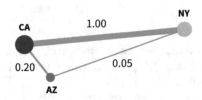

Fig. 10.12 Network diffusion model of California, New York, and Arizona. Discrete graph $\mathcal{G}$ with three nodes and three edges. The size and color of the nodes represent the degree D_{II}; the thickness and number of the edges represent the adjacency A_{IJ} estimated from weighted annual incoming and outgoing passenger air travel.

10.8 Implicit network SIR model. For the SIR model in Chapter 3, discretize the set of equations on each node of the network, connect the nodes through graph Laplacian terms, and discretize the equations in time using an implicit backward Euler method. Derive the discrete residuals for the unknowns, $R_{S,I}$, $R_{I,I}$, and $R_{R,I}$. Compare your result to the network SEIR model in equation (10.7).

10.9 Adjacency matrix, degree matrix, and graph Laplacian. For the discrete graph $\mathcal{G}$ of California, New York, and Arizona in Figure 10.12 with $n_{nd} = 3$ nodes and $n_{el} = 3$ edges, calculate the adjacency matrix A_{IJ}, the degree matrix D_{IJ}, and the graph Laplacian L_{IJ} when sorting the states by population, $I = [\,$California, New York, Arizona$\,]$. Which states have the largest and smallest degrees and what are their degrees?

10.10 Explicit network SIR model. For the discrete graph $\mathcal{G}$ of California, New York, and Arizona in Figure 10.12 with $n_{nd} = 3$ nodes and $n_{el} = 3$ edges, write the explicit update equations of the SIR model for the three unknowns $S_{I,n+1}$, $I_{I,n+1}$, and $R_{I,n+1}$ at all three nodes using the graph Laplacian from the pervious problem. Assume diffusion coefficients of $\kappa = 1$ and contact and infectious rates of $\beta = 0.50\,\text{days}^{-1}$ and $\gamma = 0.25\,\text{days}^{-1}$. Assume initial conditions of $S_{I,0} = 1$, $I_{I,0} = 0$, and $R_{I,0} = 0$ at all nodes except for New York with $I_{2,0} = 0.2$ and $S_{2,0} = 0.8$. Simulate the first time step of $\Delta t = 1$ day. Describe what happens.

10.11 Explicit network SEIR model. For the discrete graph $\mathcal{G}$ of California, New York, and Arizona in Figure 10.12 with $n_{nd} = 3$ nodes and $n_{el} = 3$ edges, write the explicit update equations of the SEIR model for the four unknowns $S_{I,n+1}$, $E_{I,n+1}$, $I_{I,n+1}$, and $R_{I,n+1}$ at all three nodes using the graph Laplacian from the pervious problem. Assume diffusion coefficients of $\kappa = 1$ and latent, contact, and infectious rates of $\alpha = 2.00\,\text{days}^{-1}$, $\beta = 0.50\,\text{days}^{-1}$, and $\gamma = 0.25\,\text{days}^{-1}$. Assume initial conditions of $S_{I,0} = 1$, $E_{I,0} = 0$, $I_{I,0} = 0$, and $R_{I,0} = 0$ at all nodes except for New York with $E_{2,0} = 0.2$ and $S_{2,0} = 0.8$. Simulate the first time step of $\Delta t = 1$ day. Describe what happens.

10.12 Network SEIR model. For the discrete graph $\mathcal{G}$ of California, Florida, New York, Washington, and Arizona in Figure 10.13 with $n_{nd} = 5$ nodes and the $n_{el} = 8$

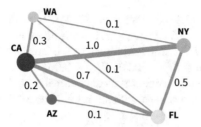

Fig. 10.13 Network diffusion model of California, Florida, New York, Washington, and Arizona. Discrete graph $\mathcal{G}$ with five nodes and the eight most traveled edges. The size and color of the nodes represent the degree D_{II}; the thickness and number of the edges represent the adjacency A_{IJ} estimated from weighted annual incoming and outgoing passenger air travel.

most traveled edges, assume an SEIR network model with four unknowns $S_{I,n+1}$, $E_{I,n+1}$, $I_{I,n+1}$, and $R_{I,n+1}$ at each node. Assume diffusion coefficients of $\kappa = 1$ and a rare disease with latent, contact, and infectious rates of $\alpha = 0 \, \text{days}^{-1}$, $\beta = 0 \, \text{days}^{-1}$ and $\gamma = 0 \, \text{days}^{-1}$. Assume the outbreak begins in New York with $S_{3,0} = 0.5$ and $I_{3,0} = 0.5$, while all other states have not observed any cases, $S_{I,0} = 1$ for $I = 1, 2, 4, 5$. Estimate the susceptible, exposed, infectious, and recovered populations $S_{I,\infty}$, $E_{I,\infty}$, $I_{I,\infty}$, and $R_{I,\infty}$ in all five states at endemic equilibrium. If you like, you can do this without a computational model. Explain you results.

10.13 Network SEIR model. For the discrete graph $\mathcal{G}$ of California, Florida, New York, Washington, and Arizona in Figure 10.13, assume an SEIR network model with diffusion coefficients of $\kappa = 1$ and a rare disease with latent, contact, and infectious rates of $\alpha = 1/3 \, \text{days}^{-1}$, $\beta = 0 \, \text{days}^{-1}$ and $\gamma = 1/7 \, \text{days}^{-1}$. Assume the outbreak begins in New York with $S_{3,0} = 0.5$ and $I_{3,0} = 0.5$, while all other states have not observed any cases, $S_{I,0} = 1$ for $I = 1, 2, 4, 5$. Estimate the susceptible, exposed, infectious, and recovered populations $S_{I,\infty}$, $E_{I,\infty}$, $I_{I,\infty}$, and $R_{I,\infty}$ in all five states at endemic equilibrium. If you like, you can do this without a computational model. Explain you results.

10.14 Network SEIR model. For the discrete graph $\mathcal{G}$ of California, Florida, New York, Washington, and Arizona in Figure 10.13, assume an SEIR network model with diffusion coefficients of $\kappa = 1000$ and latent, contact, and infectious rates of $\alpha = 1/3 \, \text{days}^{-1}$, $\beta = 2/7 \, \text{days}^{-1}$ and $\gamma = 1/7 \, \text{days}^{-1}$. Assume the outbreak begins in New York with $S_{3,0} = 0.5$ and $I_{3,0} = 0.5$, while all other states have not observed any cases, $S_{I,0} = 1$ for $I = 1, 2, 4, 5$. Estimate the susceptible, exposed, infectious, and recovered populations $S_{I,\infty}$, $E_{I,\infty}$, $I_{I,\infty}$, and $R_{I,\infty}$ in all five states at endemic equilibrium. If you like, you can do this without a computational model. Explain you results.

10.15 Network SEIR model. For the discrete graph $\mathcal{G}$ of California, Florida, New York, Washington, and Arizona in Figure 10.13, assume an SEIR network model with diffusion coefficients of $\kappa = 1$ and latent, contact, and infectious rates of

$\alpha = 1/3\,\text{days}^{-1}$, $\beta = 2/7\,\text{days}^{-1}$ and $\gamma = 1/7\,\text{days}^{-1}$. Assume the outbreak begins in New York with $S_{3,0} = 0.5$ and $I_{3,0} = 0.5$, while all other states have not observed any cases, $S_{I,0} = 1$ for $I = 1, 2, 4, 5$. Estimate by which order the five states converge to endemic equilibrium at $S_{I,\infty} = 0.22$, $E_{I,\infty} = 0.00$, $I_{I,\infty} = 0.00$, and $R_{I,\infty} = 0.78$. If you like, you can do this without a computational model. Explain you response. How would your estimate change with strict travel restrictions between New York and California?

10.16 Travel restrictions. Consider an SEIR network model on the discrete graph $\mathcal{G}$ of California, Florida, New York, Washington, and Arizona in Figure 10.13. Which parameters would you change to model strict travel restrictions between New York and California? Which parameters would you change to model temperature checks at the gates in all five locations? Discuss which effects you would expect from these measures.

References

1. Apple Mobility Trends. https://www.apple.com/covid19/mobility. accessed: June 1, 2021.
2. Balcan D, Colizzan V, Goncalves B, Hu H, Jamasco J, Vespignani A (2009) Multiscale mobility networks and the spatial spreading of infectious diseases. Proceedings of the National Academy of Sciences 106:21484-21489.
3. Chinazzi, M, Davis JT, Ajelli M, Gioanni C, Litvinova M, Merler S, Piontti A, Mu K, Rossi L, Sun K, Viboud C, Xiong X, Yu H, Halloran ME, Longini IM, Vespignani A (2020) The effect of travel restrictions on the spread of the 2019 novel coronavirus (COVID-19) outbreak. Science 368:395-400.
4. Colizza V, Barrat A, Barthelemy M, Vespignani A (2006) The role of the airline transportation network in the prediction and predictability of global epidemics. Proceedings of the National Academy of Sciences 103:2015-2020.
5. European Commission. COVID-19: Temporary restriction on non-essential travel to the EU. Communication from the Commission to the European Parliament, the European Council and the Council. Brussels, March 16, 2020.
6. European Centre for Disease Prevention and Control. Situation update worldwide. https://www.ecdc.europa.eu/en/geographical-distribution-2019-ncov-cases accessed: June 1, 2021.
7. Eurostat. Your key to European statistics. Air transport of passengers. https://ec.europa.eu/eurostat accessed: June 1, 2021.
8. Fornari S, Schafer A, Jucker M, Goriely A, Kuhl E (2019) Prion-like spreading of Alzheimer's disease within the brain's connectome. Journal of the Royal Society Interface 16:20190356.
9. Johns Hopkins University (2021) Coronavirus COVID-19 Global Cases by the Center for Systems Science and Engineering. https://coronavirus.jhu.edu/map.html, https://github.com/CSSEGISandData/covid-19 assessed: June 1, 2021.
10. Kergassner A, Burkhard C, Lippold D, Kergassner M, Pflug L, Budday D, Steinmann P, Budday S (2020) Memory-based meso-scale modeling of Covid-19. Computational Mechanics 66:1069-1079.
11. Linka K, Peirlinck M, Sahli Costabal F, Kuhl E (2020) Outbreak dynamics of COVID-19 in Europe and the effect of travel restrictions. Computer Methods in Biomechanics and Biomedical Engineering 23: 710-717.
12. Linka K, Peirlinck M, Kuhl E (2020) The reproduction number of COVID-19 and its correlation with public heath interventions. Computational Mechanics 66:1035-1050.

13. Linka K, Rahman P, Goriely A, Kuhl E (2020) Is it safe to lift COVID-19 travel restrictions? The Newfoundland story. Computational Mechanics 66:1081–1092.
14. Linka K, Goriely A, Kuhl E (2021) Global and local mobility as a barometer for COVID-19 dynamics. Biomechanics and Modeling in Mechanobiology 20:651–669.
15. Maier BF, Brockmann D (2020) Effective containment explains sub-exponential growth in confirmed cases of recent COVID-19 outbreak in mainland China. Science 368:742-746.
16. Newman MEJ (2002) Spread of epidemic disease on networks. Physical Review E 66:016128.
17. Newman M (2010) Networks: An Introduction. Oxford University Press, New York.
18. Pastor-Satorras R, Vespignani A (2001) Epidemic spreading in scale-free networks. Physical Review Letters 86:3200-3203.
19. Pastor-Satorras R, Castallano C, Van Mieghem P, Vespignani A (2015) Epidemic processes in complex networks. Reviews of Modern Physics 87:925-979.
20. Peirlinck M, Linka K, Sahli Costabal F, Kuhl E (2020) Outbreak dynamics of COVID-19 in China and the United States. Biomechanics and Modeling in Mechanobiology 19:2179-2193.
21. Viguerie A, Veneziani A, Lorenzo G, Baroli D, Aretz-Nellesen N, Patton A, Yankeelov TE, Reali A, Hughes TJR, Auricchio F (2020) Diffusion-reaction compartmental models formulated in a continuum mechanics framework: application to COVID-19, mathematical anaysis, and numerical study. Computational Mechanics 66:1131-1152.
22. World Health Organization. WHO Director-General's opening remarks at the media briefing on COVID-19., March 13, 2020 https://www.who.int/dg/speeches/detail/who-director-general-s-opening-remarks-at-the-mission-briefing-on-covid-19-13-march-2020; accessed: June 1, 2021.
23. Zlojutro A, Rey D, Gardner L (2019) A decision-support framework to optimize border control for global outbreak mitigation. Scientific Reports 9:2216.

Part IV
Data-driven epidemiology

Chapter 11
Introduction to data-driven epidemiology

Abstract Classical compartment models in epidemiology are idealized deterministic models that neglect randomness in the analysis of disease data. However, in reality disease data are inherently stochastic; they are incomplete, include noise, and contain systemic uncertainty. Here we integrate data-driven modeling and computational epidemiology to explore disease data and compartment models using a probabilistic approach and quantify the uncertainties of our analysis. To set the stage, we briefly introduce the basic concepts of data-driven modeling using Bayesian statistics and illustrate Bayes' theorem for discrete and continuous parameter values. To demonstrate the basic elements of a Bayesian analysis, we discuss the concepts of likelihood, prior, marginal likelihood, and posterior by means of the simple SIS model and apply Markov Chain Monte Carlo methods to compute them. We generate synthetic data for the SIS model, infer the posterior distribution of its parameter values, and illustrate the result using means and credible intervals. The learning objectives of this chapter on data-driven modeling are to

- think in terms of probabilistic models
- discuss the differences between a Bayesian and a frequentist approach
- explain the basic elements of Bayes' theorem: likelihood, prior, and posterior
- illustrate and discuss the importance of prior selection
- apply Bayes' theorem to calculate the discrete probability of having COVID-19
- apply Bayes' theorem to analyze the continuous SIS model
- understand the role of Markov Chain Monte Carlo methods in a Bayesian analysis
- communicate a Bayesian analysis and interpret means and credible intervals
- discuss limitations of Bayes' theorem in computational epidemiology

By the end of the chapter, you will be able to perform a Bayesian analysis that includes selecting prior probability distributions, selecting a likelihood function, estimating the maximum likelihood, calculating posterior probability distributions, and identifying the maximum a posteriori estimate for simple infectious diseases that do not provide immunity to reinfection.

11.1 Introduction to data-driven modeling

Unlike any other pandemic in the history of mankind, the COVID-19 pandemic has generated an unprecedented volume of data, well documented, continuously updated, and broadly available to the general public [12]. Yet, the precise role of physics-based modeling and machine learning in providing quantitative insight into the dynamics of COVID-19 remains a topic of ongoing debate [12]. While the previous chapters of this book have introduced deterministic models to analyze the case data of COVID-19 [3], we will now embed these models into a Bayesian analysis to quantify the uncertainty of both, the reported data and the computational models [16]. When looking at the public COVID-19 dashboards [9, 12, 25], it becomes immediately clear that the reported case data are inherently stochastic, noisy, and incomplete, and naturally contain systemic uncertainty [21]. This is why we should not just look at the plain data, but always interpret the data in the context of models [1].

Uncertainty Quantification is the science of characterizing and reducing uncertainties. Its objective is to determine the likelihood of certain outcomes when some aspects of the system are not exactly known [20]. Since disease data generally display significant variations, and small variations in the input can have a large effect on the output, uncertainty quantification is critical in computational epidemiology [26]. The two major types of uncertainties are data uncertainty and model uncertainty.

● **Data uncertainty** or *aleatoric uncertainty* characterizes the uncertainty associated with the input to our models, for example related to asymptomatic transmission, testing uncertainties, testing frequencies, and inconsistencies in recording versus reporting dates.

● **Model uncertainty** or *epistemic uncertainty* characterizes the uncertainty associated with the model itself, for example related to model selection, e.g., SIS, SIR, SEIS, SEIR, SEIIR; parameter identification, e.g., A, B, C, R_0; initial conditions, e.g., E_0, I_0, R_0; and boundary conditions, e.g., the in- and out-flux of populations, E, I, and the diffusion or mobility tensor D_{ij}.

Bayesian methods are a class of state-of-the-art statistical methods that use probability distributions as a tool to rigorously quantify parameter uncertainty by explicitly accounting for epistemic uncertainty [16]. In view of COVID-19 modeling, this becomes especially important when applying predictive models to situations not seen during training [26], e.g., in regression extrapolating or in forecasting outbreak dynamics. Uncertainty quantification in the form of credible intervals is a natural and important by-product of a Bayesian analysis [10], especially in view of forecasting and outbreak predictions [15].

Probability distributions. A probability distribution is a mathematical object that describes how likely different events are [14]. In statistics, we can think of these events as data generated from a true probability distribution with unknown parameters. Inference is the process of finding out these parameters by using just a sample, the data, from the true probability distribution. Since the true probability distribution is usually unknown, we create a model to approximate the probability distribution [16]. We can classify probability distributions into two classes: *discrete probability distributions*, which we use when the set of possible outcomes is discrete and the probabilities are encoded by a *probability function*, a discrete list of the probabilities of the outcomes, for example when rolling a die; and *continuous probability distributions*, which we use when the set of possible outcomes takes values in a continuous range and the probabilities are encoded by a *probability density function*, for example when studying COVID-19 case data. Probabilities are the essential building blocks of Bayes' theorem [11].

Fig. 11.1 The English statistician, philosopher, and Presbyterian minister Thomas Bayes is known for Bayes' theorem, a solution to a problem of inverse probability that was presented to the Royal Society and published in 1763, two years after Bayes' death, on April 7, 1761. Bayes' theorem, $P(A|B) = P(B|A)P(A)/P(B)$, describes the probability of an event, based on prior knowledge of conditions that might be related to the event. One of the many applications of Bayes' theorem is Bayesian inference, a particular approach to statistical inference, which is fundamental to Bayesian statistics.

Bayes' theorem. Bayes' theorem describes the probability of an event, based on prior knowledge of conditions that might be related to the event [3],

$$P(A|B) = \frac{P(B|A)\,P(A)}{P(B)}. \tag{11.1}$$

Here $P(B|A)$ is the *likelihood*, the conditional probability that B is true given A is true; $P(A)$ is the *prior*, the probability that A is true; $P(B)$ is the *marginal likelihood* or *evidence*, the probability that B is true; and $P(A|B)$ is the *posterior*, the conditional probability that A is true given that B is true [10]. We can derive Bayes' theorem from the product rule of probabilities, $P(A,B) = P(A|B)P(B)$, which states that the joint probability of A and B is equal to the conditional probability of A given B times the probability of B. Similarly, $P(A,B) = P(B|A)P(A)$, which states that the joint probability of A and B is equal to the conditional probability of B given A times the probability of A. By combining both equations, $P(A|B)P(B) = P(B|A)P(A)$, we obtain Bayes' theorem (11.1). An important observation of Bayes' theorem is that $P(A|B)$ is not necessarily identical to $P(B|A)$.

Bayes' theorem is central to Bayesian statistics [9]. For more than 250 years, the field of Bayesian statistics, named after the English statistician Thomas Bayes, has enjoyed as much appreciation as controversy. There are two schools of statistics,

Bayesian and frequentist [22], and we compare both in Table 11.1. Bayesian statistics is older than frequentist statistics, but it has long been ignored because of its need for priors. Before appropriate numerical methods became available, Bayesian statistics was restricted to only a small subset of problems for which conjugate priors were known. Fortunately, the development of computational techniques and the discovery of novel mathematical tools resurrected Bayesian statistics [27]. Probably the most notable of these developments is the invention of Markov Chain Monte Carlo methods, a class of algorithms that randomly sample from probability distributions to evaluate Bayesian models [23]. Throughout the past decades, Bayesian statistics has undeniably gained massive popularity and has advanced to a widely used technique in the natural sciences, computer science, and engineering. This success is a result of the rapid advancements in statistics and data science, and the birth of an entirely new field, probabilistic programming.

Table 11.1 Frequentist vs. Bayesian approach. Classification, comparison, and requirements for both approaches.

Frequentist	Bayesian

o never uses probabilities
  neither for parameters nor for data
	o uses probabilities
	  for both parameters and data

o does not require prior knowledge
  and provides an objective view
	o requires prior knowledge
	  and provides a subjective view

o provides a point estimate
  for the parameter values
  with little insight into quality of fit
	o provides a posterior distribution
	  for the parameter values
	  with high density interval, mean, mode

o confidence interval
  over infinite sample size from the population,
  95% of these contain tru population value
	o credible interval
	  95% probability that population value
	  is within the limits of this interval

o is static and cannot build in new information	o can dynamically build in new information

o computationally inexpensive
  and relatively straightforward
	o computationally expensive
	  but now feasible with Bayesian libraries

Probabilistic programming. Probabilistic programming is a programming paradigm in which the user specifies probabilistic models and the program automatically performs inference using these models [23]. Various probabilistic programming platforms exist and we can distinguish low- and high-level application programming interfaces: Low-level application programming interfaces provide a high degree of control to manipulate functions, but are limited to the basic programming features, whereas high-level application programming interfaces like PyMC3 provide less flexibility, but more functionality and are relatively easy to learn. Probabilistic programming is a powerful framework to flexibly build Bayesian statistical models into computer code. Once built, inference algorithms that work independently of the model can fit the model to the data. The unique potential of probabilistic programming is the seamless integration of flexible model selection and automatic inference analysis. Probabilistic programing makes statistical modeling broadly accessible and enables a rapid and unique insight into big data.

In the following sections, we first illustrate the concept of Bayes' theorem for a simple example with discrete parameter values, and then apply Bayes' theorem to the SIS compartment model with continuous parameter values. To illustrate the basic concepts of a Bayesian analysis, we initially perform the analyses ourselves, but then adopt the high-level application programming interface PyMC3, a Python package for Bayesian statistical modeling and probabilistic machine learning which features advanced Markov Chain Monte Carlo and variational fitting algorithms.

11.2 Bayes' theorem for discrete parameter values

In this section, we illustrate the application of Bayes' theorem to sets of separate, distinct possibilities. Sets whose members are clearly separated from each other are called discrete and the variables that represent them are *discrete variables*. The distributions of the probability values of discrete variables are called *probability functions*. Bayes' theorem for the probabilities of discrete variables,

$$\text{posterior probability} = \frac{\text{likelihood} \times \text{prior probability}}{\text{marginal likelihood}}, \quad (11.2)$$

states that the posterior parameter probability is equal to the likelihood times the prior parameter probability divided by the marginal likelihood [10]. Let's look at Bayes' theorem for a concrete example with two discrete variables, the loss of smell and no loss of smell, and two discrete parameters, COVID-19 and the common cold.

Likelihood and maximum likelihood. Imagine you feel sick and experience a loss of smell. Your doctor tells you that 86 in 100 people who have COVID-19 experience a loss of smell. In other words, the probability of experiencing a loss of smell when you have COVID-19 is 86%,

$$P(\text{loss of smell}|\text{COVID-19}) = 0.86 = 86\% .$$

Here P denotes the probability and the vertical bar | stands for "given that" or "under the condition that" and denotes a *conditional probability*. The probability of an observation or *variable*, the loss of smell, under the condition of a *parameter*, COVID-19, is called the *likelihood* of COVID-19. While this information is interesting, a probability of 86% sounds scary, and is not very useful. There could be other explanations for your symptoms. For example, the probability of experiencing a loss of smell when you have a common cold is 60%,

$$P(\text{loss of smell}|\text{common cold}) = 0.60 = 60\% .$$

In this example, the COVID-19 has a larger likelihood than the common cold. Since we only consider two diseases, of the two possible alternatives, COVID-19 has the *maximum likelihood*.

Prior probability. In fact, what you really want to know is the probability that you have COVID-19, given your experience a loss of smell. Bayes' theorem translates the *uninformative probability* of your symptoms given that you have COVID-19 into the *informative probability* that you have COVID-19 given your symptoms. To do this, we weight the likelihood with prior knowledge. In our case, this prior knowledge is the prevalence or probability that a random individual has COVID-19. As of today, March 25, 2021, the number of active COVID-19 cases is 21,378,284 in a total population of 7,794,798,739, and the *prior probability* of COVID-19 is,

$$P(\text{COVID-19}) = \frac{21,378,284}{7,794,798,739} = 0.0027 = 0.27\%,$$

compared to an estimated prior probability for the common cold of

$$P(\text{common cold}) = 0.0100 = 1.00\%.$$

It seems intuitive to weight the likelihood with the prior probability, and this is exactly what Bayes' theorem does.

Marginal likelihood. In Bayes' theorem, the denominator is called the *marginal likelihood* or *evidence* and is typically difficult to estimate. In our example, the marginal likelihood denotes the fraction of people in the general population that experience a loss of smell. Here we assume a marginal likelihood of

$$P(\text{loss of smell}) = 0.0083 = 0.83\%,$$

meaning one 83 in 10,000 people experience a loss of smell.

Posterior probability. Using Bayes' theorem (11.2), we can calculate the *posterior probability*,

$$P(\text{COVID-19}|\text{loss of smell}) = \frac{P(\text{loss of smell}|\text{COVID-19}) \cdot P(\text{COVID-19})}{P(\text{loss of smell})}$$

the probability that you have COVID-19 given you experience a loss of smell,

$$P(\text{COVID-19}|\text{loss of smell}) = \frac{0.86 \cdot 0.0027}{0.0083} = 0.28 = 28\%.$$

This is the information you were expecting to hear from your doctor. This example shows that Bayes' theorem is a mathematical method to find inverse probabilities. By using prior information, it translates the uninformative probability of observing a variable, the loss of smell, for a given parameter, COVID-19, into the informative probability that a parameter, COVID-19, explains an observable variable, the loss of smell. At this point, we can also calculate the posterior probability that a different parameter, the common cold, explains our observable variable, the loss of smell,

$$P(\text{common cold}|\text{loss of smell}) = \frac{0.60 \cdot 0.0100}{0.0083} = 0.72 = 72\%.$$

This example shows, that although the likelihood of COVID-19 is larger than the likelihood of the common cold, if you experience a loss of smell, it is more probable that you actually have a common cold than COVID-19.

Model selection, Bayes factor, prior and posterior odds. The process of identifying the most probable model is called *model selection*. It uses the *posterior odds*, the ratio of posterior probabilities,

$$R_{post} = \frac{P(\text{COVID-19}|\text{loss of smell})}{P(\text{common cold}|\text{loss of smell})}.$$

By using Bayes' theorem, the we can rewrite the posterior odds as

$$R_{post} = B \cdot R_{prior},$$

where the ratio of likelihoods B is called *Bayes factor*, and the ratio of priors R_{prior} is called the *prior odds*,

$$B = \frac{P(\text{loss of smell}|\text{COVID-19})}{P(\text{loss of smell}|\text{common cold})} \quad \text{and} \quad R_{prior} = \frac{P(\text{COVID-19})}{P(\text{common cold})}.$$

When calculating the posterior odds, the marginal likelihood, the probability of observing a variable, cancels. This is very convenient since we had difficulties calculating the marginal likelihood in the first place. In our case, we have assumed the probability of the loss of smell to $P(\text{loss of smell}) = 0.0083$. We have chosen this value such that the posterior probabilities sum up to one, $P(\text{COVID-19}|\text{loss of smell}) + P(\text{common cold}|\text{loss of smell}) = 0.28 + 0.72 = 1.00$. For most practical purposes, we are interested in ratios between two posterior probabilities and can simply ignore the marginal likelihood. For our example, the posterior odds are

$$R_{post} = B \cdot R_{prior} = \frac{0.86}{0.60} \cdot \frac{0.0027}{0.0100} = 1.433 \cdot 0.270 = 0.387.$$

While Bayes factor is larger than one, B = 1.433, and favors COVID-19, the prior odds is smaller than one, $R_{prior} = 0.270$, and favors the common cold. As a result, the posterior odds favors the common cold as an explanation for your observation, your loss of smell. In general, if the posterior odds is smaller than one third, $R_{post} < 1/3$, or larger than three, $3 < R_{post}$, this is considered a substantial difference between both models and increases our confidence in our model selection. Taken together, this example shows that prior experience is critical for interpreting observations, and Bayes' theorem provides a rigorous mathematical method for doing so. While the previous example was based on *discrete* variables, the loss of smell and no loss of smell, and *discrete* parameters, COVID-19 and the common cold, we will now turn to examples with *continuous* values.

11.3 Bayes' theorem for continuous parameter values

In this section, we introduce Bayes' theorem for *continuous variables*. The distributions of the probabilities of continuous variables are called *probability density functions*. Bayes' theorem for the probabilities of continuous variables states that the posterior probability distribution is equal to the likelihood function times the prior probability density divided by the marginal likelihood [10],

$$\frac{\text{posterior probability}}{\text{distribution}} = \frac{\text{likelihood function} \times \text{prior probability distribution}}{\text{marginal likelihood}}.$$

$$(11.3)$$

In computational epidemiology, we apply Bayes' theorem to estimate the set of parameters $\boldsymbol{\vartheta}$, such that the statistics of the model output $D(t)$ agree with the statistics of the data $\hat{D}(t)$,

$$P(\boldsymbol{\vartheta}|\hat{D}(t)) = \frac{P(\hat{D}(t)|\boldsymbol{\vartheta}) P(\boldsymbol{\vartheta})}{P(\hat{D}(t))}. \qquad (11.4)$$

Here $P(\hat{D}(t)|\boldsymbol{\vartheta})$ is the *likelihood function*, the conditional probability of the data $\hat{D}(t)$ for given fixed parameters $\boldsymbol{\vartheta}$; $P(\boldsymbol{\vartheta})$ is the *prior probability distribution* of the model parameters $\boldsymbol{\vartheta}$; $P(\hat{D}(t))$ is the *marginal likelihood* or *evidence*; and $P(\boldsymbol{\vartheta}|\hat{D}(t))$ is the *posterior probability distribution* of the parameters $\boldsymbol{\vartheta}$ for given data $\hat{D}(t)$. In the following section, we use equation (11.4) to infer the model parameters for compartment models in computational epidemiology, for example, the initial infectious population I_0, the basic reproduction number R_0, and the infectious period C, for given case data [3, 5, 17].

Likelihood function and maximum likelihood estimate. The likelihood function, $P(\hat{D}(t)|\boldsymbol{\vartheta})$, measures the *goodness of fit* between a sample of the observed data $\hat{D}(t)$ and the model output $D(\boldsymbol{\vartheta},t)$ for given values of the unknown parameters $\boldsymbol{\vartheta}$. For the example of COVID-19, the observed data $\hat{D}(t)$ are the infectious or recovered populations from publicly available COVID-19 dashboards [12], and the model output $D(\boldsymbol{\vartheta},t)$ is the temporal evolution of the populations of our SEIR model [17]. To account for the goodness of fit throughout the entire time window, from day t_0 to day t_n, we evaluate the likelihood $\mathcal{L}(\boldsymbol{\vartheta},t_i)$ at each discrete time point t_i and multiply the individual terms to calculate the overall likelihood function $P(\hat{D}(t)|\boldsymbol{\vartheta})$,

$$P(\hat{D}(t)|\boldsymbol{\vartheta}) = \prod_{i=0}^{n} \mathcal{L}(\boldsymbol{\vartheta},t_i). \qquad (11.5)$$

The product symbol, $\prod_{i=0}^{n}$, denotes the multiplication of the $i = 0, ..., n$ individual likelihood terms $\mathcal{L}(\boldsymbol{\vartheta},t_i)$. The resulting likelihood function $P(\hat{D}(t_i)|\boldsymbol{\vartheta})$ describes a hypersurface over the parameter space $\boldsymbol{\vartheta}$. Its peak defines the parameter set $\boldsymbol{\vartheta}_{\text{mle}}$ that maximizes the overall likelihood $P(\hat{I}(t)|\boldsymbol{\vartheta})$ and is known as the *maximum likelihood estimate*,

$$\boldsymbol{\vartheta}_{\text{mle}} \subset \arg \max_{\boldsymbol{\vartheta}} P(\hat{D}(t)|\boldsymbol{\vartheta}). \qquad (11.6)$$

The maximum likelihood estimate defines the set of parameters values that create the model output that best explains the observed data, in our example, the observed COVID-19 case numbers.

Prior probability distribution. The prior probability distribution, $P(\vartheta)$, or simply the prior, is the probability distribution of the parameters. It encodes our belief about the parameters before taking into account the data. We distinguish between uninformative priors, weakly informative priors, and informative priors. An *uninformative prior* encodes vague general information about a parameter, for example by using a flat distribution that assigns equal probabilities to all parameter values. A *weakly informative prior* encodes partial information about a variable, for example by using a uniform distribution that assigns equal probabilities to all possible parameter values within a defined interval or a normal or log-normal distribution around a reasonable mean with a defined standard deviation. An *informative prior* encodes specific definite information about a variable. Throughout this book, we almost exclusively use weakly informative priors.

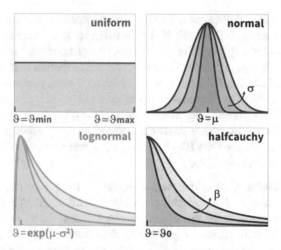

Fig. 11.2 Bayes' theorem. Prior probability density functions. Uniform distribution, normal distribution, log-normal distribution, and half Cauchy distribution. The uniform distribution is constant throughout the parameter domain $\vartheta_{\min}$ to $\vartheta_{\max}$. The normal distribution is symmetric around the mean μ with a width that increases with increasing standard deviation σ. The lognormal distribution is the probability distribution of a random variable ϑ whose logarithm $\ln(\vartheta)$ is normally distributed with mean μ, standard deviation σ, and mode $\exp(\mu - \sigma^2)$. The half Cauchy distribution is symmetric around ϑ_0 with a width that increases with increasing scale parameter β.

Figure 11.2 illustrates four popular prior probability density functions, the uniform distribution, normal distribution, log-normal distribution, and half Cauchy distribution. The uniform distribution is constant throughout the parameter domain from $\vartheta_{\min}$ to $\vartheta_{\max}$. The normal distribution is symmetric around the mean μ with a width that increases with increasing standard deviation σ. The log-normal distribution is the probability distribution of a random variable ϑ whose logarithm $\ln(\vartheta)$ is normally distributed with mean μ, standard deviation σ, and mode $\exp(\mu - \sigma^2)$. The half Cauchy distribution is symmetric around ϑ_0 with a width that increases with increasing parameter β.

Marginal likelihood. The marginal likelihood $P(\hat{D}(t))$, often also called evidence, is the probability of observing the data $\hat{D}(t)$ for all possible values the parameters $\boldsymbol{\vartheta}$ can take,

$$P(\hat{D}(t)) = \int_{\boldsymbol{\vartheta}} P(\hat{D}(t)|\boldsymbol{\vartheta})\, P(\boldsymbol{\vartheta})\, d\boldsymbol{\vartheta}\,. \tag{11.7}$$

Evaluating the marginal likelihood involves complex integrals that are generally difficult to compute. Fortunately, in Bayes' theorem, the marginal likelihood is nothing but a normalization factor, and we can rewrite Bayes' theorem as a proportionality,

$$P(\boldsymbol{\vartheta}|\hat{D}(t)) \propto P(\hat{D}(t)|\boldsymbol{\vartheta})\, P(\boldsymbol{\vartheta})\,. \tag{11.8}$$

For most of this book, we will not explicitly calculate the marginal likelihood and simply think of it as a proportionality constant.

Posterior probability distribution. The posterior probability distribution, $P(\boldsymbol{\vartheta}|\hat{D}(t))$, or simply the posterior, is the result of the Bayesian analysis in equation (11.4). Importantly, the posterior probability distribution of our parameter set $\boldsymbol{\vartheta}$ is not a set of single values. Rather, for each parameter, it is a distribution that mathematically balances the likelihood and the prior. A major strength of Bayesian inference is that we can continuously update the posterior probability distribution as new information becomes available. Conceptually, we can interpret the posterior as the updated prior in light of new data. This makes Bayesian analysis perfectly suited for studying the outbreak dynamics of the COVID-19 pandemic, in real time, with live updates, as new case data become available.

Markov Chain Monte Carlo. In Bayesian inference, we are usually not interested in the absolute value of the posterior probability distribution, $P(\boldsymbol{\vartheta}|\hat{D}(t))$, and it is often sufficient to estimate the set of parameters for which it takes a maximum value. Instead of evaluating the posterior probability distribution, $P(\boldsymbol{\vartheta}|\hat{D}(t))$, by explicitly computing the marginal likelihood, $P(\hat{D}(t))$, most methods successively compare two posterior probabilities by calculating their ratio such that the marginal likelihood cancels. A class of powerful methods that use this strategy are *Markov Chain Monte Carlo* methods. One of the most popular Markov Chain Monte Carlo methods, the Metropolis-Algorithm, was recognized as one of the ten most influential algorithms for science and engineering. The first part of the name, Markov Chain, defines the way we draw our samples and the second part, Monte Carlo, defines the sampling purpose. Briefly, Markov Chain Monte Carlo methods create samples from continuous random variables with defined probability densities. They then compare each new sample $\boldsymbol{\vartheta}_{\text{new}}$ against the pervious best sample $\boldsymbol{\vartheta}_{\text{now}}$ by evaluating the ratio of their posterior probabilities, $P(\boldsymbol{\vartheta}_{\text{new}}|\hat{D})/P(\boldsymbol{\vartheta}_{\text{new}}|\hat{D})$. If this ratio is larger than one, the new posterior probability is larger than the previous one, and the new sample explains the data better than the pervious sample. The algorithm keeps the new sample, discards the previous one, and draws the next sample. If the ratio is smaller than one, the algorithm simply keeps the pervious sample. At the end of the chain, the final remaining sample contains the set of parameters that maximize the normalized posterior, $P(\boldsymbol{\vartheta}|\hat{D}(t)|)$,

$$\boldsymbol{\vartheta}_{\text{map}} \subset \arg \max_{\boldsymbol{\vartheta}} P(\boldsymbol{\vartheta}|\hat{D}(t)). \tag{11.9}$$

This parameter estimate is called the *maximum a posteriori estimate*, $\boldsymbol{\vartheta}_{\text{map}}$. The maximum a posteriori estimate defines the set of parameters values that best explain the observed data, in our example, the observed COVID-19 case numbers, by taking into account our previous knowledge about these parameters. In the following section, we illustrate the application of Bayes' theorem for continuous parameter values and apply Markov Chain Monte Carlo sampling to analyze the classical SIS model.

11.4 Data-driven SIS model

SIS model and parameters. The SIS model characterizes infectious diseases like the common cold or influenza that do not provide immunity upon infection [2]. While the SIS model is too simplistic to explain the outbreak dynamics of complex infectious diseases, it is the only compartment model with an explicit analytical solution for the time course of its populations. This makes it ideally suited to explain and illustrate the basic principles of data-driven modeling. Figure 11.3 shows that the SIS model

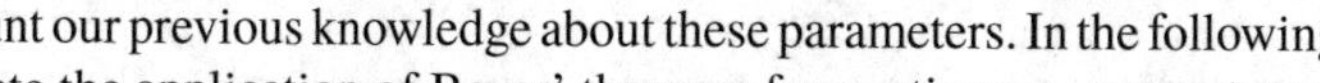

Fig. 11.3 Classical SIS model. The classical SIS model contains two compartments for the susceptible and infectious populations, S and I. The transition rates between the compartments are the contact rate $\beta = R_0/C$ and the infectious rate $\gamma = 1/C$. The dynamics of the SIS model are uniquely determined by three parameters $\boldsymbol{\vartheta} = \{I_0, R_0, C\}$, the initial infectious population I_0, the basic reproduction number R_0, and the infectious period C.

consists of only two populations, the susceptible group S and the infectious group I. As we have discussed in Chapter 2, in the SIS model, the transition between the susceptible and infectious populations is governed by a system of two coupled ordinary differential equations,

$$\begin{aligned} \dot{S} &= -\beta SI + \gamma I \\ \dot{I} &= +\beta SI - \gamma I. \end{aligned} \tag{11.10}$$

The transition rates between both compartments are the contact rate $\beta = R_0/C$ and the infectious rate $\gamma = 1/C$ in units [1/days]. The dynamics of the SIS model are uniquely determined by three parameters, the initial infectious population I_0, the basic reproduction number R_0, and the infectious period C,

$$\boldsymbol{\vartheta} = \{I_0, R_0, C\} \quad \text{such that} \quad \beta = \frac{R_0}{C} \quad \text{and} \quad \gamma = \frac{1}{C}. \tag{11.11}$$

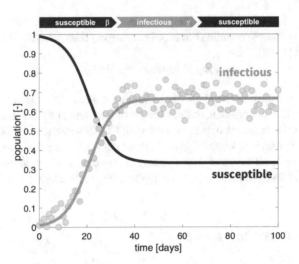

Fig. 11.4 Classical SIS model. Analytical solution $I(t)$ for given parameters $\vartheta = \{I_0, R_0, C\}$ and synthetic data $\hat{I}(t)$. The susceptible and infectious populations $S(t)$ and $I(t)$ are based on an initial infectious population $I_0 = 0.01$, a basic reproduction number $R_0 = 3.0$, and an infectious period $C = 10$ days. The synthetic data $\hat{I}(t)$ are created by adding a random noise with $\sigma = 0.05$ to the analytical solution.

In this simple format, the SIS model (11.10) neglects all vital dynamics, $\dot{S} + \dot{I} \doteq 0$ and $S + I = \text{const.} = 1$. With $S = 1 - I$, we can rephrase the system of equations of the SIS model (11.10) in terms of only one independent variable, the size of the infectious population, $\dot{I} = \beta [1 - I] I - \gamma I$, or, in terms of the basic reproduction number R_0 and the infectious period C,

$$\dot{I} = -\frac{R_0}{C} I^2 + \frac{R_0 - 1}{C} I. \tag{11.12}$$

This is a logistic differential equation,

$$\dot{I} = \frac{R_0 - 1}{C} I \left[1 - \frac{I}{1 - 1/R_0} \right], \tag{11.13}$$

which has an explicit analytical solution for the infectious population,

$$I(\vartheta, t) = \frac{[1 - 1/R_0] I_0}{I_0 + [1 - 1/R_0 - I_0] \exp([1 - R_0] t/C)}, \tag{11.14}$$

in terms of the parameters, $\vartheta = \{I_0, R_0, C\}$, the initial infectious population I_0, the basic reproduction number R_0, and infectious period C. Figure 11.4 illustrates the susceptible and infectious population of the SIS model based on an initial infectious population $I_0 = 0.01$, a basic reproduction number $R_0 = 3.0$, and an infectious period $C = 10$ days. We now use the SIS model to illustrate the use of Bayes' theorem and infer these three parameters, $\vartheta = \{I_0, R_0, C\}$, from data $\hat{I}(t)$.

Data. We have seen that real disease data contain a lot of uncertainties, for example, associated with asymptomatic transmission, testing uncertainties, and reporting inconsistencies. For our example, rather than using real disease data, we perform the analysis on *synthetic data* $\hat{I}(t)$. We assume that the synthetic data points are normally distributed around the analytical solution $I(\boldsymbol{\vartheta}, t)$ with a standard deviation of $\sigma = 0.05$, which collectively represents the data uncertainties,

$$\hat{I}(t) \sim \text{Normal}(\mu, \sigma) \quad \text{with} \quad \mu = I(\boldsymbol{\vartheta}, t) \quad \text{and} \quad \sigma = 0.05. \tag{11.15}$$

The symbol $\sim$ indicates that the synthetic data $\hat{I}(t)$ are random variables distributed as a normal distribution with parameters μ and σ. Figure 11.4 illustrates the synthetic data $\hat{I}(t)$ for for the first $t = 1, ..., 100$ days of the outbreak as orange dots.

Bayes' theorem. We now use Bayes' theorem, to estimate the posterior probability distribution of the parameters $\boldsymbol{\vartheta} = \{I_0, \mathsf{R}_0, C\}$ such that the statistics of the model output $I(t)$ agree with the statistics of the synthetic data $\hat{I}(t)$,

$$P(\boldsymbol{\vartheta}|\hat{I}(t)) = \frac{P(\hat{I}(t)|\boldsymbol{\vartheta}) \, P(\boldsymbol{\vartheta})}{P(\hat{I}(t))}. \tag{11.16}$$

Here $P(\hat{I}(t)|\boldsymbol{\vartheta})$ is the *likelihood*, the conditional probability of the data $\hat{I}(t)$ for given fixed parameters $\boldsymbol{\vartheta}$; $P(\boldsymbol{\vartheta})$ is the *prior*, the probability distribution of the model parameters $\boldsymbol{\vartheta}$; $P(\hat{I}(t))$ is the *marginal likelihood* or *evidence*; and $P(\boldsymbol{\vartheta}|\hat{I}(t))$ is the *posterior*, the conditional probability of the parameters $\boldsymbol{\vartheta}$ for the given data $\hat{I}(t)$. Bayes' theorem calculates the normalized pointwise product of the prior and the likelihood, to produce the posterior probability distribution, which is the conditional distribution of the parameters $\boldsymbol{\vartheta}$ given the data, in our case, the synthetically generated infectious population $\hat{I}$.

Prior probability distributions. The SIS model has three parameters, $\boldsymbol{\vartheta} = \{I_0, \mathsf{R}_0, C\}$, the initial infectious population I_0, the basic reproduction number R_0, and the infectious period C. For each parameter we would like to infer, we have to select a prior probability distribution. For illustrative purposes, we select three different weakly informative priors. For the initial infectious population I_0, we choose a log-normal distribution with a probability density function $P(I_0)$,

$$I_0 \sim \text{LogNormal}(\mu, \sigma) \quad \text{with} \quad P(I_0) = \frac{1}{\sqrt{2\pi}\sigma I_0} \exp\left(-\frac{(\ln(I_0) - \mu)^2}{2\sigma^2}\right), \tag{11.17}$$

with mean, $\exp(\mu + \sigma^2/2)$, median, $\exp(\mu)$, and mode, $\exp(\mu - \sigma^2)$. For the basic reproduction number R_0, we choose a uniform distribution with a probability density function $P(\mathsf{R}_0)$,

$$\mathsf{R}_0 \sim \text{Uniform}(\mathsf{R}_{0,\text{min}}, \mathsf{R}_{0,\text{max}}) \quad \text{with} \quad P(\mathsf{R}_0) = \frac{1}{\mathsf{R}_{0,\text{max}} - \mathsf{R}_{0,\text{min}}}, \tag{11.18}$$

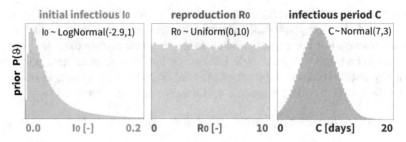

Fig. 11.5 Prior probability distribution. Histograms of the log-normal distribution of the initial infectious population I_0, the uniform distribution of the basic reproduction number R_0, and the normal distribution of the infections period C.

with lower and upper bounds $R_{0,min}$ and $R_{0,min}$. For the infectious period C, we choose a normal distribution with a probability density function $P(C)$,

$$C \sim \text{Normal}(\mu, \sigma) \quad \text{with} \quad P(C) = \frac{1}{\sqrt{2\pi}\sigma} \exp\left(-\frac{(C - \mu)^2}{2\sigma^2}\right), \tag{11.19}$$

with mean μ and standard deviation σ. Table 11.2 summarizes our prior probability distributions and their parameters. Figure 11.5 illustrates the three probability distributions as discrete histograms. We will now infer our three parameters, I, R_0, and C. First, we infer each parameter individually, while keeping the remaining two parameters fixed at $I_0 = 0.001$, $R_0 = 3$, and $C = 10$ days. Then, we infer all three parameters simultaneously.

Table 11.2 Prior probability distributions. Priors for the initial infectious population I_0, the basic reproduction number R_0, the infectious period C, and the likelihood width σ.

parameter	distribution	parameters
I_0	LogNormal (μ, σ)	$\mu = -2.9$, $\sigma = 1$
R_0	Uniform $(R_{0,min}, R_{0,max})$	$R_{0,min} = 0$, $R_{0,max} = 10$
C	Normal (μ, σ)	$\mu = 7$, $\sigma = 3$
σ	HalfCauchy (β)	$\beta = 1$

Likelihood function and maximum likelihood estimate. The likelihood function evaluates the goodness of fit between the model output $I(\vartheta, t)$, the simulated infectious population for given parameters ϑ, and the observed data $\hat{I}(t)$, the synthetically generated infectious population. For the individual likelihood functions $\mathcal{L}(\vartheta, t_i)$, at each time point t_i, we select a normal distribution, where the mean is the model result $I(\vartheta, t)$ for the given parameter set ϑ and the standard deviation σ accounts for the observation error in the data $\hat{I}(t)$. The product of $i = 0.., n$ individual likelihood functions $\mathcal{L}(\vartheta, t_i)$, evaluated at the discrete time points t_i, defines the overall likelihood $P(\hat{I}(t)|\vartheta)$,

$$P(\hat{I}(t)|\boldsymbol{\vartheta}) = \prod_{i=0}^{n} \mathcal{L}(\boldsymbol{\vartheta}, t_i) \quad \text{with} \quad \mathcal{L}(\boldsymbol{\vartheta}, t_i) = \frac{1}{\sqrt{2\pi}\sigma} \exp\left(-\frac{(\hat{I}(t_i) - I(\boldsymbol{\vartheta}, t_i))^2}{2\sigma^2}\right).$$

(11.20)

The product symbol, $\prod_{i=0}^{n}$, denotes the multiplication of the $i = 0, ..., n$ individual terms $\mathcal{L}(\boldsymbol{\vartheta}, t_i)$, in our case, at one time point per day throughout a time window of 100 days. The values of $\boldsymbol{\vartheta}$ that maximize the overall likelihood function $P(\hat{I}(t)|\boldsymbol{\vartheta})$,

$$\boldsymbol{\vartheta}_{\text{mle}} \subset \arg\max_{\boldsymbol{\vartheta}} P(\hat{I}(t)|\boldsymbol{\vartheta}),$$

(11.21)

are called the *maximum likelihood estimate* $\boldsymbol{\vartheta}_{\text{mle}}$ of the true parameter values $\boldsymbol{\vartheta}$. Intuitively, the maximum likelihood estimate defines the set of parameters values that make the observed data, in our case the synthetically generated infectious population $\hat{I}$, most probable.

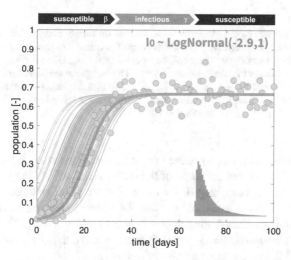

Fig. 11.6 Classical SIS model. Analytical solution for log-normally distributed initial infectious population I_0. Sampling I_0 reveals that the outbreak curve shifts leftward for increasing I_0, while the slope and final infectious population I_∞ remain unchanged. The closer the gray curves of the model output $I(I_0, t)$ to the orange dots of the observed data $\hat{I}(t)$, the larger the likelihood $P(\hat{I}(t)|I_0)$. The orange curve is closest to all orange dots. It defines the maximum likelihood estimate $I_{0,\text{mle}} = 0.11$. Infectious period $C = 10$ days, basic reproduction number $R_0 = 3.0$, and initial infectious population $I_0 \sim \text{LogNormal}(\mu, \sigma)$ with mean $\mu = -2.91$ and standard deviation $\sigma = 1$.

Figures 11.6, 11.7, and 11.8 illustrate the analytical solutions of the SIS model for a log-normally distributed initial infectious population I_0, a uniformly distributed basic reproduction number R_0, and a normally distributed infectious period C. Sampling for the three parameters, $\boldsymbol{\vartheta} = \{I_0, R_0, C\}$, of the SIS model reveals similar trends as the sensitivity analysis of the SIS model in Figures 2.3, 2.4, and 2.5 in Chapter 2: Increasing the parameter I_0 induces a leftward shift of the outbreak curve, but the slope and the final infectious population I_∞ remain unchanged; increasing the pa-

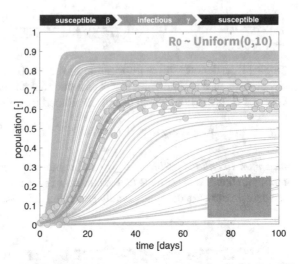

Fig. 11.7 Classical SIS model. Analytical solution for uniformly distributed basic reproduction number R_0. Sampling R_0 reveals that the slope and final infectious population I_∞ increase with increasing R_0. The closer the gray curves of the model output $I(R_0, t)$ to the orange dots of the observed data $\hat{I}(t)$, the larger the likelihood $P(\hat{I}(t)|R_0)$. The orange curve is closest to all orange dots. It defines the maximum likelihood estimate $R_{0,mle} = 3.0$. Initial infectious population $I_0 = 0.01$, infectious period $C = 10$ days, and basic reproduction number $R_0 \sim$ Uniform(R_{min}, R_{max}) with lower and upper bounds $R_{min} = 0$ and $R_{min} = 10$.

rameter R_0 increases the slope and the final infectious population I_∞; and increasing the parameter C increases the slope of the outbreak curve, but the final infectious population I_∞ remains unchanged. On average, more samples lie to the left of the solid orange curve of the analytical solution for all three parameter samples. This is intuitive since the medians of the samples $I_0 = 0.03$, $R_0 = 5.0$, and $C = 7.0$ days are larger than the true initial infectious population $I_0 = 0.01$, the true basic reproduction number $R_0 = 3.0$, and the true infectious period $C = 10$ days.

For each set of given model parameters $\boldsymbol{\vartheta} = \{I_0, R_0, C\}$, the likelihood function in equation (11.20) evaluates the proximity between the gray curves of the model output $I(\boldsymbol{\vartheta}, t)$ and the orange dots of the observed data $\hat{I}(t)$. The closer a gray curve to the orange dots, the larger its likelihood $P(\hat{I}(t)|\boldsymbol{\vartheta})$. The orange curve in each graph is the curve that is closest to all orange dots. It maximizes the likelihood function over the analyzed parameter space. The parameters associated with the orange curves are the maximum likelihood estimate $\boldsymbol{\vartheta}_{mle}$. For our example, the maximum likelihood estimates are $I_{0,mle} = 0.11$, $R_{0,mle} = 3.0$, and $C_{mle} = 10.0$ days.

Marginal likelihood. The marginal likelihood function $P(\hat{I}(t))$ is also often referred to as evidence in Bayesian statistics. It represents a likelihood function in which the parameter variables are marginalized by integrating over the entire parameter domain,

$$P(\hat{I}(t)) = \int_{\boldsymbol{\vartheta}} P(\hat{I}(t), \boldsymbol{\vartheta}) \, d\boldsymbol{\vartheta} . \tag{11.22}$$

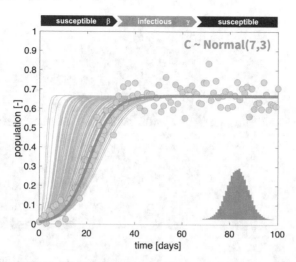

Fig. 11.8 Classical SIS model. Analytical solution for normally distributed infectious period C. Sampling C reveals that the slope of the outbreak curve increases with increasing C, while the final infectious population I_∞ remains unchanged. The closer the gray curves of the model output $I(C_0, t)$ to the orange dots of the observed data $\hat{I}(t)$, the larger the likelihood $P(\hat{I}(t)|C_0)$. The orange curve is closest to all orange dots. It defines the maximum likelihood estimate $C_{0,\text{mle}} = 10.0$ days. Initial infectious population $I_0 = 0.01$, basic reproduction number $R_0 = 3.0$, and infectious period $C \sim \text{Normal}(\mu, \sigma)$ with mean $\mu = 7$ and standard deviation $\sigma = 1$.

We have already seen that marginal likelihoods are generally difficult to compute. If needed, we could integrate equation (11.22) using numerical integration schemes such as Gaussian integration or Monte Carlo methods. Fortunately, for most practical purposes, we do not need to know the precise value of the marginal likelihood $P(\hat{I}(t))$. Instead, most methods simply compare two posterior probabilities, for example by calculating their ratio, such that the marginal likelihood cancels.

Posterior probability distributions and maximum a posteriori estimate. The posterior probability distribution $P(\boldsymbol{\vartheta}|\hat{I}(t))$ is the conditional probability of the parameters $\boldsymbol{\vartheta}$ for the given data, the synthetically generated infectious population $\hat{I}(t)$. We calculate the posterior probability from our prior knowledge, encoded through the prior probability distribution $P(\boldsymbol{\vartheta})$, and the likelihood $P(\hat{I}(t)|\boldsymbol{\vartheta})$, the conditional probability of the data for given the parameters, using Bayes' theorem,

$$P(\boldsymbol{\vartheta}|\hat{I}(t)) = \frac{P(\hat{I}(t)|\boldsymbol{\vartheta}) \, P(\boldsymbol{\vartheta})}{P(\hat{I}(t))} . \tag{11.23}$$

The values of $\boldsymbol{\vartheta}$ that maximize the posterior probability distribution $P(\boldsymbol{\vartheta}|\hat{I}(t))$,

$$\boldsymbol{\vartheta}_{\text{map}} \subset \arg\max_{\boldsymbol{\vartheta}} P(\boldsymbol{\vartheta}|\hat{I}(t)), \tag{11.24}$$

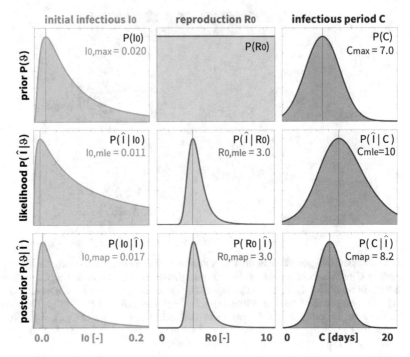

Fig. 11.9 Bayesian analysis of the SIS model. Prior probability density functions, $P(\vartheta)$, likelihood functions $P(\hat{I}|\vartheta)$, and scaled posterior probability density functions, $P(\vartheta|\hat{I})$, as the products of the two, for the initial infectious population I_0, the basic reproduction number R_0, and the infectious period C. Vertical lines highlight the maximum priors, maximum likelihood estimates, and maximum a posteriori estimates. For the uniform prior, middle column, the likelihood and posterior probability density functions display the same shape and maxima. For the log-normal and normal priors, left and right columns, the maximum a posteriori estimates are located between the maximum priors and maximum likelihood estimates.

are called the *maximum a posteriori estimate* ϑ_{map} of the true parameter values ϑ. Importantly, the marginal likelihood $P(\hat{I}(t))$ is simply a scaling factor in equation (11.23). It scales the ordinate or y−axis of the posterior plots, but does not affect the abscissa or x−axis. The maximum a posteriori estimate, which is usually the objective of a Bayesian analysis, is not affected by the value of the marginal likelihood $P(\hat{I}(t))$.

Figure 11.9 illustrates the application of Bayes' theorem (11.23) to estimate the posterior probability density functions $P(\vartheta|\hat{I})$, as the product of the prior probability density functions $P(\vartheta)$, and the likelihood functions $P(\hat{I}|\vartheta)$, from top to bottom. The individual columns represent the three parameters $\vartheta = \{I_0, R_0, C\}$, the initial infectious population I_0, the basic reproduction number R_0, and the infectious period C, from left to right. Each graph is based on a parameter sweep with $I_0 = 0.0, ..., 0.2$, $R_0 = 0, ..., 10$, and $C = 0, ..., 20$ days. The top row highlights the log-normal prior $P(I_0)$ of the initial infectious population $I_0 \sim \text{LogNormal}(-2.9, 1)$, the uniform prior $P(R_0)$ of the basic reproduction number $R_0 \sim \text{Uniform}(0, 10)$, and the normal prior $P(C)$ of the infectious period $C \sim \text{Normal}(7, 3)$ as the continuous analogue of the discrete histograms in Figure 11.5 using the parameterizations of Table 11.2. The middle row summarizes the likelihood functions $P(\hat{I}|I_0)$, $P(\hat{I}|R_0)$, and $P(\hat{I}|C)$ for all three parameters. The bottom row illustrates the product of the posterior probability densities $P(I_0|\hat{I})$, $P(R_0|\hat{I})$, and $P(C|\hat{I})$ as the product of prior and likelihood. The vertical lines and parameter values in each graph highlight the maximum of each curve, the maximum prior, the maximum likelihood estimate, and the maximum a posteriori estimate.

For the uniform prior probability density function in the middle column, the likelihood and posterior probability density functions display the same shape and the maximum likelihood estimate and maximum a posteriori estimate are identical, $R_{0,\text{mle}} = R_{0,\text{map}} = 3.0$. As such, for a uniform prior distribution, a Bayesian analysis is similar to a frequentist analysis and reduces to a maximum likelihood evaluation, which implicitly assumes a uniform prior. For the log-normal and normal prior probability density functions in the left and right columns, the maximum a posteriori estimates, $I_{0,\text{map}} = 0.017$ and $C_{\text{map}} = 8.2$, are located between the maxima of the prior probability density, $I_{0,\text{max}} = 0.020$ and $C_{\text{max}} = 7.0$, and the maximum likelihood estimates, $I_{0,\text{mle}} = 0.011$ and $C_{\text{mle}} = 10.0$. Taken together, this example emphasizes the importance of prior selection. Here we have intentionally selected priors with maxima at $I_{0,\text{max}} = 0.020$ and $C_{\text{max}} = 7.0$, instead of using the parameters we had adopted to generate the synthetic data, $I_0 = 0.010$ and $C = 10$ days. Selecting inappropriate priors can significantly bias the maximum a posteriori estimate, and with it the outcome of a Bayesian analysis.

Markov Chain Monte Carlo and credible interval. Markov Chain Monte Carlo methods are a class of algorithms that sample from probability distributions. They are based on creating Markov chains, series of samples from a continuous random variable with a probability density distribution that is proportional to a known function. These samples can be used to evaluate an integral over that variable, as its expected value or variance. Here we use Markov Chain Monte Carlo methods to draw samples from our prior distributions and smartly decide how to move towards a new, better sample. Generally, Markov Chain Monte Carlo methods continuously compare two consecutive samples and make a smart choice for drawing the next sample. They use an *acceptance ratio* a that scores the current and new samples relative to each other without the need for calculating the marginal likelihood. Specifically, assume we have a current parameter sample $\boldsymbol{\vartheta}_{\text{now}}$ and draw a new sample $\boldsymbol{\vartheta}_{\text{new}}$. By calculating the ratio of their posterior probability distributions,

$$\text{a} = \frac{P(\boldsymbol{\vartheta}_{\text{new}}|\hat{I})}{P(\boldsymbol{\vartheta}_{\text{now}}|\hat{I})} = \frac{P(\hat{I}|\boldsymbol{\vartheta}_{\text{new}})\,P(\boldsymbol{\vartheta}_{\text{new}})}{P(\hat{I}|\boldsymbol{\vartheta}_{\text{now}})\,P(\boldsymbol{\vartheta}_{\text{now}})}, \tag{11.25}$$

the marginal likelihoods $P(\hat{I})$ cancel and the acceptance ratio only depends on the priors and likelihoods of the current and new parameters $\boldsymbol{\vartheta}_{\text{now}}$ and $\boldsymbol{\vartheta}_{\text{new}}$. An acceptance ratio larger than one, a > 1, implies that the model with the new parameters $\boldsymbol{\vartheta}_{\text{new}}$ results in a curve that is closer to the data $\hat{I}$. The new parameters $\boldsymbol{\vartheta}_{\text{new}}$ explain the data better than the current parameters $\boldsymbol{\vartheta}_{\text{now}}$ and we update the current parameter set, $\boldsymbol{\vartheta}_{\text{now}} \leftarrow \boldsymbol{\vartheta}_{\text{new}}$. An acceptance ratio smaller than one, a < 1, implies that the

Table 11.3 Bayesian analysis of the SIS model. Sampling and convergence towards maximum a posteriori estimate. Sample number, posterior probability, and updated parameter values of the initial infectious population I_0, basic reproduction number R_0, and infectious period C for a total of 100,000 samples. True initial infectious population $I_0 = 0.001$, basic reproduction number $R_0 = 3.0$, and infectious period $C = 10$ days.

sample	posterior	I0	sample	posterior	R0	sample	posterior	C
0	2.861e-40	0.041225	0	4.964e-44	6.7089	0	4.696e-42	7.7629
1	4.456e-40	0.011554	2	1.882e-42	2.4317	4	4.754e-42	8.3493
3	4.912e-40	0.015714	10	2.423e-42	3.7578	46	4.763e-42	8.0346
29	4.929e-40	0.017417	12	4.265e-42	3.0790	87	4.768e-42	8.0837
165	4.930e-40	0.017378	91	4.280e-42	3.0451	204	4.770e-42	8.1588
283	4.932e-40	0.017010	697	4.281e-42	3.0157	788	4.770e-42	8.1602
694	4.932e-40	0.017005	774	4.282e-42	3.0281	991	4.770e-42	8.1621
851	4.932e-40	0.016951	20463	4.282e-42	3.0276	3691	4.770e-42	8.1623
1479	4.932e-40	0.016936	21712	4.282e-42	3.0278	39922	4.770e-42	8.1626
39368	4.932e-40	0.016944	51361	4.282e-42	3.0278	55566	4.770e-42	8.1626
map	4.932e-40	0.016944	map	4.282e-42	3.0278	map	4.770e-42	8.1626

model with the current parameters $\boldsymbol{\vartheta}_{\text{now}}$ explains the data better and we simply continue to draw the next new sample. In addition to the information which sample to keep and which to discard, the acceptance ratio also provides information about how to sample smartly by systematically visiting regions of high posterior probability more often than those of low posterior probability.

Table 11.3 illustrates the convergence towards the maximum posterior estimate, estimated individually for the initial infectious population I_0, basic reproduction number R_0, and infectious period C. For each block, the first column shows the number of the sample for which the acceptance ratio was larger than one, $a > 1$, the second column shows its posterior probability, $P(\boldsymbol{\vartheta}_{\text{new}} | \hat{I})$, and the third column shows its sampled parameter value. For all three parameters, the posterior probabilities and parameter values converge toward a reasonable degree of accuracy within 100,000 sampling steps. In agreement with Figure 11.9, the final maximum a posteriori estimates of the bottom row are $I_{0,\text{map}} = 0.016944$, $R_0 = 3.0278$, and $C_{\text{map}} = 8.1626$. The convergence table reveals several characteristics of the ad hoc sampling algorithm: The sampled parameter estimates I_0, R_0, and C improve rapidly within the first 200 samples, but then improve only marginally within the remaining 99800 samples; by design of the priors, convergence is neither exclusively from above nor exclusively from below, instead, the parameters alternate towards their maximum a posteriori estimates; since the overall likelihood is the product 100 individual likelihoods, the value of the posterior probability is notably small. This is why, in practice, most sampling algorithms replace the likelihood by the log-likelihood, the natural logarithm of the likelihood function.

In the remainder of this book, we will no longer implement the sampler manually ourselves. Instead, we adopt a probabilistic programming environment and use PyMC3, a Python library specifically developed to perform Bayesian inference with an easy-to-understand syntax to create prior probabilities and likelihood functions.

We illustrate its application by analyzing the same SIS model as before, with the priors and parameters from Figure 11.2 and Table 11.2. However, now, we infer all three parameters simultaneously. For the likelihood function, we select a half-Cauchy distribution. We can see in Figure 11.2 that the half-Cauchy distribution has a longer tail than the classical normal distribution which makes it more robust to outliers. We solve this distribution numerically using the NO-U-Turn sampler implementation of the python package PyMC3. For the Markov Chain Monte Carlo method, we use four chains. Each chain starts from a set of arbitrarily selected points at sufficient distance from one another and moves around the parameter space by smartly sampling parameter sets which are expected to have a reasonably high probability. The first 1000 samples are used to tune the sampler and are later discarded; the subsequent 2000 samples are used to estimate the set of parameters ϑ. From the converged posterior distributions, we sample multiple combinations of parameters from which we can post-process the time evolution of the infectious population $I(t)$. These posterior samples allow us to quantify the uncertainty on each parameter.

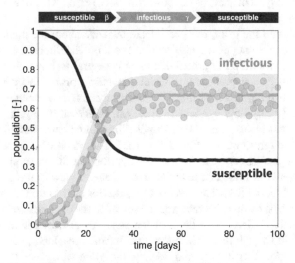

Fig. 11.10 Classical SIS model. Analytical solution $I(t)$ **for inferred parameters** ϑ = $\{I_0, R_0, C\}$ **and credible interval for synthetic data** $\hat{I}(t)$. The synthetic data $\hat{I}(t)$ are created by adding a random noise with $\sigma = 0.05$ to the true analytical solution. The susceptible and infectious populations $S(t)$ and $I(t)$ and the 95% credible interval are based on an inferred initial infectious population $I_0 = 0.012 \pm 0.004$, basic reproduction number $R_0 = 3.009 \pm 0.060$, and infectious period $C = 10.162 \pm 0.872$ days.

An important by-product of a Baysian inference are *credible intervals*. Credible intervals summarize the uncertainty related to the inferred parameters and the data. As such, they appear conceptually similar to the frequentist confidence intervals. Yet, from a statistics point, their definition and meaning is quite different: The Bayesian inference returns a posterior distribution $P(\vartheta_{new}|\hat{I})$. The credible interval is simply the range that contains a particular percentage of probable parameters

ϑ. For example, the 95% credible interval is the central portion of the posterior probability distribution that contains 95% of the parameters. In other words, for given observed data $\hat{I}(t)$, the parameters ϑ have a 95% probability of falling within the 95% credible interval. A strength of a Bayesian analysis is that credible intervals are a natural result of the analysis and fairly straightforward to compute.

Figure 11.10 illustrates the result of the Bayesian inference similar to Figures 11.6, 11.7, and 11.8. However, now, we infer all three parameters, $\vartheta = \{I_0, R_0, C\}$, simultaneously. The Markov Chain Monte Carlo method converges towards an initial infectious population of $I_0 = 0.012 \pm 0.004$, a basic reproduction number of $R_0 = 3.009 \pm 0.060$, and an infectious period of $C = 10.162 \pm 0.872$ days. From these inferred parameters, we calculate the susceptible and infectious populations, $S(t)$ and $I(t)$. The orange region around the infectious population highlights the 95% credible interval. We observe that almost all synthetically generated data points $\hat{I}(t)$ fall within this interval and conclude that our inferred model successfully captures the data.

Taken together, Bayesian analysis is a powerful tool to explore the outbreak dynamics of a disease by integrating the likelihood, the way we introduce reported case data into our model, and the prior, our previous knowledge about the model parameters [6, 5, 11, 17, 19]. We encode the information about the likelihood and prior in terms of probabilities, and thinking probabilistically is central to Bayesian analysis [27]. Fortunately, with the unprecedented developments in probabilistic programming and machine learning, much of the Bayesian analysis is now handled by libraries such as PyMC3 that allow us to learn Bayesian inference by doing, visualize the impact of different priors, manipulate the likelihood, and gain valuable insight into disease data through the lens of modeling [23]. The result of a Bayesian analysis is the posterior probability distribution, the distribution of our model parameters, which encodes all our information about an outbreak in view of the given case data and the selected model [26]. The peak of the posterior distribution defines the set of parameters that best explain the data, and the spread of the distribution is proportional to the uncertainty [16]: We can think of the posterior distribution as an update of the prior in light of new data. The more data become available, the more we can refine the posterior distribution, and, ideally, the more we reduce the uncertainty. This makes Bayesian analysis ideally suited when analyzing data in the midst of a global pandemic, which is constantly evolving and continuously generating new data. Bayesian analysis allows us to seamlessly integrate new data–in real time–with live updates, as new case data become available.

Problems

11.1 Probabilities. Which of the following expressions correspond to 'the probability of contracting COVID-19 in strict lockdown'?

○ $P(\text{COVID-19})$ ○ $P(\text{lockdown})$

○ $P(\text{COVID-19}|\text{lockdown})$ ○ $P(\text{lockdown}|\text{COVID-19})$

○ $P(\text{COVID-19,lockdown})P(\text{lockdown})$ ○ $P(\text{lockdown,COVID-19})P(\text{COVID-19})$

○ $\dfrac{P(\text{COVID-19}|\text{lockdown})P(\text{lockdown})}{P(\text{COVID-19})}$ ○ $\dfrac{P(\text{lockdown}|\text{COVID-19})P(\text{COVID-19})}{P(\text{lockdown})}$

11.2 Posterior probability. Three in five people with the common cold experiences a loss of taste, whereas 83 in 10,000 people of the general population experience a loss of taste. On average, one in 100 people has a common cold. Calculate the posterior probability that someone who experiences a loss of taste has a common cold.

11.3 Posterior probability. Four in five people with COVID-19 experiences a loss of taste, whereas 83 in 10,000 people of the general population experience a loss of taste. On April 5, 2021, the number of active COVID-19 cases was 22,662,573 in a total population of 7,794,804,703. Calculate the posterior probability that someone who experiences a loss of taste has COVID-19.

11.4 Bayes factor and prior and posterior odds. Calculate Bayes factor B = $P(\text{loss of taste}|\text{COVID-19})/P(\text{loss of taste}|\text{common cold})$, and the prior and posterior odds, $R_{\text{prior}} = P(\text{COVID-19})/P(\text{common cold})$ and $R_{\text{post}} = \text{B} \cdot R_{\text{prior}}$. If you were a doctor and a patient would present with a loss of taste, what would be your diagnosis? How confident are you in your diagnosis?

11.5 Posterior probability. Four in five people with COVID-19 experiences a loss of taste, whereas 83 in 10,000 people of the general population experience a loss of taste. In January, 2021, the number of active COVID-19 cases was 20% larger than in April, 27,195,876 in a total population of 7,794,204,001. Calculate the posterior probability that someone who experienced a loss of taste in January had COVID-19. Interpret your results.

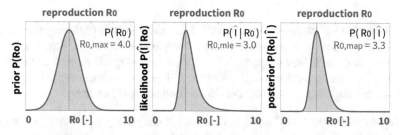

Fig. 11.11 Bayesian analysis of the SIS model. Prior probability density functions, $P(R_0)$, likelihood functions $P(\hat{I}|R_0)$, and scaled posterior probability density functions, $P(R_0|\hat{I})$, as the products of the two, for the basic reproduction number R_0. Vertical lines highlight the maximum prior $R_{0,\text{max}} = 4.0$, maximum likelihood estimate $R_{0,\text{mle}} = 3.0$, and maximum a posteriori estimate $R_{0,\text{map}} = 3.3$.

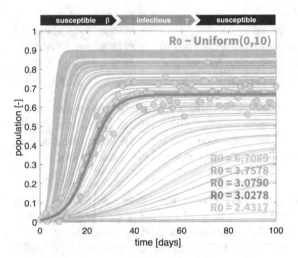

Fig. 11.12 Classical SIS model. Analytical solution for uniformly distributed basic repro-duction number R_0. The closer the sampled gray curves of the model output $I(R_0, t)$ to the orange dots of the observed data $\hat{I}(t)$, the larger the likelihood $P(\hat{I}(t)|R_0)$ and the posterior probability $P(R_0|\hat{I}(t))$. The yellow-to-orange curves are associated with the basic reproduction numbers $R_0 = [6.7089, 2.4317, 3.7578, 3.0790]$ according to Table 11.3. The darkest orange curve is closest to all orange dots and defines the maximum likelihood and a posteriori estimate $R_{0,mle} = R_{0,map} = 3.0278$. Initial infectious population $I_0 = 0.01$, infectious period $C = 10$ days, and basic reproduction number $R_0 \sim \text{Uniform}(R_{min}, R_{max})$ with lower and upper bounds $R_{min} = 0$ and $R_{min} = 10$.

11.6 Posterior probability. Fever is one of the most common symptoms of COVID-19 present in approximately 90% of cases. At the University of Notre Dame, during the peak of the pandemic, 391 students of a total of 12,607 enrolled students tested positive for COVID-19 during the second week of August in 2020. During this time, 3% of students reported a fever. What is the probability that a Notre Dame student who experienced a fever during the second week of August had COVID-19?

11.7 Bayesian vs. frequentist approach. Discuss the results of the Bayesian analysis in the previous examples in view of a comparison between a Bayesian and a frequentist approach. Which approach provides a more dynamic interpretation of the available information and can adjust automatically as data become available? Which terms in the Bayesian inference reflect these continuous adjustments?

11.8 Prior, likelihood, and posterior. Illustrate the prior probability distribution, likelihood function, and posterior probability distribution to infer the basic reproduction number R_0. Follow the same steps as in Figure 11.9. Use the parameters from Table 11.2, but now, instead of using a uniform prior for R_0, use a normal distribution $R_0 \sim \text{Normal}(4, 1)$. Perform a parameter sweep with $R_0 = 0, ..., 10$. Plot the prior, likelihood, and posterior. Discuss how your graphs differ from the middle column in Figure 11.9.

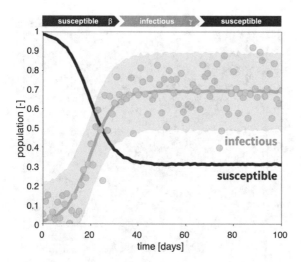

Fig. 11.13 Classical SIS model. Analytical solution $I(t)$ **for inferred parameters** ϑ =
$\{I_0, R_0, C\}$ **and credible interval for synthetic data** $\hat{I}(t)$. The synthetic data $\hat{I}(t)$ are created
by adding a random noise with $\sigma = 0.10$ to the true analytical solution. The susceptible and
infectious populations $S(t)$ and $I(t)$ and the 95% credible interval are based on an inferred initial
infectious population $I_0 = 0.015 \pm 0.006$, basic reproduction number $R_0 = 3.240 \pm 0.126$, and
infectious period $C = 11.359 \pm 1.429$ days.

11.9 Maximum likelihood and maximum a posteriori estimates. Apply Bayes'
theorem to infer the basic reproduction number R_0. Follow the same steps as in Figure
11.9. Use the parameters from Table 11.2, but now, instead of using a uniform prior
for R_0, use a normal distribution $R_0 \sim$ Normal(4, 1). Perform a parameter sweep
with $R_0 = 0, ..., 10$. Identify the maximum prior, maximum likelihood estimate, and
maximum a posteriori estimate and compare your results against Figure 11.11.

11.10 Effects of prior selection. Apply Bayes' theorem to infer the basic reproduction number R_0. Follow the same steps as in Figure 11.9. Use the parameters from
Table 11.2, but now, instead of using a uniform prior for R_0 use a normal distribution
$R_0 \sim$ Normal($R_{0,max}, 1$). Perform a parameter sweep with $R_0 = 0, ..., 10$. Plot the
prior, likelihood, and posterior for $R_{0,max} = 4.0$ and compare your result to Figure
11.11. Vary $R_{0,max}$. Discuss the effects of reducing and increasing the mean of the
prior, $R_{0,max}$, on the maximum a posteriori estimate $R_{0,map}$ in view of the importance
of appropriate prior selection.

11.11 Parameter sampling and acceptance ratio. Assume you perform a Bayesian
analysis to infer the basic reproduction number R_0 for the SIS model in Figure
11.12. Use a uniform prior distribution, $R_0 \sim$ Uniform(0, 10), with a constant prior
probability, $P(R_0) = 0.1$. According to Table 11.3, the initial sample has a basic
reproduction number $R_{0,now} = 6.7089$ and a likelihood $P(\hat{I}|R_{0,now}) = 4.964e{-}43$.
The first sample has a basic reproduction number $R_{0,new} = 1.2431$ and a likelihood

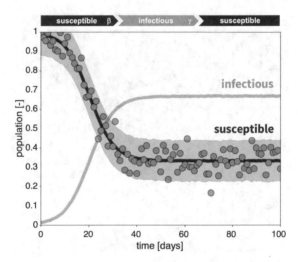

Fig. 11.14 Classical SIS model. Analytical solution $S(t)$ **for inferred parameters** ϑ = $\{S_0, R_0, C\}$ **and credible interval for synthetic data** $\hat{S}(t)$. The synthetic data $\hat{S}(t)$ are created by adding a random noise with $\sigma = 0.05$ to the true analytical solution. The susceptible and infectious populations $S(t)$ and $I(t)$ and the 95% credible interval are based on an inferred initial infectious population $S_0 = 0.988 \pm 0.004$, basic reproduction number $R_0 = 3.008 \pm 0.062$, and infectious period $C = 10.163 \pm 0.893$ days.

$P(\hat{I}|R_{0,\text{new}}) = 1.8010\text{e}{-}47$. Calculate the acceptance ratio a to decide whether the first sample explains the data better than the initial sample. Which sample do you discard and which do you keep?

11.12 Parameter sampling and acceptance ratio. Assume you perform a Bayesian analysis to infer the basic reproduction number R_0 for the SIS model in Figure 11.12. Use a uniform prior distribution, $R_0 \sim \text{Uniform}(0, 10)$, with a constant prior probability, $P(R_0) = 0.1$. According to Table 11.3, the initial sample has a basic reproduction number $R_{0,\text{now}} = 6.7089$ and a likelihood $P(\hat{I}|R_{0,\text{now}}) = 4.964\text{e}{-}43$. The second sample has a basic reproduction number $R_{0,\text{new}} = 2.4317$ and a likelihood $P(\hat{I}|R_{0,\text{new}}) = 1.882\text{e}{-}41$. Calculate the acceptance ratio a to decide whether the second sample explains the data better than the initial sample. Which sample do you discard and which do you keep?

11.13 Bayes' theorem, inferred parameters, and credible interval. Figure 11.13 shows the results of a Bayesian inference with the same true parameters, $I_0 = 0.01$, $R_0 = 3.0$, and $C = 10$ days as Figure 11.10. However, now, the synthetically generated orange data points $\hat{I}(t)$ are created by perturbing the true data by a random noise with $\sigma = 0.10$ and display a larger scatter. Compare Figures 11.10 and 11.13 and discuss the effect of a larger noise in the data on the inferred parameters $\vartheta = \{I_0, R_0, C\}$ and on the credible interval.

11.14 Data and potential sources of error. Assume that, from surveillance testing, we observe the daily susceptible population $S(t)$ rather than the infectious population

$I(t)$. To mimic the results of the surveillance tests, we create synthetic data $\hat{S}(t)$ by adding a random noise with $\sigma = 0.05$ to the true analytical solution based on the true parameters $S_0 = 0.99$, $R_0 = 3.0$, and $C = 10$ days. Figure 11.14 shows the observed susceptible population $\hat{S}(t)$ as red dots. Compare Figures 11.14 and 11.10 and discuss differences and similarities. Discuss potential sources of error that could have caused a scatter in the observed data $I(t)$ or $S(t)$.

11.15 Bayes' theorem, inferred parameters, and credible interval. Assume that, from surveillance testing, we observe the daily susceptible population $S(t)$. Figure 11.14 shows the results of a Bayesian inference using the priors from in Table 11.2. The synthetic data $\hat{S}(t)$ are created by adding a random noise with $\sigma = 0.05$ to the true analytical solution based on the parameters, $S_0 = 0.99$, $R_0 = 3.0$, and $C = 10$ days. Write the equations for the prior probability of the susceptible population $P(S_0)$, the likelihood $P(\hat{S}(t)|\vartheta)$, and the posterior probability $P(\vartheta|\hat{S}(t))$.

References

1. Alber M, Buganza Tepole A, Cannon W, De S, Dura-Bernal S, Garikipati K, Karniadakis G, Lytton WW, Perdikaris P, Petzold L, Kuhl E (2019) Integrating machine learning and multiscale modeling: Perspectives, challenges, and opportunities in the biological, biomedical, and behavioral sciences. npj Digital Medicine 2:115.
2. Anderson RM, May RM (1981) The population dynamics of microparasites and their invertebrate hosts. Philosophical Transactions of the Royal Society London B. 291:451–524.
3. Bayes T, Price R. (1763). An Essay towards Solving a Problem in the Doctrine of Chances. By the Late Rev. Mr. Bayes, F.R.S. communicated by Mr. Price, in a Letter to John Canton, A.M.F.R.S. Philosophical Transactions of the Royal Society of London. 53:370–418.
4. Bhouri MA, Sahli Costabal F, Wang H, Linka K, Peirlinck M, Kuhl E, Perdikaris P (2021) COVID-19 dynamics across the US: A deep learning study of human mobility and social behavior. Computer Methods in Applied Mechanics and Engineering 382: 113891.
5. Brauer F, Castillo-Chavez C, Feng Z (2019) Mathematical Models in Epidemiology. Springer-Verlag New York.
6. Britton T, O'Neil PD (2002) Bayesian inference for stochastic epidemics in populations with random social structure. Scandinavian Journal of Statistics 29:375-390.
7. Dehning J, Zierenberg J, Spitzner FP, Wibral M, Pinheiro Neto J, Wilczek M, Priesemann V (2020) Inferring COVID-19 spreading rates and potential change points for case number forecasts. Science 369:eabb9789.
8. European Centre for Disease Prevention and Control. Situation update worldwide. https://www.ecdc.europa.eu/en/geographical-distribution-2019-ncov-cases accessed: June 1, 2021.
9. Gelman A, Hill J (2006) Data Analysis using Regression and Multilevel/Hierarchical Models. Cambridge University Press.
10. Gelman A, Carlin JB, Stern HS, Dunson DB, Vektari A, Rubin DB (2013) Bayesian Data Analysis. Chapman and Hall/CRC, 3rd edition.
11. Hastie T, Tibshirani R, Friedman J (2017) The Elements of Statistical Learning. Data Mining, Inference, and Prediction. Springer Science+Business Media, 2nd edition.
12. Holmdahl I, Buckee C. Wrong but useful–What Covid-19 epidemiolgic models can and cannot tell us. New England Journal of Medicine 383:303-305.
13. Ioannidis JPA, Cripps S, Tanner MA (2021) Forecasting for COVID-19 has failed. International Journal of Forecasting, in press.

14. Jaynes ET (2003) Probability Theory: The Logic of Science. Cambridge University Press.
15. Jha PK, Cao L, Oden JT (2020) Bayesian-based predictions of COVID-19 evolution in Texas using multispecies mixture-theoretic continuum models. Computational Mechanics 66:1055–1068.
16. Johns Hopkins University (2021) Coronavirus COVID-19 Global Cases by the Center for Systems Science and Engineering. https://coronavirus.jhu.edu/map.html, https://github.com/CSSEGISandData/covid-19 assessed: June 1, 2021.
17. Kennedy MC, O'Hagan A (2001) Bayesian calibration of computer models. Journal of the Royal Statistical Society B 63:425-464.
18. Linka K, Peirlinck M, Kuhl E (2020) The reproduction number of COVID-19 and its correlation with public heath interventions. Computational Mechanics 66:1035-1050.
19. Lourenco J, Maia de Lima M, Rodrigues Faria N, Walker A, Kraemer MUG, Villabona-Arenas CJ, Alcantara LCJ, Recker M (2017) Epidemiological and ecological determinants of Zika virus transmission in an urban setting. eLife 6:e29820.
20. Oden JT, Moser R, Ghattas O (2010) Computer predictions with quantified uncertainty. SIAM News 43:9.
21. Oden JT, Babuska I, Faghihi D (2017) Predictive computational science: Computer predictions in the presence of uncertainty. Encyclopedia of Computational Mechanics. John Wiley & Sons.
22. Oden JT (2018) Adaptive multiscale predictive modelling. Acta Numerica 2018:353-450.
23. Osvaldo M (2018) Bayesian Analysis with Python: Introduction to Statistical Modeling and Probabilistic Programming Using PyMC3 and ArviZ. Packt Publishing, 2nd edition.
24. New York Times (2020) Coronavirus COVID-19 Data in the United States. https://github.com/nytimes/covid-19-data/blob/master/us-states.csv assessed: June 1, 2021.
25. Peirlinck M, Linka K, Sahli Costabal F, Bendavid E, Bhattacharya J, Ioannidis J, Kuhl E (2020) Visualizing the invisible: The effect of asymptomatic transmission on the outbreak dynamics of COVID-19. Computer Methods in Applied Mechanics and Engineering 372:113410.
26. Peng GCY, Alber M, Buganza Tepole A, Cannon W, De S, Dura-Bernal S, Garikipati K, Karniadakis G, Lytton WW, Perdikaris P, Petzold L, Kuhl E (2021) Multiscale modeling meets machine learning: What can we learn? Archive of Computational Methods in Engineering 28:1017-1037.
27. Stone JV (2013) Bayes' Rule. A Tutorial Introduction to Bayesian Analysis. Sebtel Press.

Chapter 12
Data-driven dynamic SEIR model

Abstract Throughout the year of 2020, no number has dominated the public media more persistently than the reproduction number. This powerful but simple concept is widely used by the public media, scientists, and political decision makers to explain and justify political strategies to control the COVID-19 pandemic. Here we explore the effectiveness of political interventions using the reproduction number of COVID-19 across Europe. We combine what we have learnt throughout this book, SEIR compartment modeling, dynamic contact rates, computational modeling, and Bayesian analysis to create a data-driven dynamic SEIR model. To illustrate the basic elements of data-driven dynamic SEIR modeling, we discuss the concepts of priors, likelihood, and posteriors and apply Markov Chain Monte Carlo methods to compute them. We draw early case data of COVID-19 and the daily air traffic, driving, walking, and transit mobility for all 27 countries of the European Union, infer the dynamic reproduction number for each country, and correlate mobility and reproduction. The learning objectives of this chapter on data-driven modeling are to

- think in terms of data-driven models for COVID-19
- explain the basic elements of Bayes' theorem: likelihood, prior, and posterior
- illustrate and discuss the importance of prior selection in view of COVID-19
- rationalize and design a Bayesian analysis using the dynamic SEIR model
- infer the dynamic reproduction number from reported COVID-19 case data
- interpret the correlation between mobility and reproduction of COVID-19
- discuss limitations of the data-driven modeling of COVID-19

By the end of the chapter, you will be able to design a Bayesian analysis including prior and likelihood selection to infer dynamic reproduction numbers, probe different exit strategies from lockdown, and correlate mobility and reproduction for infectious diseases like COVID-19 based on reported case data.

12.1 Introduction of the data-driven dynamic SEIR model

Dynamic SEIR model and parameters. We model the epidemiology of the early COVID-19 outbreak using a dynamic SEIR model with four compartments, the susceptible, exposed, infectious, and recovered populations [8]. Figure 12.1 illustrates the four populations and the latent, contact, and infectious rates α, $\beta(t)$, and γ that define the transition between them. Their inverses $A = 1/\alpha$, $B(t) = 1/\beta(t)$, and

| susceptible β=R(t)/C | E₀ exposed α=1/A | I₀ infectious γ=1/C | recovered |

Fig. 12.1 Dynamic SEIR model. The dynamic SEIR model contains four compartments for the susceptible, exposed, infectious, and recovered populations, S, E, I, and R. The transition rates between the compartments are the latent and infectious rates $\alpha = 1/A$ and $\gamma = 1/C$ and the dynamic contact rate $\beta(t) = R(t)/C$. With a hyperbolic tangent type ansatz for the dynamic reproduction number, $R(t) = R_0 - \frac{1}{2}[1 + \tanh([t - t^*]/T)][R_0 - R_t]$, the dynamics of the SEIR model are determined by eight parameters, $\vartheta = \{E_0, I_0, A, C, R_0, R_t, t^*, T\}$.

$C = 1/\gamma$ define the latent, contact, and infectious periods that are typically reported in disease reports, and the ratio $R(t) = \beta(t)/\gamma = C/B(t)$ defines the dynamic reproduction number. As we have discussed in Section 7.7, the dynamic SEIR model is governed by a system of four coupled ordinary differential equations [16],

$$
\begin{aligned}
\dot{S} &= -\beta(t)\,S\,I \\
\dot{E} &= +\beta(t)\,S\,I - \alpha\,E \\
\dot{I} &= \qquad\quad + \alpha\,E - \gamma\,I \\
\dot{R} &= \qquad\qquad\quad + \gamma\,I\,.
\end{aligned}
\tag{12.1}
$$

In contrast to the classical SEIR model with a fixed contact rate β, the dynamic SEIR model features a time-varying dynamic contact rate, $\beta(t) = R(t)/C$. This implies that the reproduction number is no longer constant, but rather a time-varying disease characteristic that characterizes the outbreak dynamics in response to behavioral changes at any point during the outbreak [19]. According to Section 7.7, we adopt the following hyperbolic tangent type ansatz[17],

$$
R(t) = R_0 - \frac{1}{2}[1 + \tanh([t - t^*]/T)][R_0 - R_t],
\tag{12.2}
$$

which ensures a smooth transition from the basic reproduction number R_0 at the beginning of the outbreak to the reduced reproduction number R_t under travel restrictions and lockdown, where t^* is the adaptation time and T is the transition time as illustrated in Figure 7.11. It is convenient to reparameterize the system (7.15) in terms of the time-dependent dynamic reproduction number $R(t) = \beta(t)/\gamma$,

$$
\begin{aligned}
\dot{S} &= -R(t)\,\gamma\,S\,I \\
\dot{E} &= +R(t)\,\gamma\,S\,I - \alpha\,E \\
\dot{I} &= \qquad\qquad + \alpha\,E - \gamma\,I \\
\dot{R} &= \qquad\qquad\qquad + \gamma\,I\,.
\end{aligned}
\tag{12.3}
$$

The dynamic SEIR model is uniquely determined by eight parameters, the initial exposed and infectious populations, E_0 and I_0, the latent and infectious periods, A and C, the initial and reduced reproduction numbers reproduction numbers at the beginning of the outbreak and under lockdown, R_0 and R_t, and the adaptation and transition times t^* and T,

$$\vartheta = \{E_0, I_0, A, C, R_0, R_t, t^*, T\}, \qquad (12.4)$$

where $\alpha = 1/A$, $\beta(t) = R(t)/C$, and $\gamma = 1/C$. We interpret the latent and infectious periods A and C as disease-specific for COVID-19, and assume that they are constant across all countries. We interpret the contact period $B(t) = R(t)/C$ as behavior specific, and assume that the reproduction number $R(t)$ that defines it is different for each country and can vary in time to reflect the effect of societal and political actions.

Data. Our objectives are to explore the evolution of the dynamic reproduction number across all 27 countries of the European Union, to probe the effects of different exit strategies from lockdown, and to correlate mobility and reproduction.

To explore the evolution of the dynamic reproduction number for each country, we infer the initial exposed and infectious populations, E_0 and I_0, and the four parameters that define the dynamic reproduction number, R_0, R_t, t^*, and T, using Bayes' theorem. We draw the COVID-19 outbreak data for all 27 countries of the European Union from public dashboards [9]. For each country, our simulation window during which we infer the model parameters begins on the day on which the total number of reported cases exceeds 100 individuals and ends on May 10, 2020. For this time window, Bayes' theorem provides the posterior distribution of the parameter set ϑ for which the modeled daily new cases $D(\vartheta, t)$ best approximate the reported daily new COVID-19 cases $\hat{D}(t)$.

To probe the effects of different exit strategies from lockdown, we use the posterior parameter distributions to analyze three possible dynamic reproduction numbers $R(t)$ and predict the outbreak dynamics for the period from May 10 to June 20, 2020. The first scenario assumes a constant dynamic reproduction number $R(t) = R_t$, the second and third scenarios simulate the effect of a linear return from R_t to the country-specific basic reproduction number R_0, either rapidly within one month, or more gradually within three months. After having made these three predictions, we return to the public COVID-19 dashboards of all 27 countries [9] and draw the reported daily new cases from May 10 to June 20, 2020. This allows us to compare our model predictions against the real case data.

To correlate mobility and reproduction, we sample all European air traffic data from the Eurocontrol dashboard, a pan-European Organization dedicated to support European aviation [11]. In addition, we approximate car, walking, and transit mobility using a database generated from cell phone data [2]. These data represent the relative volume of location requests per city, subregion, region, and country, scaled by the baseline volume on January 13, 2020. We smoothen the weekday-weekend fluctuations in outbreak and mobility data by applying a moving average window of seven days.

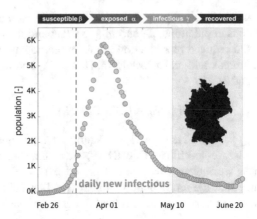

Fig. 12.2 Daily new COVID-19 infections in Germany. Reported new infectious cases from February 26 to June 20, 2020. The time window during which we infer the model parameters begins on March 4 and ends on May 10, 2020. The time window during which we predict the effects of three different exit strategies from lockdown begins on May 10 and ends on June 20, 2020, highlighted through the gray box.

Figure 12.2 illustrates the daily new COVID-19 infections in Germany. As we had discussed in Section 7.9, the first two cases in Germany were reported on February 26, 2020. From Table 7.2, we conclude that on March 4, the country reported more than 100 cases in total, and we begin our simulation. Our time window to infer the model parameters runs from March 4 to May 10, 2020, [9]. With the inferred parameters, we perform a predictive simulation from May 10 to June 20, 2020, highlighted through the gray box.

Figure 12.3 illustrates the hyperbolic tangent type dynamic reproduction number, $R(t) = R_0 - \frac{1}{2}[1 + \tanh([t - t^*]/T)][R_0 - R_t]$, and the physical interpretation of its four parameters, R_0, R_t, t^*, and T. The inference part of the analysis is similar to the estimates of the reproduction number in Chapter 7 in Figure 7.20. However, now, we apply Bayesian statistics [16], which not only return a single set of parameters, but set of probability distributions, along with means and credible intervals for each parameter. The forward analysis from May 10, 2020, onward is based on the three dashed lines in the gray box which illustrate the three possible exit strategies: a constant reproduction number R_t, and a linear return from to the basic reproduction number R_0, either rapidly within one month, or more gradually within three months.

Figure 12.4 illustrates the reduction in mobility in Germany from February 26 to June 20, 2020 compared to the same period in 2019. The dots represent the percentage reduction in daily air traffic [10], and the curve represents a hyperbolic-tangent type fit, $M(t) = M_0 - \frac{1}{2}[1 + \tanh([t - t^*]/T)][M_0 - M_t]$. Daily air traffic in Germany dropped by up to 89.2% from late February to early April compared to the previous year. By comparing the adaptation times in mobility $t^* = 20$ days in Figure 12.4 and in reproduction $t^* = 25$ days in Figure 12.3, we obtain the time delay Δt, a quantitative measure of the responsiveness to political interventions.

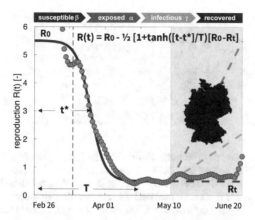

Fig. 12.3 Dynamic reproduction number in Germany. Hyperbolic-tangent type dynamic repro-duction number, $R(t) = R_0 - \frac{1}{2}[1 + \tanh([t - t^*]/T)][R_0 - R_t]$, inferred from daily new cases from March 4 to May 10, 2020, and linearly increasing reproduction number R_t to predict the effects of three different exit strategies from lockdown from May 10 to June 20, 2020, shown as dashed lines in the gray box; latent and infectious periods $A = 2.5$ days and $C = 6.5$ days, initial an reduced reproduction numbers, $R_0 = 5.5$ and $R_t = 0.5$, and the adaptation and transition times $t^* = 25$ days and $T = 50$ days.

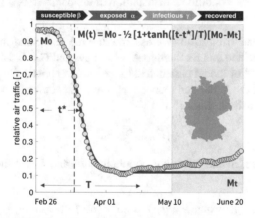

Fig. 12.4 Mobility by daily air traffic variation in Germany. Hyperbolic-tangent type mobility, $M(t) = M_0 - \frac{1}{2}[1 + \tanh([t - t^*]/T)][M_0 - M_t]$, from reduction in air traffic from February 26 to June 20, 2020. Daily air traffic dropped by up to 89.2% in Germany from late February to early April compared to previous year. Initial an reduced mobility, $M_0 = 96.5$ and $M_t = 12.0$, and the adaptation and transition times $t^* = 20$ days and $T = 50$ days.

Bayes' theorem. We use Bayes' theorem to estimate the posterior probability dis-tribution of the parameters $\boldsymbol{\vartheta} = \{E_0, I_0, A, C, R_0, R_t, t^*, T\}$, such that the statistics of the simulated daily new cases $D(\boldsymbol{\vartheta}, t)$ agree with the reported daily new cases $\hat{D}(t)$,

$$P(\boldsymbol{\vartheta}|\hat{D}(t)) = \frac{P(\hat{D}(t)|\boldsymbol{\vartheta}) P(\boldsymbol{\vartheta})}{P(\hat{D}(t))}, \tag{12.5}$$

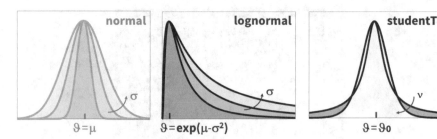

Fig. 12.5 Prior probability distribution. Normal distribution, log-normal distribution, and Student's t-distribution. The normal distribution is symmetric around the mean μ with a width that increases with increasing standard deviation σ. The log-normal distribution is the probability distribution of a random variable ϑ whose logarithm $\ln(\vartheta)$ is normally distributed with mean μ, standard deviation σ, and mode $\exp(\mu - \sigma^2)$. Student's t-distribution is symmetric around the mean ϑ_0 with a width that decreases with increasing parameter ν.

where $P(\hat{D}(t)|\vartheta)$ is the likelihood, $P(\vartheta)$ is the prior, $P(\hat{D}(t))$ is the marginal likelihood, and $P(\vartheta|\hat{D}(t))$ is the posterior. Bayes' theorem calculates the pointwise product of the prior and the likelihood, to produce the normalized posterior probability distribution, which is the conditional distribution of the parameters ϑ given the data, in our case, the daily new COVID-19 cases $\hat{D}(t)$.

Prior probability distributions. We assume that the latent and infectious periods are disease specific and and fix them to $A = 2.5$ days and $C = 6.5$ days [14, 15, 28]. This reduces the set of model parameters to the initial exposed and infectious populations, E_0 and I_0, the initial and reduced reproduction numbers, R_0 and R_t, and the adaptation and transition times t^* and T,

$$\vartheta = \{E_0, I_0, R_0, R_t, t^*, T\}. \tag{12.6}$$

For the initial exposed and infectious populations E_0 and I_0, we choose log-normal distributions,

$$E_0 \sim \text{LogNormal}(\mu_E, \sigma_E) \quad \text{and} \quad I_0 \sim \text{LogNormal}(\mu_I, \sigma_I) \tag{12.7}$$

Table 12.1 Prior probability distributions. Priors for the initial exposed and infectious populations E_0 and I_0, the initial and reduced reproduction number R_0 and R_t, the adaptation and transition times t^* and T, and the likelihood width σ.

parameter	distribution
E_0	LogNormal(log($\hat{D}(t = A)$),1.5)
I_0	LogNormal(log($\hat{D}(t = 0)$),1.5)
R_0	Normal(2.5,2)
R_t	Normal(2.5,2)
t^*	Normal(10,10)
T	LogNormal(log(3),1.5)
σ	HalfCauchy($\beta = 1$)

with means $\mu_E = \log(\hat{D}(t = A))$ and $\mu_I = \log(\hat{D}(t = 0))$ derived from the logarithm of the daily new cases on day $t = A$ and $t = 0$ of the outbreak and standard deviations $\sigma_E = 1.5$ and $\sigma_I = 1.5$. For the initial and reduced reproduction numbers, R_0 and R_t, we choose normal distributions,

$$R_0 \sim \text{Normal}(\mu_R, \sigma_R) \quad \text{and} \quad R_t \sim \text{Normal}(\mu_R, \sigma_R) \tag{12.8}$$

with means $\mu_R = 2.5$ and standard deviations $\sigma_R = 2$. For the adaptation and transition times t^* and T, we choose normal and log-normal distributions,

$$t^* \sim \text{Normal}(\mu_t, \sigma_t) \quad \text{and} \quad T \sim \text{LogNormal}(\mu_T, \sigma_T) \tag{12.9}$$

with means $\mu_t = 10$ and $\mu_T = \log(3)$ and standard deviations $\sigma_t = 10$ and $\sigma_T = 1.5$. Figure 12.5 illustrates the normal distribution, log-normal distribution, and Student's t-distribution of our priors and Table 12.1 summarizes both our prior selection and their parameters.

Likelihood function. The likelihood function, $P(\hat{D}(t)|\vartheta)$, measures the goodness of fit between a sample of the observed data $\hat{D}(t)$ and the model output $D(\vartheta, t)$ for given values of the unknown parameters ϑ. The observed data $\hat{D}(t)$ are the reported daily new COVID-19 cases and the model output $D(\vartheta, t)$ is the number of individuals that transition daily from the exposed to the infectious group. For the daily likelihood between the data $\hat{D}(t)$ and the model predictions $D(\vartheta, t)$, we select Student's t-distribution with a case number-dependent width [5], which results in a variance proportional to the mean,

$$\mathcal{L}(\hat{D}(t)|\vartheta) \sim \text{studentT}_{\nu=4}(\text{mean} = D(\vartheta, t), \text{width} = \sigma\sqrt{D(\vartheta, t)}). \tag{12.10}$$

We choose Student's t-distribution because it resembles a normal distribution around the mean, but, as Figure 12.5 suggests, has heavy tails that make the Markov Chain Monte Carlo method more robust to outliers. For the likelihood width σ between the reported daily new cases $\hat{D}(t)$ and the simulated daily new infections $D(\vartheta, t)$, we select a half Cauchy distribution,

$$\sigma \sim \text{HalfCauchy}(\sigma_0, \beta) \quad \text{with} \quad P(\sigma) = \frac{1}{\pi\beta\left[1 + [(\sigma - \sigma_0)/\beta]^2, \right]} \tag{12.11}$$

with mean $\sigma_0 = 0$ and scaling parameter β. To compare the data and the model throughout the entire time window, from day t_0 to day t_n, we evaluate the daily likelihood $\mathcal{L}(\vartheta, t_i)$ from equation (12.10) at each discrete time point t_i and multiply the individual terms to calculate the overall likelihood function $P(\hat{D}(t)|\vartheta)$,

$$P(\hat{D}(t)|\vartheta) = \prod_{i=0}^{n}\mathcal{L}(\vartheta, t_i), \tag{12.12}$$

where the product symbol, $\prod_{i=0}^{n}$, denotes the multiplication of the $i = 0, ..., n$ daily likelihoods $\mathcal{L}(\vartheta, t_i)$.

Posterior probability distributions. We apply Bayes' theorem (12.5) to obtain the posterior distribution of the parameters using the likelihood $P(\hat{D}(t)|\boldsymbol{\vartheta})$ and priors $P(\boldsymbol{\vartheta})$. As we have discussed in the previous chapter, the denominator in Bayes' theorem, the marginal likelihood $P(\hat{D}(t))$, is generally difficult to compute. Since it is essentially a normalization factor, we can rewrite Bayes' theorem as a proportionality,

$$P(\boldsymbol{\vartheta}|\hat{D}(t)) \propto P(\hat{D}(t)|\boldsymbol{\vartheta})\,P(\boldsymbol{\vartheta}), \qquad (12.13)$$

and simply calculate the normalized posterior distribution as the product of the likelihood and priors.

Markov Chain Monte Carlo. We solve this distribution numerically using the NO-U-Turn sampler [9] implementation of the Python package PyMC3 [27]. We use two chains. The first 1000 samples are used to tune the sampler, and are later discarded; the subsequent 1000 samples are used to infer the set of parameters $\boldsymbol{\vartheta}$. Chain convergence requires a geometric ergodicity between the Markov transition and the target distribution. In PyMC3, this is detected by split $\hat{R}$ statistics, which identify convergence by comparing the variance between the chains [23]. From the converged posterior distributions, we sample multiple combinations of parameters $\boldsymbol{\vartheta}$. From these posterior samples, we quantify the means and standard deviations of each parameter and plot the means and credible intervals of the reproduction number $R(t)$ and the daily new cases $D(\boldsymbol{\vartheta}, t)$ for all 27 countries in the European Union [17].

12.2 Example: Inferring reproduction of COVID-19 in Europe

Figure 12.6 illustrates the outbreak dynamics of COVID-19 for all 27 countries of the European Union. The dots represent daily new cases. The brown and red curves illustrate the fit of the SEIR model and the dynamic reproduction number for the time window until May 10, 2020. The gray shaded area highlights the model predictions for the 40-day period of gradual reopening, from May 10 until June 20, 2020. The dashed brown, orange, and red curves illustrate the projections for three possible exit strategies: a constant continuation at the dynamic reproduction number R_t from May 10, 2020, a gradual return to the basic reproduction number R_0 within three months, and a rapid to R_0 within one months. Naturally, the case numbers increase in all three cases, with the steepest increase for the most rapid return to baseline. Interestingly, the Bayesian analysis reveals notably different credible intervals for different countries suggesting that a controlled return will be more predictable in some countries like Austria and less in others like Hungary. The projections suggest that in Sweden, were policy makers had encouraged each individual to take responsibility for their own health rather than enforcing political constraints, the projected case numbers will follow the current curve, without major deviations. Strikingly, in most countries, the newly reported case numbers upon gradual reopening, from May 10 to June 20, 2020, follow the dashed brown curves of the prediction with a constant reproduction number.

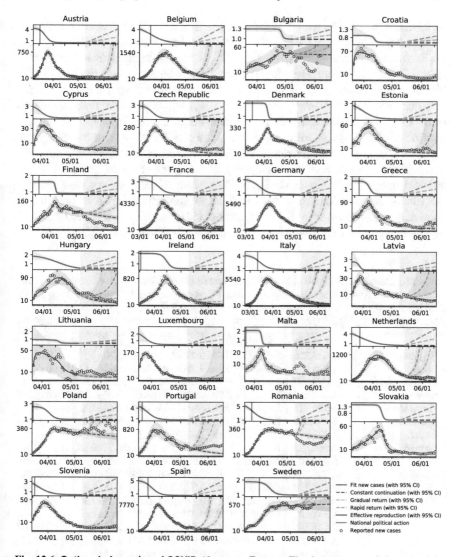

Fig. 12.6 Outbreak dynamics of COVID-19 across Europe. The dots represent daily new cases. The brown and red curves illustrate the fit of the SEIR model and the dynamic reproduction number for the time period until May 10, 2020. The gray shaded area highlights the model predictions from May 10 until June 20, 2020. The dashed brown, orange, and red curves illustrate the projections for three possible exit strategies: a constant continuation at the reproduction number R_t from May 10, 2020, a gradual return to the basic reproduction number R_0 within three months, and a rapid to R_0 within one months [17].

Table 12.2 and Figure 12.7 summarize the basic reproduction number R_0 at the beginning of the outbreak and the reduced reproduction number R_t on May 10, 2020. The basic reproduction number R_0 has maximum values in Germany, the

Netherlands, and Spain, with 6.33, 5.88, and 5.19 and minimum values in Bulgaria, Croatia, and Lithuania with 1.29, 0.93, and 0.91. Its population weighted mean across the European Union is $R_0 = 4.22 \pm 1.69$ [17]. For the basic reproduction number of $R_0 = 4.22 \pm 1.69$, the herd immunity threshold would be at $H = 1 - 1/R_0 = 60 - 83\%$ [6]. Compared to Table 1.3, this value is lower than 92-95% for the measles, 83-86% for rubella, and 80-86% for polio, but significantly higher than the values of 33% to 44% for the seasonal flu [1]. Knowing the precise basic reproduction number of COVID-19 is critical to estimate the conditions for herd immunity and predict the success of vaccination strategies. The reduced reproduction number R_t is significantly lower than the basic reproduction number R_0 suggesting that the political mitigation strategies during the first wave of the outbreak were broadly successful. In most countries, R_t is well below the critical value of one. It has maximum values in Sweden, Bulgaria, and Poland all with 1.01, 0.99, and 0.96 and minimum values in Lithuania, Hungary, and Slovakia with 0.41, 0.37, and 0.28. The

Table 12.2 Outbreak parameters of COVID-19 across Europe. Basic reproduction number R_0, reduced reproduction number R_t, adaptation time t^*, adaptation speed T, and time delay Δt for fixed latent period $A = 2.5$ days and infectious period $C = 6.5$ days.

Country	Population	R_0	R_t	t^*	T	Δt
Austria	8.840.521	4.38±0.36	0.45±0.01	13.37±0.68	6.49±0.47	8.33±1.70
Belgium	11.433.256	5.00±0.73	0.54±0.03	13.31±2.84	19.30±1.57	4.00±2.35
Bulgaria	7.025.037	1.29±0.04	0.99±0.07	37.04±1.99	1.64±1.56	43.00±2.83
Croatia	4.087.843	0.93±0.22	0.49±0.03	22.36±2.90	2.46±3.61	27.33±2.36
Cyprus	1.189.265	3.35±1.14	0.50±0.02	6.60±2.87	8.02±1.47	14.00±0.00
Czech Republic	10.629.928	2.92±0.47	0.60±0.01	14.04±1.71	8.44±1.12	14.00±1.73
Denmark	5.793.636	2.00±0.05	0.81±0.01	24.74±0.29	1.72±0.45	24.00±2.92
Estonia	1.321.977	3.12±0.78	0.45±0.04	10.72±3.80	14.19±2.87	12.50±2.06
Finland	5.515.525	1.62±0.05	0.92±0.01	25.05±0.51	1.20±0.68	24.25±2.49
France	66.977.107	3.46±0.29	0.62±0.02	24.79±1.30	10.58±1.17	10.50±1.50
Germany	82.905.782	6.33±0.64	0.58±0.01	17.06±1.39	12.41±0.71	3.25±1.92
Greece	10.731.726	1.66±0.12	0.61±0.02	18.93±0.87	4.38±1.22	17.33±3.09
Hungary	9.775.564	1.97±0.55	0.37±0.15	25.62±6.55	20.23±7.33	31.67±1.89
Ireland	4.867.309	1.94±0.06	0.57±0.03	30.78±0.53	5.94±1.28	30.00±3.46
Italy	60.421.760	4.25±0.42	0.74±0.01	19.24±1.57	12.06±1.13	5.00±0.71
Latvia	1.927.174	2.50±0.89	0.76±0.01	6.99±1.32	2.70±0.90	14.67±1.89
Lithuania	2.801.543	0.91±0.88	0.41±0.09	26.23±9.88	2.25±6.25	34.67±1.89
Luxembourg	607.950	2.42±1.21	0.46±0.01	5.77±4.20	8.78±1.88	10.00±2.35
Malta	484.630	2.08±0.14	0.51±0.03	16.24±0.42	1.21±0.51	23.00±0.00
Netherlands	17.231.624	5.88±0.88	0.49±0.03	7.61±3.12	23.25±1.92	0.75±2.77
Poland	37.974.750	2.62±0.26	0.96±0.01	18.15±1.47	7.38±1.47	20.33±0.94
Portugal	10.283.822	5.10±0.86	0.73±0.02	8.93±1.86	10.40±1.22	8.67±2.62
Romania	19.466.145	6.06±0.84	0.95±0.01	8.12±1.60	11.55±0.70	8.33±2.36
Slovakia	5.446.771	1.46±0.04	0.28±0.03	31.80±0.47	2.65±0.70	40.25±0.43
Slovenia	2.073.894	3.83±0.96	0.44±0.03	4.65±3.50	15.33±1.96	6.33±2.36
Spain	46.796.540	5.19±0.50	0.57±0.01	15.90±1.17	10.70±0.69	5.50±2.60
Sweden	10.175.214	1.89±0.09	1.01±0.03	29.70±1.30	7.99±2.65	23.75±2.77
European Union	446.786.293	4.22±1.69	0.67±0.18	18.61±6.43	10.82±4.65	17.24±2.00

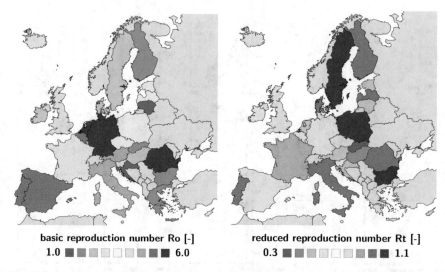

basic reproduction number Ro [-]
1.0 ■■■□□□□■■■ 6.0

reduced reproduction number Rt [-]
0.3 ■■■□□□□■■■ 1.1

Fig. 12.7 Basic and reduced reproduction numbers R_0 **and** R_t **of the COVID-19 outbreak across Europe.** The basic and reduced reproduction numbers characterize the initial and modified numbers of new infectious created by one infectious individual. R_0 has maximum values in Germany, the Netherlands, and Spain, with 6.33, 5.88, and 5.19 and minimum values in Bulgaria, Croatia, and Lithuania with 1.29, 0.93, and 0.91. R_t has maximum values in Sweden, Bulgaria, and Poland all with 1.01, 0.99, and 0.96 and minimum values in Lithuania, Hungary, and Slovakia with 0.41, 0.37, and 0.28 as of May 10, 2020 [17].

population weighted mean of the reduced reproduction number across the European Union is $R_t = 0.67 \pm 0.18$.

12.3 Example: Correlating mobility and reproduction of COVID-19

Figures 12.8 and 12.9 provide a direct correlation between the reduction in mobility and the dynamic reproduction number of the COVID-19 outbreak across Europe. The black dots in Figure 12.8 represent the reduction in air traffic, and the purple, blue, grey, and black dots in Figure 12.9 represent the reduction in air traffic, driving, walking, and transit mobility. The red curves in both Figures show the inferred dynamic reproduction number with 95% credible interval. The time delay Δt highlights the temporal delay between reduction in mobility and dynamic reproduction number. The reported Spearman's rank correlation ρ and p-value measure of the statistical dependency between mobility and reproduction. For all four types of mobility in Figure 12.9, Spearman's rank correlation reveals the strongest correlation in the Netherlands, Germany, Ireland, Spain, and Sweden with 0.99 and 0.98. Only in Slovakia, Slovenia and Lithuania, where the number of cases has not yet plateaued and the dynamic reproduction number does not show a clear smoothly decaying trend, there is no significant correlation between mobility and reproduction.

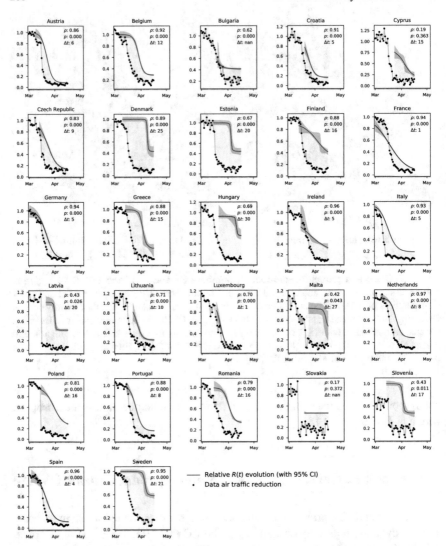

Fig. 12.8 Correlation between reduction in mobility and reproduction during the early COVID-19 outbreak across Europe. Black dots represent reduction in air traffic; red curves show the dynamic reproduction number R(t) with 95% credible intervals. The mean time delay Δt indicates the difference between the reduction in mobility and the dynamic reproduction number. Spearman's rank correlation ρ and p-value measure of the statistical dependency between mobility and reproduction [17].

Table 12.2 and Figure 12.10 summarize the time-varying dynamic reproduction number R(t) and highlight the adaptation time t^* and time delay Δt of its reduction with respect to the European travel restrictions. The *adaptation time t^** is the time between the beginning of the outbreak at 100 confirmed cases and the inflection

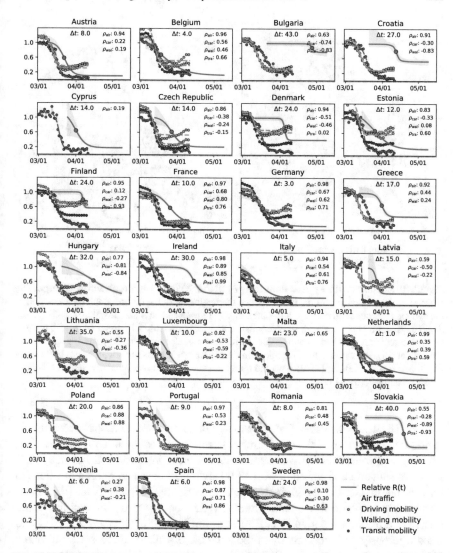

Fig. 12.9 Correlation between reduction in mobility and reproduction during the early COVID-19 outbreak across Europe. Purple, blue, grey, and black dots represent reduction in air traffic, driving, walking, and transit mobility; red curves show the dynamic reproduction number R(t) with 95% credible intervals. The mean time delay Δt highlights the temporal delay between reduction in mobility and dynamic reproduction number. Spearman's rank correlation ρ, measures of the statistical dependency between mobility and reproduction, and reveals the strongest correlation in the Netherlands, Germany, Ireland, Spain, and Sweden with 0.99 and 0.98 [17].

point of the dynamic reproduction curve. It has maximum values in Bulgaria and Slovakia with 37.04 and 31.80 days and minimum values in Luxembourg and Slovenia with 5.77 and 5.64 days. The mean adaptation time across the European Union

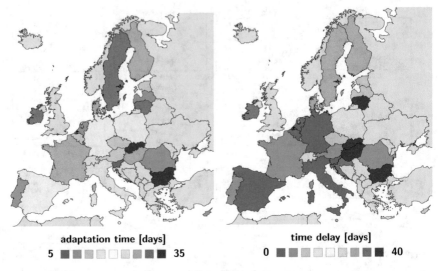

adaptation time [days]

5 ■■■■□■■■■ 35

time delay [days]

0 ■■■■□■■■■ 40

Fig. 12.10 Adaptation time t^* **and time delay** Δt **of the COVID-19 outbreak across Europe.** The adaptation time characterizes the time between the beginning of the outbreak and the reduction in the dynamic reproduction number. It has maximum values in Bulgaria and Slovakia with 37.04 and 31.80 days and minimum values in Luxembourg and Slovenia with 5.77 and 5.64 days. The time delay characterizes the mean time between the reduction in air travel, driving, walking, and transit mobility and the reduction in the dynamic reproduction number. It has maximum values in Bulgaria and Slovakia with 43.00 and 40.25 days and minimum values in Germany and the Netherlands both with 3.25 and 0.75 days [17].

is $t^* = 18.61 \pm 6.43$ days. The *time delay* Δt is an important socio-economical metric that characterizes the mean time between the reduction in air travel, driving, walking, and transit mobility and the reduction in reproduction. It has maximum values in Bulgaria and Slovakia with 43.00 and 40.25 days and minimum values in Italy, Belgium, Germany, and the Netherlands with 5.00, 4.00, 3.25, and 0.75 days. These rapid response times naturally also reflect decisions on the national level. France had the first reported COVID-19 case in Europe on January 24, 2020 and acted rigorously and promptly by introducing the first national measures on March 16. Similarly, Italy, Spain, and Germany had introduced their national measures on March 9, March 9, and March 13, 2020 . Figures 12.6 and 12.9 clearly highlight the special role of Sweden, where the government encouraged the right behavior and created social norms rather than mandatory restrictions: The time delay of 23.75 days is above the European Union average of 17.24 days, and Sweden is one of the few countries where the dynamic reproduction number has not yet decreased below one. Taken together, these drastic political actions to control the COVID-19 pandemic have triggered an ongoing debate about the effectiveness of different outbreak strategies and the appropriate level of constraints [22]. Our results confirm that, especially during the early stages of an outbreak, mobility plays a critical role in spreading a disease [3, 4, 5, 19]. However, drastic political measures to control it have stimulated an active debate when and how it would be safe to lift these restrictions [18].

Limitations. Just like any infectious disease model, our model inherently faces limitations associated with data uncertainties from differences in testing, inconsistent diagnostics, incomplete counting, and delayed reporting. For the specific examples on the early outbreak dynamics of COVID-19 in Europe, we encounter a few additional limitations [19]: First, although a massive amount of data are freely available through numerous well-documented public databases, the selection of the model naturally limits what we can predict and it remains challenging to map the available information into the particular format required for the SEIR model. Second, the initial conditions for our exposed and infectious populations will always remain unknown and many new first cases have been discovered since the beginning of the outbreak. To reduce the influence of unknown initial conditions, our Bayesian inference learns the initial conditions on these populations alongside the parameters that define the dynamic reproduction number [26]. Third, in its current state, our model does not distinguish between community mitigation strategies, local public health recommendations, and global political actions [6]. In Chapter 14, we will integrate the current approach with a global network model that will provide more granularity to include other community mitigation strategies in addition to mobility [3]. Fourth, our current model is not directly informed by mobility data. We have recently proposed a new method that uses a stochastic process to directly incorporate mobility as a latent variable into the present SEIR model framework [19]. Fifth, and probably most importantly, our current knowledge limits our ability to make firm predictions about the recovered group, which is critical to estimate the return to normal. Recent studies have shown that the dimension of unreported asymptomatic transmission is enormous, up to an order of magnitude larger than the reported symptomatic transmission traced in our study [26]. In Chapter 13, we will expand our current approach by including asymptomatic transmission to visualize the effects of this invisible but important subgroup of the population. A related challenge is that the number of reported cases strongly depends on the testing strategy of each country. A possibility to eliminate testing bias could be to use death counts rather than case counts [12]; however, this would also require a consistent Europe-wide definition of death with versus death caused by COVID-19. In general, broad testing can be help identify the size of the asymptomatic population and explore whether it behaves differently in terms of contact rate and infectious period, which would both radically change the overall reproduction number. The more data become available, the more we will become confident in our data-driven models, and the more we will learn how to quickly respond to changes in disease dynamics.

Taken together, this example shows that the dynamic SEIR model with a hyperbolic tangent type contact rate is well-suited to model the interplay between the early outbreak dynamics and outbreak control of COVID-19. We have learnt the dynamic dynamic reproduction number for all 27 countries of the European Union from their individual reported cases using machine learning and uncertainty quantification. During the early stages of the outbreak, the dynamic reproduction number across Europe was 4.2 ± 1.7. By May 10, 2020, massive public health interventions successfully reduced the dynamic reproduction number to $= 0.7 \pm 0.2$. Strikingly, this reduction displays a strong correlation with air traffic, driving, walking, and transit mobility, with a time delay of 17 ± 2 days. This time delay is an important metric that characterizes the response time of the population to public health interventions. The data-driven dynamic epidemiology model provides the flexibility to simulate the effects and timelines of various outbreak control and exit strategies to inform political decision making and identify solutions that minimize the impact of COVID-19 on global health.

Problems

12.1 Data-driven SIS vs. SEIR model. Which elements of the Bayesian analysis make the data-driven SEIR model from this chapter computationally more intense than the data-driven SIS model from the previous chapter?

- prior selection
- SEIR model has no analytical solution
- more unknown parameters
- product of daily likelihoods

- synthetic data generation
- explicit time integration
- Student's t-distribution
- marginal likelihood calculation

12.2 Data-driven SEIR model. Design an Bayesian analysis to infer the dynamic reproduction numbers of COVID-19 during the early stages of the outbreak. Draw a flow chart that includes at least the five most important elements of the Bayesian analysis. Describe the input and output to each element.

12.3 Prior selection. Plot the prior probability distribution for the initial reproduction number. Assume a normal distribution, $R_0 \sim \text{Normal}(\mu_R, \sigma_R)$, with mean $\mu = 2.5$ and standard deviation $\sigma = 2$. From what you know now about the initial reproduction number, would you have made a different choice? How would this affect your posterior probability distribution?

12.4 Prior selection. Throughout this chapter, we have used similar prior probability distribution for the initial and reduced reproduction numbers $R_0 \sim \text{Normal}(\mu_R, \sigma_R)$ and $R_t \sim \text{Normal}(\mu_R, \sigma_R)$ with similar means $\mu = 2.5$ and standard deviations $\sigma = 2$. From what you know now about the initial and reduced reproduction numbers, would you have made a different choice? Discuss the effect of different priors in view of the posterior probability distributions.

12.5 Prior selection. We have selected log normal distributions for the initial exposed and infectious populations, E_0 and I_0, and normal distributions for the initial and reduced reproduction numbers, R_0 and R_t, see Table 12.1. Compare both distributions in Figure 12.5 and justify these selection.

12.6 Data for SEIR model. Find a public database for COVID-19 for your own country, state, county, or city. What data are reported? Assume you want to do a Bayesian analysis of your location, how do these data map into the susceptible, exposed, infectious, and recovered populations of the SEIR model? Discuss challenges when aligning reported data to the output of epidemiological compartment models.

12.7 Data for mobility correlation. Find a public database for the mobility during COVID-19 in your own country, state, county, or city. What data are reported? Can you see a drop in mobility during the first wave of the pandemic? Plot the data and try to manually fit a hyperbolic-tangent type mobility function, $M(t) = M_0 - \frac{1}{2}[1 + \tanh([t - t^*]/T)][M_0 - M_t]$. Do your data display weekday-weekend alterations? Discuss challenges when analyzing mobility data.

12.8 Likelihood function. In this Chapter, the likelihood function, $\mathcal{L}(\hat{D}(t)|\vartheta)$, uses the reported daily new cases of COVID-19 as illustrated in Figures 12.2 and 12.6. Describe how you would calculate the reported daily new cases from the output of the SEIR model. What other daily data could you use instead in the likelihood function?

12.9 Inferred reproduction dynamics. Early in the pandemic, policy makers in Sweden consciously decided against political interventions and implemented only moderate guidelines to minimize the spread of COVID-19. Compare the reproduction dynamics of Sweden in Table 12.2 against the rest of the European Union. How do Sweden's initial and basic reproduction numbers, R_0 and R_t, differ from the European mean? Discuss how this difference reflects on the effectiveness of political interventions.

12.10 Credible intervals. For the predictions of upon gradual reopening for all 27 countries of the European Union, identify five countries with narrow 95% credible intervals for the projected new cases, the orange shaded regions in Figure 12.6. What do the countries with narrow 95% credible intervals have in common? Identify several features that could help narrow the credible intervals in the forward projection.

12.11 Credible intervals. For the predictions of upon gradual reopening for all 27 countries of the European Union, identify five countries with wide 95% credible intervals for the projected new cases, the orange shaded regions in Figure 12.6. What do the countries with wide 95% credible interval have in common? Identify several features that could cause wide credible intervals in the forward projection.

12.12 Time delay. Using a hyperbolic tangent-type function for both the reproduction number $R(t)$ and the mobility $M(t)$ allows us to quantitatively compare the inflection points t^* of both curves. The time delay Δt, the difference between both inflection points, measures the time delay between reduced mobility and reduced reproduction. Compare at the different adaptation times t^* and time delays Δt across the European Union in Figure 12.10 and Table 12.2. Discuss what policy makers can learn, for example, from the mean time delay of $\Delta t = 17.24 \pm 2.00$ days.

References

1. Anderson RM, May RM (1982) Directly transmitted infectious diseases: control by vaccination. Science 215:1053-1060.
2. Apple Mobility Trends. https://www.apple.com/covid19/mobility. accessed: June 1, 2021.
3. Bhouri MA, Sahli Costabal F, Wang H, Linka K, Peirlinck M, Kuhl E, Perdikaris P (2021) COVID-19 dynamics across the US: A deep learning study of human mobility and social behavior. Computer Methods in Applied Mechanics and Engineering 382: 113891.
4. Chinazzi, M, Davis JT, Ajelli M, Gioanni C, Litvinova M, Merler S, Piontti A, Mu K, Rossi L, Sun K, Viboud C, Xiong X, Yu H, Halloran ME, Longini IM, Vespignani A (2020) The effect of travel restrictions on the spread of the 2019 novel coronavirus (COVID-19) outbreak. Science 368:395-400.
5. Colizza V, Barrat A, Barthelemy M, Vespignani A (2006) The role of the airline transportation network in the prediction and predictability of global epidemics. Proceedings of the National Academy of Sciences 103:2015-2020.
6. Chu DK, Akl EA, Duda S, Solo K, Yaacoub S, Schünemann HJ (2020) Physical distancing, face masks, and eye protection to prevent person-to-person transmission of SARS-CoV-2 and COVID-19: A systematic review and meta-analysis. Lancet 395:1973-1987.
7. Dehning J, Zierenberg J, Spitzner FP, Wibral M, Pinheiro Neto J, Wilczek M, Priesemann V (2020) Inferring COVID-19 spreading rates and potential change points for case number forecasts. Science 369:eabb9789.
8. Dietz K (1993) The estimation of the basic reproduction number for infectious diseases. Statistical Methods in Medical Research 2:23-41.
9. European Centre for Disease Prevention and Control. Situation update worldwide. https://www.ecdc.europa.eu/en/geographical-distribution-2019-ncov-cases. accessed: June 1, 2021.
10. Eurostat. Your key to European statistics. Air transport of passengers. https://ec.europa.eu/eurostat accessed: June 1, 2021.
11. Eurocontrol. Flights 2020. Daily traffic variation. http://eurocontrol.int. accessed: June 1, 2021.
12. Flaxman S, Mishra S, Gandy A, Unwin HJT, Mellan TA, Coupland H, Whittaker C, Zhu H, Berah T, Eaton JW, Monod M, Imperial College COVID-19 Response Team, Ghani AC, Donnelly CA, Riley S, Vollmer MAC, Ferguson NM, Okell LC, Bhatt S (2020) Estimating the effects of non-pharmaceutical interventions on COVID-19 in Europe. Nature 584:257–261.
13. Hethcote HW (2000) The mathematics of infectious diseases. SIAM Review 42:599-653.
14. Hoffman MD, Gelman A (2014) The No-U-Turn sampler: adaptively setting path lengths in Hamiltonian Monte Carlo. Journal of Machine Learning Research 15:1593-1623.
15. Ioannidis JPA, Cripps S, Tanner MA (2021) Forecasting for COVID-19 has failed. International Journal of Forecasting, in press.
16. Kennedy MC, O'Hagan A (2001) Bayesian calibration of computer models. Journal of the Royal Statistical Society B 63:425-464.
17. Lauer SA, Grantz KH, Bi Q, Jones FK, Zheng Q, Meredith HR, Azman AS, Reich NG, Lessler J (2020) The incubation period of coronavirus disease 2019 (COVID-19) from publicly reported confirmed cases: estimation and application. Annals of Internal Medicine 172:577-582.
18. Li Q, Guan X, Wu P, Wang X, ... Feng Z (2020) Early transmission dynamics in Wuhan, China, of novel coronavirus-infected pneumonia. New England Journal of Medicine 382:1199-1207.
19. Linka K, Peirlinck M, Kuhl E (2020) The reproduction number of COVID-19 and its correlation with public heath interventions. Computational Mechanics 66:1035-1050.
20. Linka K, Rahman P, Goriely A, Kuhl E (2020) Is it safe to lift COVID-19 travel restrictions? The Newfoundland story. Computational Mechanics 66:1081–1092.
21. Linka K, Goriely A, Kuhl E (2021) Global and local mobility as a barometer for COVID-19 dynamics. Biomechanics and Modeling in Mechanobiology 20:651–669.

22. Maier BF, Brockmann D (2020) Effective containment explains sub-exponential growth in confirmed cases of recent COVID-19 outbreak in mainland China. Science 368:742-746.
23. Osvaldo M (2018) Bayesian Analysis with Python: Introduction to Statistical Modeling and Probabilistic Programming Using PyMC3 and ArviZ. Packt Publishing, 2nd edition.
24. Peirlinck M, Linka K, Sahli Costabal F, Bendavid E, Bhattacharya J, Ioannidis J, Kuhl E (2020) Visualizing the invisible: The effect of asymptomatic transmission on the outbreak dynamics of COVID-19. Computer Methods in Applied Mechanics and Engineering 372:113410.
25. Salvatier J, Wiecki TV, Fonnesbeck C (2016) Probabilistic programming in Python using PyMC3. PeerJ Computer Science 2:e55.
26. Sanche S, Lin Y, Xu C, Romero-Severson E, Hengartner N, Ke R (2020) High contagiousness and rapid spread of severe acute respiratory syndrome coronavirus 2. Emerging Infectious Diseases 2020;26(7):1470-1477.

Chapter 13
Data-driven dynamic SEIIR model

Abstract In the early stages of a pandemic, doctors, researchers, and political decision makers mainly focus on symptomatic individuals and address those who require
the most urgent medical attention. In the more advanced stages, the interest shifts
towards mildly symptomatic and asymptomatic individuals who–by definition–are
difficult to trace and likely to retain normal social and travel patterns. In the case
of COVID-19, early antibody seroprevalence studies suggested that the number of
asymptomatic cases outnumbered the symptomatic cases by an order of magnitude
or more. Estimating the prevalence and contagiousness of these asymptomatic cases
is critical to our understanding of the overall dimension and management of a disease. However, without broad seroprevalence testing, the effects of the asymptomatic
population, its size, and its outbreak dynamics remain largely unknown. Mathematical modeling can illustrate the potential effects of asymptomatic transmission and
visualize the dynamics of the asymptomatic population for various different scenarios. Knowing the exact dimension of the asymptomatic transmission is critical
to estimate the true severity of the outbreak, its hospitalizations, and true mortality
rates, and to reliably predict the success of surveillance and control strategies, contact tracing, and vaccination. The learning objectives of this chapter on data-driven
modeling of asymptomatic transmission are to

- interpret seroprevalence studies and explain the notion of undercount
- recognize and discuss the importance of asymptomatic transmission
- rationalize and design a Bayesian analysis using the dynamic SEIIR model
- propagate uncertainty through the model and interpret credible intervals
- infer the reproduction number from COVID-19 case and seroprevalence data
- discuss limitations of seroprevalence and asymptomatic transmission studies

By the end of the chapter, you will be able to design a Bayesian analysis including
prior and likelihood selection to infer the outbreak dynamics of symptomatic and
asymptomatic transmission for infectious diseases like COVID-19 using reported
case and seroprevalence data.

13.1 Introduction of the data-driven dynamic SEIIR model

Dynamic SEIIR model and parameters. We model the epidemiology of COVID-19 using an SEIIR model that accounts for both symptomatic and asymptomatic transmission [3]. The SEIIR model consists of five compartments, the susceptible, exposed, symptomatic infectious, asymptomatic infectious, and recovered populations [26]. Figure 13.1 shows the five populations and the latent, contact, and infectious

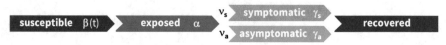

Fig. 13.1 Dynamic SEIIR model. The dynamic SEIIR model contains five compartments for the susceptible, exposed, symptomatic infectious, asymptomatic infectious, and recovered populations. The transition rates between the compartments, $\beta(t)$, α, and γ, are inverses of the dynamic contact period, $B(t) = 1/\beta(t)$, the latent period, $A = 1/\alpha$, and the infectious periods, $C_s = 1/\gamma_s$ and $C_a = 1/C_a$. Upon exposure, only a fraction, $\nu_s = 1 - \nu_a$, becomes symptomatic, and a fraction, ν_a, remains asymptomatic. The symptomatic and asymptomatic groups have the same dynamic contact and latent periods, $B(t)$, and A, but individual infectious periods, C_s and C_a.

rates α, β, and γ that define the transition between them. Their inverses $B = 1/\beta$, $A = 1/\alpha$, and $C = 1/\gamma$ define the contact, latent, and infectious periods. As we have discussed in Chapter 8, the dynamic SEIIR model is governed by a system of five coupled ordinary differential equations,

$$
\begin{aligned}
\dot{S} &= -S[\beta_s I_s + \beta_a I_a] \\
\dot{E} &= +S[\beta_s I_s + \beta_a I_a] - \alpha E \\
\dot{I}_s &= \qquad\qquad\qquad + \nu_s \alpha E - \gamma_s I_s \\
\dot{I}_a &= \qquad\qquad\qquad + \nu_a \alpha E - \gamma_a I_a \\
\dot{R} &= \qquad\qquad\qquad\qquad\quad + \gamma_s I_s + \gamma_a I_a .
\end{aligned}
\tag{13.1}
$$

We postulate that upon exposure, only a fraction, ν_s, becomes symptomatic, and a fraction, ν_a, remains asymptomatic, such that $\nu_s + \nu_a = 1$ [12]. The ratio between the asymptomatic and symptomatic fractions defines the undercount, $U = \nu_a/\nu_s = [1 - \nu_s]/\nu_s$, that characterizes the number of asymptomatic individuals per single symptomatic case. We assume that the symptomatic and asymptomatic groups have the same contact and latent rates, β and α, but can have individual infectious rates, γ_s and γ_a, to mimic a different infectiousness. The symptomatic and asymptomatic infectious rates define the overall infectious rate, $\gamma = \nu_s\gamma_s + \nu_a\gamma_a$, and with it the dynamic reproduction number,

$$
R(t) = \frac{\beta(t)}{\gamma} = \frac{\beta(t)}{\nu_s\gamma_s + \nu_a\gamma_a} .
\tag{13.2}
$$

Here, in contrast to the classical SEIIR model in Chapter 8 with a fixed contact rate β [3], we assume a dynamic SEIIR model with a time-varying dynamic contact rate

$\beta(t)$ to account for behavioral changes during the course of the pandemic [26]. With these considerations, we can rewrite the SEIIR model (13.1) as follows,

$$
\begin{aligned}
\dot{S} &= -S\,\beta(t)\,[\,I_s + I_a\,] \\
\dot{E} &= +S\,\beta(t)\,[\,I_s + I_a\,] - \qquad\qquad \alpha\,E \\
\dot{I}_s &= \qquad\qquad\qquad + \quad v_s\,\alpha\,E - \gamma_s\,I_s \\
\dot{I}_a &= \qquad\qquad\qquad + [1 - v_s]\,\alpha\,E - \gamma_a\,I_a \\
\dot{R} &= \qquad\qquad\qquad\qquad\qquad + \gamma_s\,I_s + \gamma_a\,I_a\,.
\end{aligned}
\tag{13.3}
$$

With $\alpha = 1/A$, $\gamma_s = 1/C_s$, and $\gamma_a = 1/C_a$, our dynamic SEIIR model (13.3) uses the following set of parameters,

$$
\vartheta = \{\, A, C_s, C_a, v_s, \beta(t), E_0, I_{s0}, I_{a0}\,\},
\tag{13.4}
$$

the latent period A, the symptomatic and asymptomatic infectious periods C_s and C_a, the symptomatic fraction v_s, the dynamic contact rate $\beta(t)$, and the initial exposed and infectious populations E_0, I_{s0}, and I_{a0}.

Discrete dynamic SEIIR model. We discretize the SEIIR model (13.3) in time using a finite difference approximation, $(\dot{\circ}) = [\,(\circ)_{n+1} - (\circ)_n\,]/\Delta t$, for the evolution of all five populations $(\circ) = S, E, I_s, I_a, R$. Here $\Delta t = t_{n+1} - t_n$ denotes the discrete time increment, usually $\Delta t = 1$ day, and $(\circ)_{n+1}$ and $(\circ)_n$ denote the populations of the new and previous time steps. We apply an explicit time integration scheme to obtain the following discrete system of equations[22],

$$
\begin{aligned}
S_{n+1} &= [1 - \beta(t)]\,[I_{s,n} + I_{a,n}]\,\Delta t]\,S_n \\
E_{n+1} &= \quad +\beta(t)\,[I_{s,n} + I_{a,n}]\,\Delta t]\,S_n + \quad [1 - \alpha\Delta t]E_n \\
I_{s,n+1} &= \qquad\qquad\qquad + \quad v_s\,\alpha\Delta t\,E_n + [1 - \gamma_s\Delta t]I_{s,n} \\
I_{a,n+1} &= \qquad\qquad\qquad + [1 - v_s]\alpha\Delta t\,E_n + [1 - \gamma_a\Delta t]I_{a,n} \\
R_{n+1} &= \qquad\qquad\qquad\qquad\qquad + [\gamma_s\,I_{s,n} + \gamma_a\,I_{a,n}]\Delta t + R_n.
\end{aligned}
\tag{13.5}
$$

We begin our simulation on day t_0, the first day of the local lockdown, with initial conditions, S_0, E_0, I_{s0}, I_{a0}, R_0. We further assume that $S_0 = 1 - E_0 - I_{s0} - I_{a0}$ and $E_0 = [A/C_s]\,[I_{s0} + I_{a0}]$ [7] and $v_s\,I_{a0} = [1 - v_s]\,I_{s0}$. This allows us to express all five initial conditions exclusively in terms of the initial symptomatic infectious population I_{s0}, such that $S_0 = 1 - [1 + A/C_s]/v_s\,I_{s0}$ and $E_0 = [A/C_s]/v_s\,I_{s0}$ and $I_{a0} = [1 - v_s]/v_s\,I_{s0}$ and $R_0 = 0$. From the solution of the SEIIR model (13.5), for each point in time, we calculate the of detected population,

$$
D_{n+1} = I_{s,n+1} + R_{s,n+1} = D_n + v_s\,\alpha\,E_n\,\Delta t,
\tag{13.6}
$$

as the discrete sum of the symptomatic infectious and recovered populations $I_{s,n+1}$ and $R_{s,n+1}$ with $\dot{R}_s = \gamma_s I_s$, or, equivalently, as the explicit update of the previous detected population D_n with the new symptomatic infectious population, $v_s\,\alpha\,E_n$, that transitions from the exposed to the symptomatic infectious group. We compare the simulated detected population $D(t)$ against the reported detected population $\hat{D}(t)$, which we obtain by scaling the daily detected cases by the total population N.

Fig. 13.2 Representative seroprevalence studies and reported undercount. Location, time of sampling, and undercount for nine seroprevalence studies with a representative population larger than 0.02% by mid June 2020 as summarized in Table 13.1 [26].

Data. Our objectives are to visualize the effects of asymptomatic transmission, to compare the asymptomatic COVID-19 transmission worldwide, and to estimate the outbreak date of COVID-19 in a specific location.

The first evidence of asymptomatic individuals in a family cluster of three was reported in late January of 2020, where one individual was mildly symptomatic and two remained asymptomatic, with normal lymphocyte counts and chest computer tomography images, but positive quantitative reverse transcription polymerase chain reaction tests [19]. By mid June of 2020, more than 50 studies have reported an asymptomatic population, twenty-three of them with a sample size of at least 500, with a median undercount of 20 across all studies [10], suggesting that only one in twenty COVID-19 cases is noticed and reported. These studies are based on polymerase chain reaction or antibody seroprevalence tests in different subgroups of the population, at different locations, at different points in time [2, 24]. Naturally, the reported undercount varies significantly, ranging from 3.5 and 5.0 for studies in Luxembourg and Germany to 543 and 627 for studies in Iran and Japan. Yet, the reported trend across all studies is strikingly consistent: Six months after the outbreak of the pandemic, a much larger number of individuals displayed antibody prevalence than we would expect from the reported symptomatic case numbers [26].

For the seroprevalence data, we select all seroprevalence studies published until mid June 2020 for which the representation ratio, the ratio between the amount of seroprevalence samples and the total population, was larger than 0.02% [26]. For each location, we estimate the symptomatic fraction ν_s as the ratio between the reported

Table 13.1 Representative seroprevalence studies and reported symptomatic fraction ν_s. Location, time of sampling, number of samples, representation ratio, total population, symptomatic fraction ν_s, and reference for nine seroprevalence studies with a representative population larger than 0.02% by mid June 2020 as illustrated in Figure 13.2.

location study	time of sampling	number of samples	total population	represent ratio	symptomatic fraction
Heinsberg, Germany	Mar 30 - Apr 6	919	41,946	2.191%	20.00%
Ada, Idaho	late April	4,856	481,587	1.008%	7.90%
New York City, New York	May 5 - Jun 5	28,523	8,398,748	0.340%	5.76%
Santa Clara, California	Apr 2 - 3	3,330	1,781,642	0.187%	1.77%
Denmark	Apr 6 - 17	9,496	5,824,857	0.163%	6.95%
Geneva, Switzerland	Apr 20-27	576	504,031	0.114%	10.34%
The Netherlands	Apr 1-15	7,361	17,282,163	0.043%	17.31%
Rio Grande do Sul, Brazil	Apr 25 - 27	4,188	11,377,239	0.037%	8.10%
Belgium	Apr 14 - May 13	2,700	11,492,641	0.023%	10.21%

confirmed cases and the amount of seroprevalence-estimated cases on the last day of the study. Figure 13.2 and Table 13.1 summarize the location, time of sampling, number of samples, representation ratio, total population, and symptomatic fraction ν_s for all nine locations.

For the COVID-19 epidemiology data, we draw the COVID-19 history of all nine locations, Heinsberg, Ada County, New York City, Santa Clara County, Denmark, Geneva Canton, Netherlands, Rio Grande do Sul, and Belgium [12] from the beginning of the pandemic until mid June, 2020. From these data, we extract the number of cumulative cases $\hat{D}(t)$ as the total number of reported COVID-19 cases up until day (t) and scale it by the local population N.

Bayes' theorem. We use Bayes' theorem to estimate the posterior probability distribution of the parameters ϑ, such that the statistics of the simulated detected population $D(\vartheta,t) = I(\vartheta,t) + R(\vartheta,t)$ agrees with the reported detected population $\hat{D}(t)$,

$$P(\vartheta|\hat{D}(t)) = \frac{P(\hat{D}(t)|\vartheta)\,P(\vartheta)}{P(\hat{D}(t))}, \qquad (13.7)$$

where $P(\hat{D}(t)|\vartheta)$ is the likelihood, $P(\vartheta)$ is the prior, $P(\hat{D}(t))$ is the marginal likelihood, and $P(\vartheta|\hat{D}(t))$ is the posterior. Bayes' theorem calculates the point wise product of the likelihood and the prior to produce the normalized posterior probability distribution. The posterior probability distribution is the conditional distribution of the parameters ϑ given the data, in our case, the total number of detected COVID-19 cases $\hat{D}(t)$ including both, currently active and recovered cases.

Prior probability distributions. To reduce the number of free parameters, we assume that the latent and symptomatic infectious periods are disease specific and fix them to $A = 2.5$ days and $C_s = 6.5$ days [14, 15, 28]. Since the asymptomatic infectious period C_a is unknown, we explore three cases with $C_a = 0.5, 1.0, 2.0\,C_s$ resulting in infectious periods of $C_a = 3.25, 6.5$, and 13.0 days. We approximate the

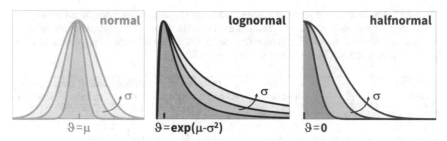

Fig. 13.3 Prior probability distribution. Normal distribution, log-normal distribution, and half-normal distribution. The normal distribution is symmetric around the mean μ with a width that increases with increasing standard deviation σ. The log-normal distribution is the probability distribution of a random variable ϑ whose logarithm $\ln(\vartheta)$ is normally distributed with mean μ, standard deviation σ, and mode $\exp(\mu - \sigma^2)$. The half-normal distribution is symmetric around the mean $\mu = 0$ with a standard deviation σ.

initial exposed and asymptomatic populations as functions of the initial symptomatic population I_{s0}, as $E_0 = [A/C_s][I_{s0} + I_{a0}]$ [7] and $I_{a0} = [1 - v_s]/v_s\, I_{s0}$. This reduces the set of model parameters (13.4) to the symptomatic fraction v_s, the dynamic contact rate $\beta(t)$, and the initial symptomatic population I_{s0},

$$\boldsymbol{\vartheta} = \{\, v_s, \beta(t), I_{s0} \,\}. \tag{13.8}$$

For the symptomatic fraction v_s, we adopt the normal distributions from the individual local antibody seroprevalence studies of all nine locations [26],

Table 13.2 Prior probability distributions and fixed material parameters. Priors for the dynamic contact rate $\beta(t)$ in terms of the initial contact rate $\mu_{\beta 0}$, the overall drift μ_β, and the daily step width σ_β, the initial symptomatic population I_{s0}, and the likelihood width σ; with fixed latent and infectious periods A, C_s, and C_a, and deterministic initial exposed and initial asymptomatic populations E_0 and I_{a0}.

parameter	interpretation	distribution
v_s	symptomatic fraction	see Table 13.1 [26]
A	latent period	constant ($A = 2.5$ days) [14, 15, 20]
C_s	symptomatic infectious period	constant ($C_s = 6.5$ days) [8, 15, 28]
C_a	asymptomatic infectious period	constat ($C_a = 3.25, 6.5, 13.0$ days) [8, 15, 28]
$\log(\beta(t))$	dynamic contact rate	GaussianRandomWalk ($\mu = \mu_\beta, \sigma = \sigma_\beta$)
$\mu_{\beta 0}$	initial contact rate	Normal ($\mu = 0.0, \sigma = 1.0$)
η	overal drift	Normal ($\mu_\eta = 0.0, \sigma_\eta = 1.0$)
σ_β	daily step width	HalfNormal ($\sigma = 0.02$)
E_0	initial exposed	deterministic ($[\,[A/C_s]\,[I_{s0} + I_{a0}\,]\,)$ [7]
I_{s0}	initial symptomatic	LogNormal ($\mu = \hat{D}(t_0), \sigma = 1.0$)
I_{a0}	initial asymptomatic	deterministic ($[1 - v_s\,]/v_s\, I_{s0}$)
σ	likelihood width	HalfCauchy ($\beta = 1.0$)

$$v_s \sim \text{Normal}(\mu_{v_s}, \sigma_{v_s}) \quad \text{with} \quad P(v_s) = \frac{1}{\sqrt{2\pi}\sigma_{v_s}} \exp\left(-\frac{(v_s - \mu_{v_s})^2}{2\sigma_{v_s}^2}\right), \quad (13.9)$$

where μ_{v_s} and σ_{v_s} are the mean and standard deviation of the symptomatic fraction based on the ratio of the reported and seroprevalence-estimated cases for each location. For the dynamic contact rate $\beta(t)$, we postulate that its logarithm follows a Gaussian random walk,

$$\log(\beta(t)) \sim \text{GaussianRandomWalk}(t; \mu_\beta, \sigma_\beta). \quad (13.10)$$

This random walk assumes a sequence of inter-dependent Gaussian distributions where the contact rate $\beta(t_{n+1})$ on day t_{n+1} depends the contact rate $\beta(t_n)$ of the previous day t_n. We update the contact rate on a daily basis assuming a normal distribution around the contact rate of the previous day,

$$\log(\beta(t_{n+1})) \sim \text{Normal}(\log(\beta(t_n)), \sigma_\beta) \quad \text{with}$$
$$P(\log(\beta(t_{n+1}))) = \frac{1}{\sqrt{2\pi}\sigma_\beta} \exp\left(-\frac{(\log(\beta(t_{n+1})) - \log(\beta(t_n)))^2}{2\sigma_\beta^2}\right). \quad (13.11)$$

The standard deviation σ_β defines the daily step width for which we assume a half-normal distribution,

$$\sigma_\beta \sim \text{HalfNormal}(\sigma_{\sigma\beta}) \quad \text{with} \quad P(\sigma_\beta) = \frac{1}{\sqrt{2\pi}\sigma_{\sigma\beta}} \exp\left(-\frac{\sigma_\beta^2}{2\sigma_{\sigma\beta}^2}\right), \quad (13.12)$$

with standard deviation $\sigma_{\sigma\beta}$. For the logarithm of the initial contact rate, we assume a normal distribution,

$$\log(\beta_0) \sim \text{Normal}(\mu_{\beta0}, \sigma_{\beta0}) \text{ with } P(\beta_0) = \frac{1}{\sqrt{2\pi}\sigma_{\beta0}} \exp\left(-\frac{(\log(\beta_0) - \mu_{\beta0})^2}{2\sigma_{\beta0}^2}\right),$$
$$(13.13)$$

with mean and standard deviation $\mu_{\beta0}$ and $\sigma_{\beta0}$. For the overall drift between the initial value $\log(\beta(t_0))$ and the final value $\log(\beta(t_{n+1}))$, we assume a normal distribution,

$$\eta \sim \text{Normal}(\mu_\eta, \sigma_\eta) \text{ with } P(\eta) = \frac{1}{\sqrt{2\pi}\sigma_\eta} \exp\left(-\frac{(\eta - \mu_\eta)^2}{2\sigma_\eta^2}\right), \quad (13.14)$$

with mean and standard deviation μ_η and σ_η. For the initial symptomatic infectious population I_{s0}, we postulate a log-normal distribution,

$$I_{s0} \sim \text{LogNormal}(\mu_{Is0}, \sigma_{Is0}) \text{ with } P(I_{s0}) = \frac{1}{\sqrt{2\pi}\sigma I_{s0}} \exp\left(-\frac{(\ln(I_{s0}) - \mu_{Is0})^2}{2\sigma_{Is0}^2}\right),$$
$$(13.15)$$

with median $\exp(\mu_{Is0})$ and standard deviation σ_{Is0}. We express the median in terms of the reported detected population on day t_0 as $\mu_{Is0} = \hat{D}(t_0)$.

Table 13.2 summarizes our prior distributions and the SEIIR model parameters.

Likelihood function. For the likelihood $P(\hat{D}(t)|\boldsymbol{\vartheta})$, we introduce a likelihood function $\mathcal{L}(\boldsymbol{\vartheta}, t_i)$ that evaluates the proximity between a sample of the observed data $\hat{D}(t_i)$ and the model output $D(\boldsymbol{\vartheta}, t_i)$ for given values of the unknown parameters $\boldsymbol{\vartheta}$ at evert time point of interest t_i. The observed data $\hat{D}(t_i)$ are the cumulative reported COVID-19 cases and the model output $D(\boldsymbol{\vartheta}, t_i)$ are the cumulative simulated symptomatic cases. We choose a Student's t-distribution for the likelihood $\mathcal{L}$,

$$\mathcal{L}(\boldsymbol{\vartheta}, t_i) \sim \text{studentT}_{\nu=4}(\hat{D}(t_i) - D(\boldsymbol{\vartheta}, t_i), \sigma). \tag{13.16}$$

because it resembles a Gaussian distribution around the mean, but has heavy tails, which make it robust with respect to outliers [5, 13]. For the likelihood width σ between the cumulative reported and simulated symptomatic COVID-19 cases $\hat{D}(t_i)$ and $D(\boldsymbol{\vartheta}, t_i)$, we select a half Cauchy distribution,

$$\sigma \sim \text{HalfCauchy}(\beta) \quad \text{with} \quad P(\sigma) = \frac{1}{\pi\beta\left[1 + [\sigma/\beta]^2,\right]} \tag{13.17}$$

with scaling parameter β, see Table 13.2. To compare the reported and simulated cases throughout the entire time window, from day t_0 to day t_n, we evaluate the likelihood function $\mathcal{L}(\boldsymbol{\vartheta}, t_i)$ from equation (13.16) at each discrete time point t_i and multiply the individual terms to calculate the overall likelihood $P(\hat{D}(t)|\boldsymbol{\vartheta})$,

$$P(\hat{D}(t)|\boldsymbol{\vartheta}) = \prod_{i=0}^{n} \mathcal{L}(\boldsymbol{\vartheta}, t_i). \tag{13.18}$$

Here the product symbol, $\prod_{i=0}^{n}$, denotes the multiplication of the $i = 0, ..., n$ individual likelihood functions $\mathcal{L}(\boldsymbol{\vartheta}, t_i)$.

Posterior probability distributions. We apply Bayes' theorem (13.7) to obtain the posterior distribution of the parameters [26] using the likelihood $P(\hat{D}(t)|\boldsymbol{\vartheta})$ and priors $P(\boldsymbol{\vartheta})$. As we have discussed in the previous chapters, the denominator in Bayes' theorem, the marginal likelihood $P(\hat{D}(t))$, is generally difficult to compute. It essentially acts as a normalization factor and we can rewrite Bayes' theorem as a proportionality,

$$P(\boldsymbol{\vartheta}|\hat{D}(t)) \propto P(\hat{D}(t)|\boldsymbol{\vartheta}) P(\boldsymbol{\vartheta}), \tag{13.19}$$

and simply calculate the normalized posterior distribution as the product of the likelihood and priors. Since we cannot describe the posterior distribution over the model parameters $\boldsymbol{\vartheta}$ analytically, we adopt approximate-inference techniques to calibrate our model on the available data.

Markov Chain Monte Carlo. We numerically approximate the posterior probability distribution, using the NO-U-Turn sampler [9] implementation of the Python package PyMC3 [23, 27]. We use four chains. The first four times 500 samples serve to tune the sampler and are later discarded; the subsequent four times 1000 samples define the posterior distribution of our model parameters. From the converged posterior distribution $P(\boldsymbol{\vartheta} \mid \hat{D}(t))$, we sample multiple combinations of parameters $\boldsymbol{\vartheta}$, from

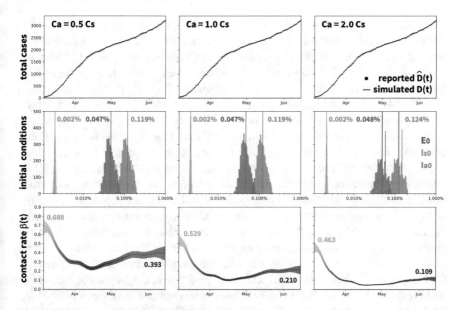

Fig. 13.4 Outbreak dynamics of COVID-19 in Santa Clara County. The simulation learns the dynamic contact rate $\beta(t)$ for fixed latent and symptomatic infectious periods $A = 2.5$ days and $C_s = 6.5$ days, and for three asymptomatic infectious periods $C_a = 3.25$ days, 6.5 days, and 13.0 days (from left to right). Reported cases $\hat{D}(t)$ and simulated cases $D(t) = I_s(t) + R_s(t)$ (top row); initial exposed and infectious populations, E_0, I_{s0}, and I_{a0} (middle row); and dynamic contact rate, $\beta(t)$ (bottom row). Colored regions highlight the 95% credible intervals [26].

which we quantify means and credible intervals as a measure of the uncertainty on each parameter. Each parameter set ϑ provides a set of values for the symptomatic fraction ν_s, the dynamic contact rate $\beta(t)$, and the initial symptomatic populations I_{s0}. From these values, we quantify the effective reproductive number $R(t)$ using equation (13.2) and the time evolution of the susceptible, exposed, symptomatic infectious, asymptomatic infectious, and recovered populations, $S(t)$, $E(t)$, $I_s(t)$, $I_a(t)$, and $R(t)$ using equations (13.5) and report their means and 95% credible intervals.

13.2 Example: Visualizing asymptomatic COVID-19 transmission

To visualize the effects of asymptomatic transmission, we combine the deterministic SEIIR model with a dynamic reproduction number, machine learning, and uncertainty quantification to learn the reproduction number–in real time–and quantify and propagate uncertainties in the symptomatic-to-asymptomatic ratio and in the initial exposed and infectious populations [26]. We focus on Santa Clara County, the first county in the United States to introduce stay-at-home orders and the first to publish a broad seroprevalence study [2].

Figure 13.4 illustrates the outbreak dynamics of COVID-19 in Santa Clara County. To explore the effect of the asymptomatic infectious period C_a on the initial dynamic reproduction number, we begin the simulations on March 6, 2020, ten days before the local lockdown date [25]. The black dots highlight the reported detected population $\hat{D}(t)$ from which we learn the posterior distributions of our SEIIR model parameters for fixed latent and symptomatic infectious periods $A = 2.5$ days and $C_s = 6.5$ days, and for three asymptomatic infectious periods, $C_a = 3.25$, 6.5, and 13.0 days. The gray and green-purple regions highlights the 95% credible intervals on the simulated cases $D(t)$ and the contact rate $\beta(t)$ based on the reported cases $\hat{D}(t)$, while taking into account uncertainties on the symptomatic infectious fraction ν_s, and the initial exposed and infectious populations E_0, I_{s0}, and I_{a0}. The red, orange, and green histograms display the inferred initial exposed and infectious populations, E_0, I_{s0}, and I_{a0}. The graphs confirm that, irrespective of the asymptomatic infectious period C_a, our dynamic SEIIR epidemiology model is capable of correctly capturing the gradual flattening and re-steepening of the case curve. The consistent downward trend of the contact rate $\beta(t)$ after the lockdown date on March 16, 2020 quantifies the efficiency of public health interventions. The different magnitudes in the contact rate highlight the effect of the three different asymptomatic infectious periods C_a: For larger asymptomatic infectious periods C_a, from left to right, to explain the same number of confirmed cases $D(t) = I_s(t) + R_s(t)$, the contact rate $\beta(t)$ has to decrease. On March 6, 2020, the mean contact rate $\beta(t)$ was 0.688 (95% CI: 0.612 - 0.764) for an infectious period of C_a=3.25 days, 0.529 (95% CI: 0.464 - 0.593) for C_a= 6.5 days, and 0.463 (95% CI: 0.403 - 0.523) for C_a=13.0 days. By March 16, 2020, the day Santa Clara County announced the first county-wide shelter-in-place order in the entire United States, these contact rates $\beta(t)$ were 0.491 (95% CI: 0.462 - 0.523) for an infectious period of C_a=3.25 days, 0.328 (95% CI: 0.305 - 0.352) for C_a= 6.5 days, and 0.252 (95% CI: 0.234 - 0.271) for C_a=13.0 days.

Figure 13.5 visualizes the effect of asymptomatic transmission in Santa Clara County. The simulation learns the dynamic contact rate $\beta(t)$, and with it the time-varying dynamic reproduction number $R(t)$, for three asymptomatic infectious periods $C_a = 3.25$ days, 6.5 days, and 13.0 days. The dynamic reproduction number $R(t)$ follows a similar downward trend as the contact rate $\beta(t)$. For larger asymptomatic infectious periods C_a, since $R(t) = C_s \beta(t) / [\nu_s + \nu_a C_s / C_a]$, as C_a increases, C_s / C_a decreases, and $R(t)$ increases. Since $R(t)$ represents the number of new infections from a single case, a decrease below $R(t) < 1$ implies that a single infectious individual infects less than one new individual, which indicates that the outbreak decays [6]. The dashed vertical lines mark the date at which it drops below one, $R(t) < 1$, meaning that one infectious individual, either symptomatic or asymptomatic, infects less than one other individual. For an asymptomatic infectious period of C_a=3.25 days, it took until March 28 before Santa Clara County managed to reduce $R(t)$ below one. For C_a=6.5 days, this only occurred by April 1, and for C_a=13.0 days, this occurred on April 8, 2020. This confirms our intuition that, the larger the asymptomatic infectious period C_a, for example because asymptomatic individuals will not isolate as strictly as symptomatic individuals, the higher the dynamic reproduction number $R(t)$, and the more difficult it will be to control $R(t)$ by public health interventions.

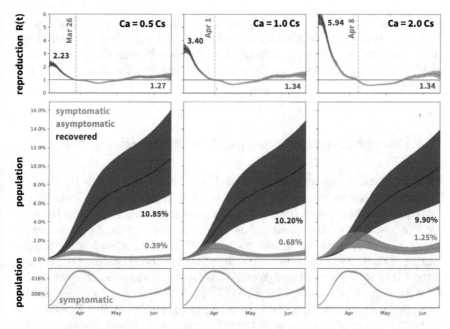

Fig. 13.5 Outbreak dynamics of COVID-19 in Santa Clara County. The simulation learns the dynamic contact rate $\beta(t)$ for fixed latent and symptomatic infectious periods $A = 2.5$ days and $C_s = 6.5$ days, and for three asymptomatic infectious periods $C_a = 3.25$ days, 6.5 days, and 13.0 days (from left to right). Dynamic reproduction number $R(t)$ (top row) and symptomatic, asymptomatic, and recovered populations I_s, I_a, and R (middle and bottom rows). Colored regions highlight the 95% credible intervals [26].

For larger asymptomatic infectious periods C_a, the total infectious population I increases and its maximum occurs later in time. Specifically, the model predicts a maximum infectious population of 0.70% (95% CI: 0.43%-0.97%) on March 28, 2020 for $C_a = 3.25$ days, 1.23% (95% CI: 0.72%-1.75%) on April 2, 2020 for $C_a = 6.5$ days, and 2.10% (95% CI: 1.25%-2.94%) on April 7, 2020 for $C_a = 13.0$ days. For the date of this study, June 15, 2020, the model predicts a total infectious population, $I = I_s + I_a$, of 0.39% (95% CI: 0.24%-0.58%) for $C_a = 3.25$ days, 0.68% (95% CI: 0.40%-1.01%) for $C_a = 6.5$ days, and 1.25% (95% CI: 0.73%-1.77%) for $C_a = 13.0$ days. For larger asymptomatic infectious periods C_a, the recovered population R decreases. For June 15, 2020, the model predicts a recovered population R of 10.85% (95% CI: 7.06%-16.09%) for an infectious period of $C_a = 3.25$ days, 10.20% (95% CI: 6.07%-14.90%) for $C_a = 6.5$ days, and 9.90% (95% CI: 6.07%-13.93%) for $C_a = 13.0$ days.

This example shows that the overall dynamic reproduction number, $R(t) = [C_a C_s]/[v_s C_a + v_a C_s]\, \beta(t)$, and with it the outbreak dynamics, depend critically on the fractions of the symptomatic and asymptomatic populations v_s and v_a but, to a lesser extent, on the ratio of their infectious periods C_s and C_a. This ratio reflects a combined effect of asymptomatic individuals being less contagious but, at the same

time, having more contacts, since, by definition, they do not realize that they are
spreading the disease. We conclude that neither a smaller nor larger infectiousness
of the asymptomatic group significantly alters the overall outbreak dynamics compared to a similar infectiousness, $C_a = C_s$, in the middle columns of Figures 13.4
and 13.5.

13.3 Example: Asymptomatic COVID-19 transmission worldwide

We now expand the analysis beyond Santa Clara County and explore the effects of
asymptomatic transmission in all nine locations in Figure 13.2 and Table 13.1 to
show how the observations from a single location generalize to different locations
across the world at different stages of the pandemic [26]. Figure 13.6 visualizes the
effects of asymptomatic transmission worldwide. The simulation learns the time-
varying dynamic contact rate $\beta(t)$, and with it the time-varying dynamic reproduction
number $R(t)$, for fixed latent, symptomatic and asymptomatic infectious periods $A =$
2.5 days, $C_s = 6.5$ days, and $C_a = 5.76$ (95%CI: 3.59-8.09) days, a period that we
have previously inferred in a hierarchical analysis [26]. For all nine locations, the
symptomatic, asymptomatic, and recovered populations I_s, I_a, and R display the fine
balance between the dynamics of the reproduction number and the control of the
epidemic outbreak. The downward trend of the effective reproductive number $R(t)$
quantifies how fast each location managed to control the spreading of COVID-19.
The dashed vertical line marks the date at which each location reduced the effective
reproduction below one, $R(t) < 1$. This critical transition occurred on March 20
for Heinsberg, April 4 for Ada County, April 28 for New York City, March 30 for
Santa Clara County, April 8 for Denmark, April 2 for Geneva, April 14 for the
Netherlands, June 8 for Rio Grande do Sul and April 11 for Belgium. Based on our
simulations, the maximum infectious population for this first wave of the outbreak
was 3.54% (95% CI: 2.96%-4.11%) on March 19 in Heinsberg, 0.47% (95% CI:
0.41%-0.55%) on May 5 in Ada County, 6.11% (95% CI: 5.99%-6.21%) on April
13 in New York City, 1.11% (95% CI: 0.69%-1.65%) on March 29 in Santa Clara
County, 0.38% (95% CI: 0.27%-0.50%) on April 9 in Denmark, 2.10% (95% CI:
1.65%-2.69%) on April 2 in Geneva, 0.62% (95% CI: 0.56%-0.68%) on April 15
in the Netherlands, 0.28% (95% CI: 0.14%-0.48%) on June 8 in Rio Grande do
Sul, and 0.70% (95% CI: 0.62%-0.78%) on April 11 in Belgium. On Jun 15, 2020,
the estimated recovered population reached 24.15% (95% CI: 20.48%-28.14%) in
Heinsberg, NRW, Germany 2.40% (95% CI: 2.09%-2.76%) in Ada County, ID, USA
46.19% (95% CI: 45.81%-46.60%) in New York City, NY, USA 11.26% (95% CI:
7.21%-16.03%) in Santa Clara County, CA, USA 3.09% (95% CI: 2.27%-4.03%)
in Denmark 12.35% (95% CI: 10.03%-15.18%) in Geneva Canton, Switzerland
5.24% (95% CI: 4.84%-5.70%) in Netherlands 1.53% (95% CI: 0.76%-2.62%) in
Rio Grande do Sul, Brazil 5.32% (95% CI: 4.77%-5.93%) in Belgium. Despite
these differences, the dynamic reproduction numbers $R(t)$ and the infectious and
recovered populations I_s, I_a, and R in Figure 13.6 display remarkably similar trends:

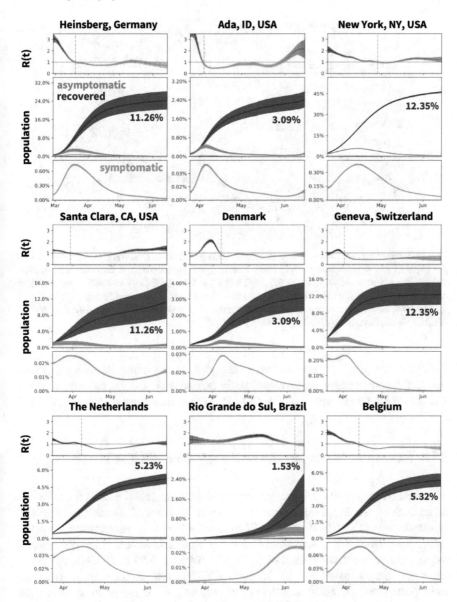

Fig. 13.6 Outbreak dynamics of COVID-19 worldwide. The simulation learns the time-varying dynamic contact rate $\beta(t)$ for fixed latent, symptomatic, and asymptomatic infectious periods $A = 2.5$ days, $C_s = 6.5$ days, and $C_a = 5.76$ days, to predict the dynamic reproduction number $R(t)$ and the symptomatic, asymptomatic, and recovered populations I_s, I_a, and R. Colored regions highlight the 95% credible intervals [26].

In most locations, the dynamic reproduction number $R(t)$ drops rapidly to values below one within a window of about three weeks after the lockdown date. However,

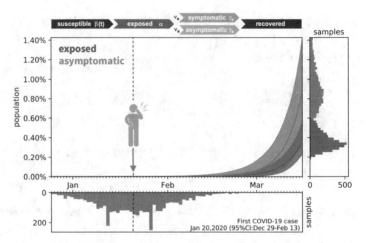

Fig. 13.7 Inferring the outbreak date of COVID-19 in Santa Clara County. The simulation learns the exposed and asymptomatic populations E and I_a for fixed latent, symptomatic, and asymptomatic periods $A = 2.5$ days, $C_s = 6.5$ days, and $C_a = 5.76$ (95%CI: 3.59-8.09) days. The right histograms highlight the credible intervals on the exposed and asymptomatic populations E and I_a. The bottom histogram shows the distribution around the most probable origin date, January 20, 2020 (95% CI: December 29, 2019 - February 13, 2020) [26].

the maximum infectious population, a value that is closely monitored by hospitals and health care systems, varies significantly ranging from 0.28% and 0.38% in Rio Grande do Sul and Denmark to 3.54% and 6.11% in Heinsberg and New York City.

13.4 Example: Inferring the outbreak date of COVID-19

A critical question early in a pandemic is to identify the initial date of the outbreak to understand the impact of community spread. Figure 13.7 shows the estimated outbreak date of COVID-19 in Santa Clara County. For fixed latent and symptomatic infectious periods $A = 2.5$ days and $C_s = 6.5$ days, and for a hierarchically estimated asymptomatic infectious period of $C_a = 5.76$ (95%CI: 3.59-8.09) days, the graph highlights the estimated date of the first COVID-19 case in the county. Based on the reported case $\hat{D}(t)$ [25], and the uncertainty on the fraction of the symptomatic infectious population ν_s [2], and the initial exposed, symptomatic, and asymptomatic populations E_0, I_{s0}, and I_{a0}, we can systematically backtrack the date of the first undetected infectious individual. Our results suggest that the first case of COVID-19 in Santa Clara County dates back to January 20, 2020 (95% CI: December 29, 2019 - February 13, 2020). Santa Clara County was home to the first individual who died with COVID-19 in the United States. Although this happened as early as February 6, the case remained unnoticed until April 22, 2020. This suggests that the new coronavirus was circulating in the Bay Area as early as January. The estimated uncertainty on the exposed, symptomatic, and asymptomatic populations allows us to infer the initial outbreak date dates back to January 20, 2020. This back-calculated early outbreak date is in line with our intuition that COVID-19 is often present in a population long before the first official case is reported. Interestingly, our analysis comes to this conclusion purely based on a local serology antibody study and the number of reported cases after lockdown.

Taken together, these examples reveal strikingly consistent trends across different locations and different time points: A much larger number of individuals displays antibody prevalence than we would expect from the reported symptomatic case numbers. Knowing the exact dimension of asymptomatic transmission is critical for two reasons: first, to truly estimate the severity of the outbreak, e.g., hospitalization or mortality rates, and second, to reliably predict the success of surveillance and control strategies, e.g., contact tracing or vaccination. While it is difficult to measure the effects of the unreported asymptomatic group directly, mathematical modeling, in conjunction with reported symptomatic case data, antibody seroprevalence studies, and machine learning allows us to infer–in real time–the epidemiology characteristics of COVID-19. We can visualize the invisible asymptomatic population, estimate its role in disease transmission, and quantify the confidence in these predictions. A better understanding of asymptomatic transmission will help us evaluate strategies to manage the impact of COVID-19 on both our economy and our health care system. A large asymptomatic population is associated with a high risk of community spread and could require a conscious shift from containment to mitigation induced by behavior changes. Our study suggests that, without vaccination and treatment, increasing population awareness, encouraging increased hygiene, mandating the use of face masks, restricting travel, and promoting physical distancing could be the most successful strategies to manage the impact of COVID-19 on both our economy and our health care system.

Problems

13.1 Limitations of seroprevalence studies. On April 17, 2020, the first study on "COVID-19 Antibody Seroprevalence in Santa Clara County, California" was published on medRxiv. It estimated an undercount larger than 50 and generated a wave of criticism in public media. Find and read some of the comments on the study. What are the major points of criticism? Discuss the potential limitations of seroprevalence studies in general and of the COVID-19 antibody seroprevalence study in Santa Clara County in particular.

13.2 Credible intervals. The Bayesian analysis provides posterior probability distributions of the parameter values. These not only contain information about the means of the parameters, but also about their credible intervals. Compare the 95% credible intervals on the basic reproduction number $R(t)$ and on the symptomatic, asymptomatic, and recovered populations I_s, I_a, and R. Explain, in your own words,

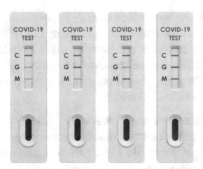

Fig. 13.8 COVID-19 seroprevalence test. The rapid test cassette of this COVID-19 seroprevalence test has three lines: one for the control, one for IgG, and one for IgM. IgM begins to rise about one week after the initial infection and IgG appears about a weeks later. However, IgG can last for months or even years, and serves as a long-term indicator for a previous infection.

why the credible intervals change in time and why some credible intervals are larger than others. Explain what this implies.

13.3 Antibody isotypes. The accurate interpretation of seroprevalence tests depends on antigen specificity and on the type of antibody. Humans have five classes of antibodies, IgM, IgG, IgA, IgD, and IgE. COVID-19 seroprevalence tests target IgM, IgG, IgA and the total antibody count. Of the nine locations we analyzed here, Heinsberg tested IgG and IgA, Ada County tested IgG, New York City tested IgG, Santa Clara County tested IgG and IgM, Denmark tested IgG and IgM, Geneva tested IgG, the Netherlands tested IgG, IgM, and IgA, Rio Grande do Sul tested IgG and IgM, and Belgium tested IgG. Discuss how the results in Figure 13.2 would have changed, if all locations had tested for all three antibodies. How would this have affected the dynamics in Figure 13.6.

13.4 COVID-19 seroprevalence. COVID-19 seroprevalence tests target three antibodies, IgM, IgG, and IgA. Figure 13.8 shows a rapid test cassette of a COVID-19 seroprevalence test with three lines, one for the control, one for IgG, and one for IgM. IgM is a ring-type pentamer of five antibodies, the largest antibody in size, and the first to be produced during infection. IgG is the smallest and most abundant antibody, it appears later in the infection, is the most specific and long-lasting isotype, and a key player in establishing post-infection immunity. Ada County, New York City, Geneva, and Belgium only tested tested IgG, whereas Santa Clara County, Denmark, and Rio Grande do Sul tested IgG and IgM. Discuss how the type of antibody test affects the dynamics in Figure 13.6.

13.5 Seroprevalence. After mid June, numerous more sophisticated seroprevalence studies have been performed, some of them even longitudinally. Figure 13.9 shows the results of a seroprevalence study across the United States based on 47,909 samples collected between September 7-24, 2020 [1]. Interpret the results in view

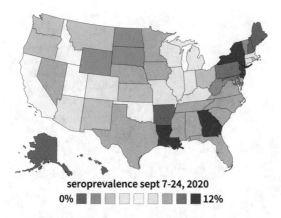

seroprevalence sept 7-24, 2020
0% ■ ■ ■ ■ □ □ ■ ■ ■ ■12%

Fig. 13.9 Estimated seroprevalence across the United States. Seroprevalence was largest in New York with 17.0%, New Jersey with 15.1%, and Georgia with 13.0% and smallest in New Hampshire with 0.7%, Maine with 0.5%, and Alaska with 0.4%, from September 7-24, 2020.

of seroprevalence heterogeneity and try to find explanations for the large and small prevalence.

13.6 Seroprevalence. Figure 13.9 shows the results of a seroprevalence study across the United States based on 47,909 samples collected between September 7-24, 2020 [1]. Compare its seroprevalence against the early COVID-19 spreading across the United States in Figure 10.5. Comment on regions with high seroprevalence and regions of early COVID-19 spread.

13.7 Seroprevalence. Figure 13.9 shows the results of a seroprevalence study across the United States based on 47,909 samples collected between September 7-24, 2020 [1]. Compare the seroprevalence in the state of New York to the estimated total recovered population in New York city in Figure 13.6. Comment on the history of COVID-19 in New York and find explanations for the large seroprevalence.

13.8 Asymptomatic transmission. On June 8, 2020, the World Health Organization announced that the asymptomatic spread of COVID-19 is "very rare". After a wave of criticism, this statement was corrected only a day later to asymptomatic spread "is a really complex question". Discuss how a large fraction of asymptomatic transmission would change mitigation strategies to manage the spread of COVID-19. Identify at least five strategies, and comment on whether or not asymptomatic transmission influences their efficiency.

13.9 Asymptomatic transmission. Discuss how the asymptomatic fraction v_a or, similarly, the undercount $U = v_a/[1 - v_a]$, may vary during the course of a pandemic. Think of the testing frequency during the beginning of the COVID-19 pandemic when testing was expensive and rare and during later stages, when regular surveillance testing had become standard in many subgroups of the population. How do these changes affect the uncertainty and credible intervals of our prediction.

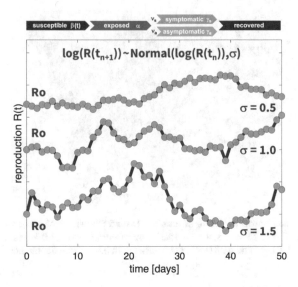

Fig. 13.10 Dynamic reproduction number of logarithmic Gaussian random walk type. The logarithmic Gaussian random walk assumes a sequence of inter-dependent log-normal distributions where the reproduction number $R(t_{n+1})$ on day t_{n+1} depends the reproduction number $R(t_n)$ of the previous day t_n, such that the standard deviation σ defines the daily step width. With increasing daily step width, $\sigma = 0.5, 1.0, 1.5$, the fluctuations around the basic reproduction number R_0 increaese.

13.10 Logarithmic Gaussian random walk prior. In our dynamic SEIIR model, in equations (13.10) and (13.11), we have assumed a logarithmic Gaussian random walk prior for the dynamic contact rate $\beta(t)$, or equivalently, for the dynamic reproduction number $R(t) = \beta(t)/\gamma$. Compare the hyperbolic tangent type ansatz (12.2) in Figure 12.3 and the logarithmic Gaussian random walk ansatz (13.10) and (13.11) in Figure 13.10. What are the advantages and disadvantages of each approach. Why, in your opinion, is the logarithmic Gaussian random walk prior a reasonable choice for the dynamic SEIIR model in this Chapter?

13.11 Logarithmic Gaussian random walk prior. In our dynamic SEIIR model, in equations (13.10) and (13.11), we have assumed a logarithmic Gaussian random walk prior for the dynamic contact rate $\beta(t)$, or equivalently, for the dynamic reproduction number $R(t) = \beta(t)/\gamma$. The logarithmic Gaussian random walk assumes a sequence of inter-dependent log-normal distributions where the reproduction number $R(t_{n+1})$ on day t_{n+1} depends the reproduction number $R(t_n)$ of the previous day t_n, such that the standard deviation σ defines the daily step width. Implement and plot the logarithmic Gaussian random walk for $\sigma = 0.5, 1.0, 1.5$. Compare your graphs against Figure 13.10. Comment on the role of the σ parameter in view of COVID-19 case data.

References

1. Bajema KL, Wiegand RE, Cuffe K, Patel SV, Iachan R, Lim T, Lee A, Moyse D, Havers FP, Harding L, Fry AM, Hall AJ, Martin K, Biel M, Deng Y, Meyer WA, Mathur M Kyle T, Gundlapalli AV, Thornburg NJ, Petersen LR, Edens C (2021) Estimated SARS-CoV-2 seroprevalence in the US as of September 2020. JAMA Internal Medicine 181:450-460.
2. Bendavid E, Mulaney B, Sood N, Shah S, Ling E, Bromley-Dulfano R, Lai C, Weissberg Z, Saavedra-Walker R, Tedrow J, Tversky D, Bogan A, Kupiec T, Eichner D, Gupta R, Ioannidis JPA, Bhattacharya J (2020). COVID-19 antibody seroprevalence in Santa Clara County, California. medRxiv doi:10.1101/2020.04.14.20062463.
3. Brauer F (2006) Some simple epidemic models. Mathematical Biosciences and Engineering 3:1-15.
4. Brauer, F, Castillo-Chavez C, Feng Z (2019) Mathematical Models in Epidemiology. Springer-Verlag New York.
5. Dehning J, Zierenberg J, Spitzner FP, Wibral M, Pinheiro Neto J, Wilczek M, Priesemann V (2020) Inferring COVID-19 spreading rates and potential change points for case number forecasts. Science 369:eabb9789.
6. Dietz K (1993) The estimation of the basic reproduction number for infectious diseases. Statistical Methods in Medical Research 2:23-41.
7. Engbert R, Drepper FR (1994) Chance and chaos in population biology-Models of recurrent epidemics and food chain dynamics. Chaos, Solutions & Fractals 4:1147-1169.
8. He X, Lau EH, Wu P, Deng X, Wang J, Hao X, Lau YC, Wong JY, Guan Y, Tan X, X. Mo X (2020) Temporal dynamics in viral shedding and transmissibility of COVID-19. Nature Medicine 26:672-675.
9. Hoffman MD, Gelman A (2014) The No-U-Turn sampler: adaptively setting path lengths in Hamiltonian Monte Carlo. Journal of Machine Learning Research 15:1593-1623.
10. Ioannidis JPA (2021) The infection fatality rate of COVID-19 inferred from seroprevalence data. Bulletin of the World Health Organization 99:19-33.
11. Johns Hopkins University (2021) Coronavirus COVID-19 Global Cases by the Center for Systems Science and Engineering. https://coronavirus.jhu.edu/map.html, https://github.com/CSSEGISandData/covid-19 assessed: June 1, 2021.
12. Kuhl E (2020) Data-driven modeling of COVID-19 – Lessons learned. Extreme Mechanics Letters 40:100921.
13. Lange KL, Little RJA, Taylor MG (1989) Robust statistical modeling using the t distribution. Journal of the American Statistical Association 84:881-896.
14. Lauer SA, Grantz KH, Bi Q, Jones FK, Zheng Q, Meredith HR, Azman AS, Reich NG, Lessler J (2020) The incubation period of coronavirus disease 2019 (COVID-19) from publicly reported confirmed cases: estimation and application. Annals of Internal Medicine 172:577-582.
15. Li Q, Guan X, Wu P, Wang X, ... Feng Z (2020) Early transmission dynamics in Wuhan, China, of novel coronavirus-infected pneumonia. New England Journal of Medicine 382:1199-1207.
16. Moin P (2000) Fundamentals of Engineering Numerical Analysis. Cambridge University Press.
17. Osvaldo M (2018) Bayesian Analysis with Python: Introduction to Statistical Modeling and Probabilistic Programming Using PyMC3 and ArviZ. Packt Publishing, 2nd edition.
18. New York Times (2020) Coronavirus COVID-19 Data in the United States. https://github.com/nytimes/covid-19-data/blob/master/us-states.csv assessed: June 1, 2021.
19. Pan X, Chen D, Xia Y, Wu X, Li T, Ou X, Zhou L, Liu J (2020) Asymptomatic cases in a family cluster with SARS-CoV-2 infection. Lancet Infectious Diseases 20:410-411.
20. Peirlinck M, Linka K, Sahli Costabal F, Kuhl E (2020) Outbreak dynamics of COVID-19 in China and the United States. Biomechanics and Modeling in Mechanobiology 19:2179-2193.
21. Peirlinck M, Linka K, Sahli Costabal F, Bendavid E, Bhattacharya J, Ioannidis J, Kuhl E (2020) Visualizing the invisible: The effect of asymptomatic transmission on the outbreak dynamics of COVID-19. Computer Methods in Applied Mechanics and Engineering 372:113410.
22. Salvatier J, Wiecki TV, Fonnesbeck C (2016) Probabilistic programming in Python using PyMC3. PeerJ Computer Science 2:e55.

23. Sanche S, Lin Y, Xu C, Romero-Severson E, Hengartner N, Ke R (2020) High contagiousness and rapid spread of severe acute respiratory syndrome coronavirus 2. Emerging Infectious Diseases 2020;26(7):1470-1477.
24. Streeck H, Schulte B, Kümmerer BM, Richter E, Höller T, Fuhrmann C, Bartok E, Dolscheid R, Berger M, Wessendorf L, Eschbach-Bludau M, Kellings A, Schwaiger A, Coenen M, Hoffmann P, Stoffel-Wagner B, Nöthen MM, Eis-Hübinger AM, Exner M, Schmithausen RM, Schmid M, Hartmann G (2020) Infection fatality rate of SARS-CoV-2 infection in a German community with a super-spreading event. Nature Communication 11:5829.

Chapter 14
Data-driven network SEIR model

Abstract During a global pandemic, a key strategy to prevent a local outbreak is to restrict incoming travel. Once a region has successfully contained the disease, a critical question is to decide when and how to reopen the borders. Here we explore the impact of border reopening for the example of Newfoundland and Labrador, a Canadian province that has been under a strict travel ban and COVID-free for almost for almost two months. We combine what we have learned throughout this book, SEIR compartment modeling, dynamic contact rates, network modeling, and Bayesian analysis to explore different reopening strategies. We draw the case data of COVID-19 across North America and the daily incoming air traffic to Newfoundland to infer our model parameters, and predict the COVID-19 dynamics upon partial and total airport reopening, with perfect and imperfect quarantine conditions. We demonstrate that banning air travel from outside Canada is more efficient in managing the pandemic than fully reopening and quarantining 95% of the incoming population. The learning objectives of this chapter on data-driven network modeling are to

- analyze epidemiological case data using data-driven physics-based models
- design a Bayesian analysis using network SEIR models
- explain the risk of mixing populations with different disease prevalence
- quantify the effects of quarantine and travel restrictions
- design what-if studies to answer relevant questions and guide decision making
- understand and discuss the limitations of data-driven disease models

By the end of the chapter, you will be able to to infer dynamic reproduction numbers, design studies that answer important questions, and inform political decision making to manage infectious diseases like COVID-19 based on reported case data.

14.1 Introduction of the data-driven network SEIR model

Dynamic SEIR model and parameters. We model the local epidemiology of the COVID-19 outbreak using a dynamic SEIR model (7.15) as introduced in Section 7.7.

Fig. 14.1 Network SEIR model. Each node of the network SEIR model contains four compartments for the susceptible, exposed, infectious, and recovered populations, S, E, I, and R. The transition rates between the compartments of each node are the dynamic contact rate $\beta(t)$, and the latent, and infectious rates, α and γ. The diffusion between the four compartments of connected nodes I and J is a product of the mobility coefficients κ_S, κ_E, κ_I, and κ_R and the graph Laplacian L_{IJ}.

As indicated in Figure 14.1, the dynamic SEIR consists of four populations, the susceptible, exposed, infectious, and recovered groups, governed by the following system of four coupled ordinary differential equations [8],

$$
\begin{aligned}
\dot{S} &= -\beta(t)\,S\,I \\
\dot{E} &= +\beta(t)\,S\,I - \alpha\,E \\
\dot{I} &= \qquad\quad + \alpha\,E - \gamma\,I \\
\dot{R} &= \qquad\qquad\qquad + \gamma\,I\,.
\end{aligned}
\tag{14.1}
$$

The local transition rates between the compartments of each node are the dynamic contact rate, $\beta(t)$, and the latent and infectious rates, α and γ, in units [1/days]. They are inverses of the dynamic contact period $B(t)$, and the the latent and infectious periods, $A = 1/\alpha$ and $C = 1/\gamma$, in units [days]. We assume that the latent and infectious periods, $A = 2.5$ days and $C = 6.5$ days, are disease-specific for COVID-19, and constant in space and time. To account for societal and political interventions, we introduce a behavior-dependent dynamic contact rate, $\beta(t)$, that varies both in space and time [20]. For easier interpretation, we express the contact rate,

$$
\beta(t) = \mathsf{R}(t)\,\gamma,
\tag{14.2}
$$

as the product of the dynamic reproduction number $\mathsf{R}(t)$ and the infectious rate γ. For the dynamic reproduction number $\mathsf{R}(t)$, we choose an ansatz of Gaussian random walk type [18],

$$
\mathsf{R}(t) \sim \mathrm{Normal}(t; \mu, \tau) \quad \text{with} \quad P(\mathsf{R}(t)) = \sqrt{\frac{\tau}{2\pi}}\exp\left(-\frac{\tau(\mathsf{R}(t) - \eta)^2}{2}\right),
\tag{14.3}
$$

where $P(\mathsf{R}(t))$ is the time-varying Gaussian distribution, η is the drift, and $\tau = \tau^*/[\,1.0 - s\,]$ is the daily step width, parameterized in terms of the daily step width precision τ^* and the smoothing parameter s.

Dynamic network SEIR model and parameters. Similar to the network SIS and SEIR models in Chapters 11 and 10, we discretize the set of equations (14.1) in space using a weighted graph $\mathcal{G}$ that represents the mobility between different provinces, states, or countries [1]. We summarize the connectivity of the graph in terms of the adjacency matrix A_{IJ}, the weighted connection between two nodes I and J, and the degree matrix $D_{II} = \text{diag} \sum_{J=1, J \neq I}^{n_{nd}} A_{IJ}$, the weighted number of incoming and outgoing connections of node I [6]. The difference between the degree matrix D_{IJ} and the adjacency matrix A_{IJ} defines the weighted graph Laplacian L_{IJ} [24],

$$L_{IJ} = D_{IJ} - A_{IJ} \quad \text{with} \quad D_{II} = \text{diag} \sum_{J=1, J \neq I}^{n_{nd}} A_{IJ} . \tag{14.4}$$

To discretize the set of equations (14.1) in space, we introduce the unknown populations, $\mathbf{P}_I = [S_I, E_I, I_I, R_I]$, as global unknowns at all $I = 1, ..., n_{nd}$ nodes of the graph G [16]. This results in a network SEIR model, a discrete set of $4\,n_{nd}$ equations with $4\,n_{nd}$ unknowns at the $I = 1, ..., n_{nd}$ nodes,

$$
\begin{aligned}
\dot{S}_I &= -\kappa_S \sum_{J=1}^{n_{nd}} L_{IJ} S_J - \beta(t) S_I I_I \\
\dot{E}_I &= -\kappa_E \sum_{J=1}^{n_{nd}} L_{IJ} E_J + \beta(t) S_I I_I - \alpha E_I \\
\dot{I}_I &= -\kappa_I \sum_{J=1}^{n_{nd}} L_{IJ} I_J \qquad\qquad + \alpha E_I - \gamma I_I \\
\dot{R}_I &= -\kappa_R \sum_{J=1}^{n_{nd}} L_{IJ} R_J \qquad\qquad\qquad + \gamma I_I .
\end{aligned}
\tag{14.5}
$$

Figure 14.1 illustrates the susceptible, exposed, infectious, and recovered populations, S, E, I, and R, at two connected nodes I and J. The mobility between connected nodes is a product of the mobility coefficients κ_S, κ_E, κ_I, and κ_R and the graph Laplacian L_{IJ} in units [1/day]. We discretize the system of equations (10.3) in time using an explicit Euler forward time integration scheme according to Section 5.2 and approximate the time derivatives as $(\dot{\circ}) = [(\circ)_{n+1} - (\circ)_n]/\Delta t$, where Δt denotes that discrete time step size [22]. Our dynamic network SEIR model (14.5) introduces five parameters for each region,

$$\vartheta_I = \{ A_I, C_I, \eta_I, \tau_I^*, s_I \}, \tag{14.6}$$

the regional latent and infectious rates A_I and C_I, and the three parameters that define the local time-dependent effective reproduction number $\mathsf{R}(t)$, the regional drift η_I, and the regional daily step width precision τ_I^* and smoothing parameter s_I that define the regional daily step width, $\tau_I = \tau_I^*/[\,1.0 - s_I\,]$.

Data. Our objective is to explore whether and when it would be safe to lift COVID-19 travel restrictions [18]. Our model problem is the Canadian province of Newfoundland and Labrador, that, at the time of this study, was under a strict travel ban and had not seen any new cases for almost two months [2]. Our goal is to make informed recommendations about the effects of partial and total airport reopening, with perfect and imperfect quarantine conditions. We study three reopening scenarios by gradually adding air traffic from the Atlantic Provinces, all of Canada, and all of North America. We model each province, territory, and state as a homogeneous population with its own local SEIR dynamics and connect them to the province of Newfoundland and Labrador through a global mobility network [4]. From this network, we create a weighted graph G in which the n_{nd} nodes represent the individual provinces, territories, and states and the n_{el} weighted edges represent the mobility between them. Figure 14.2 illustrates three discrete graphs of the Atlantic Provinces, Canada, and North America and the average daily air traffic to and from Newfoundland. The

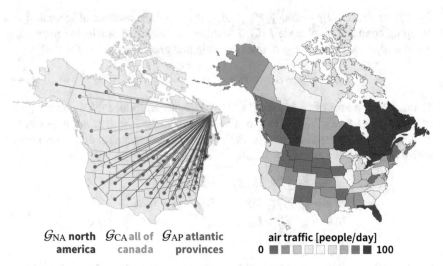

$\mathcal{G}_{\text{NA}}$ **north** $\quad$ $\mathcal{G}_{\text{CA}}$ **all of** $\quad$ $\mathcal{G}_{\text{AP}}$ **atlantic** $\qquad$ **air traffic [people/day]**
america $\qquad$ **canada** $\qquad$ **provinces** $\qquad$ 0 ▪▪▪▪▫▫▪▪▪▪ 100

Fig. 14.2 Network diffusion models of Newfoundland. Discrete graphs $\mathcal{G}_{\text{AP}}$ of the Atlantic Provinces in dark blue, $\mathcal{G}_{\text{CA}}$ of Canada in dark and light blue, and $\mathcal{G}_{\text{NA}}$ of North America in dark and light blue and red, with $n_{\text{nd}} = 4$, $n_{\text{nd}} = 13$, and $n_{\text{nd}} = 64$ nodes and $n_{\text{el}} = n_{\text{nd}} - 1$ edges that represent the main travel routes to Newfoundland; the edges are weighted by the average number of daily incoming and outgoing air passengers from the Atlantic provinces, Canadian and North America [18].

graph of the Atlantic Provinces $\mathcal{G}_{\text{AP}}$ consists of $n_{\text{nd}} = 4$ nodes and $n_{\text{el}} = 3$ edges, shown in dark blue; the graph of Canada $\mathcal{G}_{\text{CA}}$ consists of $n_{\text{nd}} = 13$ nodes and $n_{\text{el}} = 12$ edges shown in dark and light blue, and the graph of North America $\mathcal{G}_{\text{NA}}$ consists of $n_{\text{nd}} = 64$ nodes and $n_{\text{el}} = 63$ edges shown in dark and light blue and red. We approximate the weights of the edges using the average daily passenger air travel statistics.

For the COVID-19 epidemiology data, we draw the COVID-19 history of all 13 Canadian provinces and territories [2] and all 51 United States [25] from the beginning of the COVID-19 pandemic until July 1, 2020. From these data, we extract the reported daily new cases $\hat{I}(t)$ and the cumulative confirmed cases $\hat{D}(t)$ as the sum of all reported cases to date.

For the mobility data, we sample all North American air traffic data to and from the province of Newfoundland and Labrador as reported by the International Air Transport Association [10] and use a 15-month period before the COVID-19 outbreak, from January 1, 2019 to March 31, 2020, to estimate the daily air traffic. For the province of Newfoundland, the incoming and outgoing passenger air travel from and to each region is relatively similar and we can simply average both. This results in an undirected graph $\mathcal{G}$ and symmetric adjacency and Laplacian matrices, $A_{IJ} = A_{JI}$ and $L_{IJ} = L_{JI}$. Since we focus on the early phase of a resurgence, we only simulate the mobility into and out of Newfoundland and neglect the intraregional mobility between all other regions. This implies that only a single row and column of the adjacency matrix and degree matrix are populated, $A_{I1} = A_{1J} \neq 0$ and $D_{11} \neq 0$, while all other entries are zero, $A_{IJ} = 0$ for all $I, J \neq 1$ and $D_{II} = 0$ for all $I \neq 1$. Figure 14.2 illustrates the average daily air traffic to Newfoundland from all of North America. The locations with the largest mobility to the province are Ontario with 188/day, Quebec with 146/day, Alberta with 143/day, and Florida with 102/day, followed by New Jersey and British Columbia both with 67/day, Nova Scotia with 41/day, Manitoba and Texas both with 20/day.

Bayes' theorem. We apply Bayes' theorem to estimate the posterior probability distribution of the parameters $\vartheta_I = \{ \mu_I, \tau_I^*, s_I \}$, such that the statistics of the cumulative cases $D(\vartheta, t)$ agree with the reported cumulative cases $\hat{D}(t)$ [11, 19, 26],

$$P(\vartheta | \hat{D}(t)) = \frac{P(\hat{D}(t) | \vartheta) \, P(\vartheta)}{P(\hat{D}(t))}. \tag{14.7}$$

Here $P(\hat{D}(t) | \vartheta)$ is the likelihood, $P(\vartheta)$ is the prior, $P(\hat{D}(t))$ is the marginal likelihood, and $P(\vartheta | \hat{D}(t))$ is the posterior. Bayes' theorem calculates the pointwise product of the prior and the likelihood, to obtain the normalized posterior probability distribution, which is the conditional distribution of the parameters ϑ given the data, in our case, the cumulative COVID-19 cases $\hat{D}(t)$.

Prior probability distributions. We assume that the latent and infectious periods are disease specific and and fix them to $A = 2.5$ days and $C = 6.5$ days [14, 15, 28]. This reduces the set of model parameters to the three parameters that define the regional drift η_I, and the regional daily step width precision τ_I^* and smoothing parameter s_I that define the regional daily step width, $\tau_I = \tau_I^* / [\, 1.0 - s_I \,]$,

$$\vartheta_I = \{ \eta_I, \tau_I^*, s_I \}. \tag{14.8}$$

For the drift η, we choose a normal distribution with a probability density function $P(\eta)$,

$$\eta \sim \text{Normal}(\mu, \sigma) \quad \text{with} \quad P(\eta) = \frac{1}{\sqrt{2\pi}\sigma} \exp\left(-\frac{(\eta - \mu)^2}{2\sigma^2} \right), \tag{14.9}$$

with mean $\mu = 0$ and standard deviation $\sigma = 2$. For the step width precision τ^*, we choose an exponential distribution with a probability density function $P(\tau^*)$,

$$\tau^* \sim \text{Exponential}(\lambda) \quad \text{with} \quad P(\tau^*) = \lambda \exp(-\lambda \tau^*), \tag{14.10}$$

with scaling parameter $\lambda = 0.5$. For the smoothing parameter s, we choose a uniform distribution with a probability density function $P(s)$,

$$s \sim \text{Uniform}(s_{\min}, s_{\max}) \quad \text{with} \quad P(s) = \frac{1}{s_{\max} - s_{\min}}, \tag{14.11}$$

with lower and upper bounds $s_{\min} = 0$ and $s_{\max} = 1$. Figure 14.3 illustrates our priors and Table 14.1 summarizes both our prior selection and their parameterization.

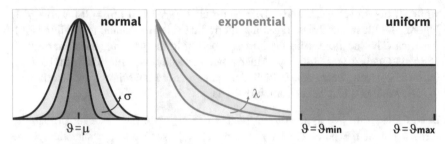

Fig. 14.3 Prior probability distributions. Normal distribution, exponential distribution, and uniform distribution. The normal distribution is symmetric around the mean μ with a width that increases with increasing standard deviation σ. The exponential distribution is symmetric around zero with a width that increases with increasing rate parameter λ. The uniform distribution is constant within the interval ϑ_{min} to ϑ_{max}.

Table 14.1 Prior probability distributions. Priors for the drift η, the step width precision τ^*, the scaling parameter s, and the likelihood width σ.

parameter	distribution	parameters
η	Normal (μ, σ)	$\mu = 0, \sigma = 2$
τ^*	Exponential(λ)	$\lambda = 0.5$
s	Uniform (s_{min}, s_{max})	$s_{min} = 0, s_{max} = 1$
σ	HalfCauchy (β)	$\beta = 1$

Likelihood. The likelihood function $P(\hat{D}(t)|\boldsymbol{\vartheta})$ measures the goodness of fit between a sample of the observed cumulative cases $\hat{D}(t)$ and the simulated cumulative cases $D(\boldsymbol{\vartheta}, t) = I(\boldsymbol{\vartheta}, t) + R(\boldsymbol{\vartheta}, t)$ for given parameters $\boldsymbol{\vartheta}$. For the daily likelihood $\mathcal{L}$, we select Student's t-distribution with a case number-dependent width [5, 13], which results in a variance proportional to the mean,

$$\mathcal{L}(\hat{D}(t)|\boldsymbol{\vartheta}) \sim \text{studentT}_{\nu=4}(\text{mean} = D(\boldsymbol{\vartheta}, t), \text{width} = \sigma\sqrt{D(\boldsymbol{\vartheta}, t)}). \qquad (14.12)$$

With its heavy tails, Student's t-distribution is more robust to outliers than the classical normal distribution. For the likelihood width σ between the reported daily new cases $\hat{D}(t)$ and the simulated daily new infections $D(\boldsymbol{\vartheta}, t)$, we select a half Cauchy distribution,

$$\sigma \sim \text{HalfCauchy}(\sigma_0, \beta) \quad \text{with} \quad P(\sigma) = \frac{1}{\pi\beta\left[1 + [(\sigma - \sigma_0)/\beta]^2,\right]} \qquad (14.13)$$

with mean $\sigma_0 = 0$ and scaling parameter $\beta = 1$. To compare the data and the model throughout the entire time window, from the beginning of the COVID-19 pandemic until July 1, 2020, we evaluate the daily likelihood $\mathcal{L}(\boldsymbol{\vartheta}, t_i)$ from equation (14.12) at each discrete time point t_i and multiply the individual terms to calculate the overall likelihood function $P(\hat{D}(t)|\boldsymbol{\vartheta})$,

$$P(\hat{D}(t)|\boldsymbol{\vartheta}) = \prod_{i=0}^{n}\mathcal{L}(\boldsymbol{\vartheta},t_i), \tag{14.14}$$

where the product symbol, $\prod_{i=0}^{n}$, denotes the multiplication of the $i = 0, ..., n$ daily likelihoods $\mathcal{L}(\boldsymbol{\vartheta},t_i)$.

Posterior probability distribution. We apply Bayes' theorem (14.7) to calculate the posterior distribution of the parameters using the likelihood $P(\hat{D}(t)|\boldsymbol{\vartheta})$ and priors $P(\boldsymbol{\vartheta})$. As we have discussed in the previous chapters, the denominator in Bayes' theorem, the marginal likelihood $P(\hat{D}(t))$ is generally difficult to compute. We ignore the marginal likelihood $P(\hat{D}(t))$, and simply evaluate the normalized probability density distribution,

$$P(\boldsymbol{\vartheta}|\hat{D}(t)) \propto P(\hat{D}(t)|\boldsymbol{\vartheta})\, P(\boldsymbol{\vartheta}), \tag{14.15}$$

as the product of the likelihood and priors.

Markov Chain Monte Carlo. Similar to the previous chapters, we solve this distribution numerically using the NO-U-Turn sampler [9] implementation of the Python package PyMC3 [23, 27]. From the converged posterior distributions, we sample multiple combinations of parameters $\boldsymbol{\vartheta}$. From these parameters, we calculate and report the means and credible intervals of the dynamic reproduction number $R(t)$ and the reported and simulated cumulative cases $\hat{D}(t)$ and $D(\boldsymbol{\vartheta},t) = I(\boldsymbol{\vartheta},t) + R(\boldsymbol{\vartheta},t)$ for each territory, province, and state from the early outbreak on March 15 until July 1, 2020. With these inferred disease dynamics, we compare the effects of partial and total airport reopening, with perfect and imperfect quarantine conditions.

14.2 Example: The Newfoundland story

On July 3, 2020, the Canadian province of Newfoundland and Labrador enjoyed the rather exceptional and enviable position of having the coronavirus pandemic under control with the total number of 261 cases, with 258 recovered, 3 deaths, no active cases for 16 consecutive days, and no new cases for 36 days [2]. On the same day, after a two-months long local travel ban, the Atlantic Bubble opened to allow air travel between the four Atlantic Provinces, Newfoundland and Labrador, Nova Scotia, New Brunswick, and Prince Edward Island, with no quarantine requirements for travelers. Under the increasing pressure to fully reopen, health officials, and political decision makers now seek to understand the risk of gradual and full reopening under perfect quarantine conditions and quarantine violation [21].

A relaxation of the travel ban naturally induces anxiety and fear of a new outbreak. From a public health perspective, the major challenges are to understand the effect of the travel bubble; to predict the effect of a wider opening to the rest of Canada and the United States; and to estimate the effect of imperfect quarantine assuming that a fraction of travelers would ignore the guidelines for self-isolation. Answering these questions will help to understand the short- and long-term effects of reopening.

To explore whether and when it would be safe to lift the travel ban, we model the dynamics of COVID-19 using a network approach that links a local epidemiological model with air-traffic mobility [18]. Local epidemiological modeling is now a well-accepted approach to follow the dynamics of a homogeneous population during an epidemic [3]. The extra network layers allow us to capture the mobility between different local populations [4]. A unique feature of this approach is that we can dynamically infer the parameters of the epidemiology model using reported case data and update it in real time during the progression of the disease [11, 17]. In addition, we can easily extract mobility data from passenger air travel statistic between the different locations [10]. Here use this approach to study three reopening scenarios by gradually adding air traffic from the Atlantic Provinces; all of Canada; and all of North America.

Outbreak dynamics of COVID-19 across North America. Figure 14.4 illustrates the outbreak dynamics of COVID-19 in the Atlantic Provinces, the other Canadian provinces and territories, and the United States from the beginning of the outbreak until July 1, 2020. The top graphs show the time-dependent dynamic reproduction number $R(t)$ of our dynamic SEIR model according to equation (14.3) as red curve. The bottom graphs show the confirmed cases $\hat{D}(t)$ as dots and the model fit $D(t) = I(t) + R(t)$ according to equations (14.5.3) and (14.5.4) as orange curve. The solid lines of each graph represent the median values, the shaded areas highlight the 95% credible intervals. From the outbreak dynamics on the last day of the Bayesian analysis, on July 1, 2020, we estimate the incoming exposed and infectious COVID-19 travelers from all 64 locations.

Estimated incoming exposed and infectious COVID-19 travelers. Table 14.2 summarizes the outbreak dynamics on July 1, 2020, the susceptible, exposed, infectious, and recovered populations S, E, I, R and the effective reproduction number R_t for all 64 locations. From these data, we calculate how many daily exposed and infectious individuals would travel to the province of Newfoundland from each location per day. Figure 14.5 illustrates the daily number of incoming exposed and infectious travelers to the province of Newfoundland. The number of exposed and infectious air passengers from the Atlantic Provinces, all of Canada, and all of North America is a reflection of both the frequency of air travel from Figure 14.2 and the local outbreak dynamics as of July 1, 2020. Most exposed travelers come from Florida with 0.137/day, Texas with 0.013/day, Nevada with 0.011/day, Alberta and Ontario both with 0.005/day. Most infectious travelers come from Florida with 0.192/day, Texas with 0.023/day, Nevada, Quebec, and Ontario all with 0.016/day. If air traffic would fully resume with the outbreak dynamics of July 1, 2020, the estimated number of incoming exposed and infectious travelers would be 0.203/day and 0.329/day, suggesting that every five days, an exposed traveler, and every three days, an infectious traveler would enter the province of Newfoundland via air travel.

Outbreak dynamics of COVID-19 in Newfoundland and Labrador. Figures 14.6 and 14.7 highlight the timeline of the COVID-19 outbreak in Newfoundland. The graphs distinguish between three time intervals: the first period without travel restrictions from March 15 until May 4, 2020, the second period with travel restrictions from May 4 until July 1, 2020, and the third period upon gradual reopening from

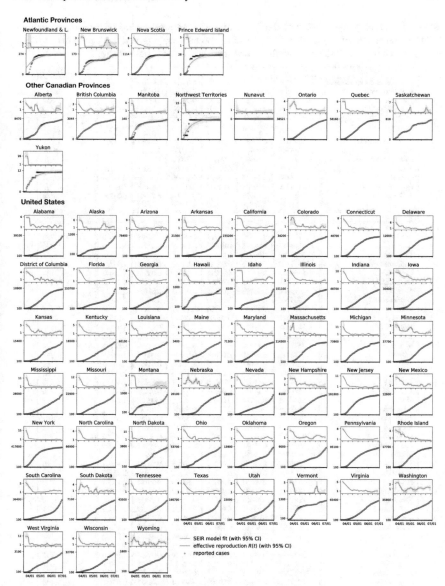

Fig. 14.4 Outbreak dynamics of COVID-19 across North America. Dynamic reproduction number (red), confirmed cases (dots), and model fit (orange) from the beginning of the outbreak until July 1, 2020. Solid lines represent the median values, shaded areas highlight the 95% credible intervals [18].

July 1, 2020 forward. The graphs report the daily new cases and total cases as dots, compared to the fit of the SEIR model as solid orange curves, from the early outbreak on from March 15 until July 1, 2020. From then on, the dashed lines highlight the

Table 14.2 Outbreak dynamics of COVID-19 across North America. Susceptible, exposed, infectious, and recovered populations S, E, I and R, and effective reproduction number R on July 1, 2020 inferred from reported case numbers and air passengers; n.d. not defined.

Atlantic Provinces	Population	S	E	I	R	R	Air Passengers
Newfoundland and Labrador	519716	0.99948	0.00000	0.00000	0.00052	n.d.	–
New Brunswick	747101	0.99979	0.00000	0.00000	0.00021	n.d.	30621
Nova Scotia	923598	0.99880	0.00000	0.00001	0.00119	n.d.	18523
Prince Edward Island	142907	0.99980	0.00000	0.00000	0.00020	n.d.	6509
Other Canadian Provinces	**Population**	**S**	**E**	**I**	**R**	**R**	**Air Passengers**
Alberta	4067175	0.99797	0.00004	0.00007	0.00192	1.31198	65048
British Columbia	4648055	0.99937	0.00001	0.00002	0.00061	1.08524	30608
Manitoba	1278365	0.99975	0.00000	0.00000	0.00025	n.d.	9098
Northwest Territories	41786	0.99987	0.00000	0.00000	0.00013	n.d.	2277
Nunavut	35944	1.00000	0.00000	0.00000	0.00000	n.d.	1020
Ontario	13448494	0.99725	0.00003	0.00008	0.00264	0.32863	85504
Quebec	8164361	0.99308	0.00003	0.00011	0.00678	n.d.	66545
Saskatchewan	1098352	0.99927	0.00001	0.00004	0.00068	0.43724	6180
Yukon	35874	0.99968	0.00000	0.00000	0.00032	n.d.	406
United States	**Population**	**S**	**E**	**I**	**R**	**R**	**Air Passengers**
Alabama	4903185	0.99164	0.00059	0.00116	0.00660	1.26937	98
Alaska	731545	0.99833	0.00013	0.00025	0.00129	1.27325	252
Arizona	7278717	0.98796	0.00115	0.00252	0.00836	1.08640	1753
Arkansas	3017825	0.99273	0.00038	0.00115	0.00574	0.57118	12
California	39512223	0.99366	0.00049	0.00089	0.00496	1.44335	4441
Colorado	5758736	0.99418	0.00014	0.00028	0.00541	1.22445	634
Connecticut	3565287	0.98688	0.00005	0.00017	0.01289	0.11798	195
Delaware	973764	0.98790	0.00035	0.00055	0.01121	1.80054	0
District of Columbia	705749	0.98531	0.00010	0.00030	0.01429	0.10039	1592
Florida	21477737	0.99145	0.00134	0.00187	0.00534	2.15466	46656
Georgia	10617423	0.99214	0.00060	0.00099	0.00627	1.63627	1347
Hawaii	1415872	0.99934	0.00003	0.00005	0.00058	1.33847	1073
Idaho	1792065	0.99611	0.00048	0.00074	0.00267	1.81372	37
Illinois	12671821	0.98848	0.00013	0.00035	0.01104	0.79747	1374
Indiana	6732219	0.99290	0.00016	0.00036	0.00659	1.03362	354
Iowa	3155070	0.99048	0.00027	0.00071	0.00854	0.79137	60
Kansas	2913314	0.99462	0.00033	0.00055	0.00450	1.68635	83
Kentucky	4467673	0.99629	0.00015	0.00031	0.00325	1.15865	510
Louisiana	4648794	0.98688	0.00061	0.00116	0.01136	1.33347	1790
Maine	1344212	0.99750	0.00008	0.00016	0.00226	1.29048	72
Maryland	6045680	0.98859	0.00014	0.00040	0.01086	0.76133	427
Massachusetts	6949503	0.98420	0.00007	0.00023	0.01550	n.d.	5534
Michigan	9986857	0.99290	0.00005	0.00017	0.00688	n.d.	946
Minnesota	5639632	0.99338	0.00019	0.00044	0.00599	0.96952	725
Mississippi	2976149	0.99033	0.00052	0.00117	0.00797	1.01498	87
Missouri	6137428	0.99614	0.00023	0.00043	0.00320	1.43212	726
Montana	1068778	0.99900	0.00011	0.00016	0.00073	1.87814	60
Nebraska	1934408	0.98986	0.00021	0.00052	0.00940	0.92765	61
Nevada	3080156	0.99304	0.00088	0.00124	0.00484	2.10005	5779
New Hampshire	1359711	0.99573	0.00004	0.00012	0.00411	0.62427	6
New Jersey	8882190	0.98026	0.00017	0.00035	0.01922	1.20830	1989
New Mexico	2096829	0.99391	0.00028	0.00054	0.00527	1.31363	71
New York	19453561	0.97946	0.00007	0.00021	0.02026	0.78300	7562
North Carolina	10488084	0.99340	0.00040	0.00085	0.00535	1.12729	1515
North Dakota	762062	0.99515	0.00015	0.00029	0.00441	1.30264	88
Ohio	11689100	0.99536	0.00020	0.00041	0.00403	1.20717	548
Oklahoma	3956971	0.99628	0.00026	0.00057	0.00290	1.06800	282
Oregon	4217737	0.99779	0.00014	0.00030	0.00177	1.17868	327
Pennsylvania	12801989	0.99276	0.00013	0.00027	0.00685	1.19329	1814
Rhode Island	1059361	0.98407	0.00009	0.00027	0.01557	0.10155	145
South Carolina	5148714	0.99202	0.00091	0.00158	0.00548	1.55259	920
South Dakota	884659	0.99219	0.00015	0.00042	0.00724	0.80479	65
Tennessee	6833174	0.99344	0.00036	0.00078	0.00542	1.06544	2807
Texas	28995881	0.99362	0.00065	0.00118	0.00455	1.41288	8915
Utah	3205958	0.99250	0.00052	0.00106	0.00592	1.21308	260
Vermont	623989	0.99810	0.00001	0.00003	0.00186	n.d.	10
Virginia	8535519	0.99250	0.00016	0.00041	0.00694	0.85318	483
Washington	7614893	0.99557	0.00011	0.00030	0.00401	0.90306	767
West Virginia	1787147	0.99831	0.00006	0.00015	0.00148	1.03586	10
Wisconsin	5822434	0.99428	0.00027	0.00052	0.00493	1.31016	353
Wyoming	578759	0.99727	0.00016	0.00033	0.00224	1.17731	3

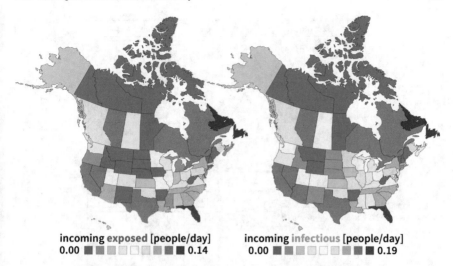

incoming **exposed** [people/day]
0.00 ▪▪▫▫▫▫▪▪▪ 0.14

incoming **infectious** [people/day]
0.00 ▪▪▫▫▫▫▪▪▪ 0.19

Fig. 14.5 Estimated exposed and infectious travelers to Newfoundland. Number of daily incoming air passengers from the Atlantic Provinces, all of Canada, and all of North America that have been exposed to or infectious with COVID-19. The estimate assumes average daily travel from Figure 14.2 and the outbreak dynamics as of July 1, 2020 [18].

reopening forecast for a 60-day period. The solid and dashed lines represent the median values, the shaded areas highlight their 95% credible intervals. Figure 14.6 explores the effect of quarantine upon reopening with all incoming travelers quarantining to 100% in dark blue, to 50% in light blue, and 0% in red. While a 100% quarantine does not result in any new cases, 50% quarantine results in a mild but notable increase of new cases, and 0% quarantine triggers a rapid exponential outbreak. Figure 14.7 explores the effect of restricted travel under the assumption of no quarantine, air traffic only between the Atlantic Provinces in dark blue, between all of Canada in light blue, and between all of North America in red. While an opening to the Atlantic Provinces and to all of Canada does not result in a notable number of new cases, under the current conditions, an opening to all of North America would trigger a rapid exponential outbreak.

Effects of quarantine and restricted travel. Figures 14.8 and 14.9 summarize the effects of quarantine and restricted travel on the number of COVID-19 cases in Newfoundland and Labrador for a 150-day period as predicted by our reopening forecast. The solid and dashed lines show the reopening forecast with incoming travelers quarantining at varying percentages and with travel restricted to the Atlantic Provinces, Canada, and all of North America. The predictions use the mean effective reproduction numbers of $R = 1.35$ for all of North America and $R = 1.16$ for Canada. The black horizontal lines mark 520 predicted cases corresponding to 0.1% of the population of Newfoundland and Labrador. Table 14.3 summarizes the dates and number of days by which the curves cross this line and the number of COVID-19 cases reaches 0.1% of the population. Our simulation predicts that this

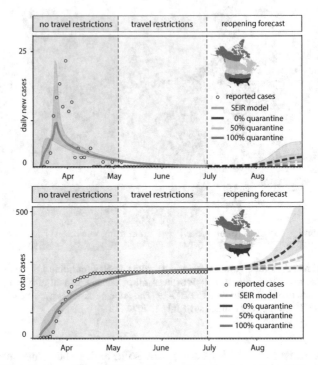

Fig. 14.6 Outbreak dynamics of COVID-19 in Newfoundland and the effect of quarantine.
Daily new cases and total cases with model fit for periods without travel restrictions and with travel
restrictions. Reopening forecast for 60-day period with all incoming travelers quarantining to 100%
in dark blue, 50% in light blue, and 0% in red. Solid and dashed lines represent the median values,
shaded areas highlight the 95% credible intervals [18].

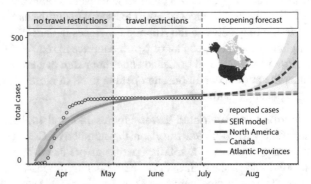

Fig. 14.7 Outbreak dynamics of COVID-19 in Newfoundland and the effect of restricted travel.
Total cases with model fit for periods without travel restrictions and with travel restrictions. Reopening forecast for 60-day period with mobility only within the Atlantic Provinces in dark blue,
all of Canada in light blue, and all of North America in red. Solid and dashed lines represent the
median values, shaded areas highlight the 95% credible intervals [18].

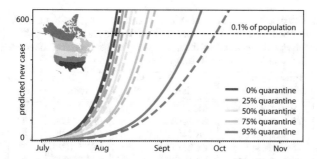

Fig. 14.8 Effect of quarantine on COVID-19 dynamics. Reopening forecast for 150-day period with incoming travelers quarantining from 95% (blue) to 0% (red). Predictions are based on a local SEIR model with incoming air traffic from all of North America using the mean effective reproduction number of R = 1.35 for all of North America (solid lines) and R = 1.16 for Canada (dashed lines) on July 1, 2020. The black horizontal line marks 0.1% of the population of Newfoundland and Labrador [18].

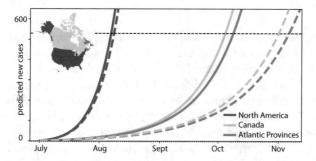

Fig. 14.9 Effect of restricted travel on COVID-19 dynamics. Reopening forecast for 150-day period with incoming travelers from Atlantic Provinces, Canada, and all of North America with no quarantine requirements. Predictions are based on a local SEIR model with incoming air traffic from all of North America using the mean effective reproduction numbers of R = 1.35 for all of North America (solid lines) and R = 1.16 for Canada (dashed lines) on July 1, 2020. The black horizontal line marks 0.1% of the population of Newfoundland and Labrador [18].

will occur after only 38 or 39 days without travel restrictions and no quarantine. Quarantining only half of the incoming population increases this time period to 46 or 48 days, quarantining 95% of the incoming population to 80 and 92 days. Limiting travel within Canada increases this time period to 97 and 125 days, limiting it within only the Atlantic Provinces to 102 and 132 days.

Two interacting features determine the outbreak dynamics of the COVID-19 pandemic: the *local epidemiology* of the disease and the *global mobility* of diseased individuals [16]. Reducing mobility is a controversial but highly effective measure to manage the outbreak dynamics [30]. Some regions, mostly smaller states or provinces, have successfully managed to reduce the number of current cases to zero [12]. One of those provinces is the Canadian province of Newfoundland [2]. An important piece of information in predicting the effects of partial and total reopening of Newfoundland is the daily count of incoming exposed and infectious travelers.

Table 14.3 Effects of quarantine and restricted travel on COVID-19 dynamics. The dates and days mark the time point at which the number of cases would reach 0.1% of the population. The forward simulations use the mean effective reproduction numbers of North America, R = 1.35, and of Canada, R = 1.16, on July 1, 2020 for the prediction.

travel from	R = 1.35		R = 1.16	
North America	date	days	date	days
0% quarantine	08/07/2020	37	08/08/2020	38
25% quarantine	08/10/2020	40	08/11/2020	41
50% quarantine	08/14/2020	44	08/16/2020	46
75% quarantine	08/22/2020	52	08/25/2020	55
95% quarantine	09/16/2020	77	09/28/2020	89
travel at	R = 1.35		R = 1.16	
0% quarantine	date	days	date	days
North America	08/07/2020	37	08/08/2020	38
Canada	10/04/2020	95	11/01/2020	123
Atlantic Provinces	10/08/2020	99	11/06/2020	128

We estimate these populations using our network epidemiology model for North America, calibrated with the reported case data for all Canadian provinces and territories [2] and all United States [25], and multiply these numbers with the daily travel data for each region [10]. Figure 14.5 illustrates the estimated number of incoming exposed and infectious travelers per day. These contour plots allow us to quickly identify critical regions like Florida, from which 0.137 exposed and 0.192 infectious individuals would enter Newfoundland every day. They also suggest that it is safe to open travel within the Atlantic Provinces, New Brunswick, Nova Scotia, and Prince Edward Island, and reasonably safe to open travel within Canada. While the travel frequency within the Atlantic Provinces and Canada is high as we can see in Figure 14.2, the case numbers in these regions are low and keep the risk of opening low. The opposite is true for states like California, with a low travel frequency but high case numbers. This suggests that opening travel within all of North America would expose Newfoundland to a disproportionally high risk.

An interesting metric is the estimated number of incoming exposed and infectious travelers upon full reopening, 0.203/day and 0.329/day. This suggests that every five days and every three days, an exposed and an infectious traveler would enter the province of Newfoundland. In other words every other day, a new COVID-19 case would enter the Newfoundland and Labrador via air travel. Since the exposed and early infectious individuals are still pre-symptomatic, it is impossible to identify and isolate them without strict quarantine requirements [7]. To estimate the precise effects of quarantine and restricted travel, Figures 14.6 and 14.7 show our reopening forecast for a 60-day window starting on July 1, 2020. The forecast confirms our intuition that there are two strategies to prevent a new outbreak, either mandating strict quarantine requirements or limited travel within the Atlantic Bubble. The increasingly wide 95% credible intervals for more relaxed conditions suggest that reliable predictions become difficult beyond this time window and case numbers could very well increase beyond control within only a few weeks.

Taken together, relaxing travel restrictions is a highly contentious political decision, especially towards regions with a higher case prevalence. Our study shows that–especially for smaller provinces or states–tight border control is often easier and more effective than quarantine. Partial reopening, for example within local travel bubbles, is an effective compromise and a reasonable first step. Our results suggest that relaxing travel restrictions entirely is possible, but would require strict quarantine conditions. Voluntary quarantine, even at an overall rate of 95%, is not enough to entirely prevent future outbreaks. These outbreaks can quickly grow out of control and require expensive lockdowns and other stringent non-pharmaceutical interventions. It is important to understand, closely monitor, and predict the local and global outbreak dynamics and be aware of the dangers associated with uncontrolled reopening.

Problems

14.1 Classical SEIQR model. In this chapter, we have modeled the effects of quarantine by modulating the initial conditions through the reducing the number of exposed and infectious incoming travelers. If we need to account for quarantine throughout the entire outbreak, we can extend the SEIR model to an SEIQR model with an additional quarantined population Q. Derive the governing equations for the SEIQR model in Figure 14.10 with five compartments for the susceptible, exposed, infectious, quarantined, and recovered populations, S, E, I, Q, and R. Assume that, similar to the classical SEIR model (4.17), individuals transition between the compartments by the contact, latent, and infectious rates, β, α, and γ, however, now, a fraction ν_q of the exposed population will go straight into quarantine, and only a fraction $\nu_i = 1 - \nu_q$ will become infectious.

Fig. 14.10 Classical SEIQR model. The classical SEIQR model contains five compartments for the susceptible, exposed, infectious, quarantined, and recovered populations, S, E, I, Q, and R. The transition rates between the compartments, the contact, latent, and infectious rates, β, α, and γ, are inverses of the contact, latent, and infectious periods, $B = 1/\beta$, $A = 1/\alpha$, and $C = 1/\gamma$. A fraction, ν_q, transitions from the exposed state straight into quarantine, and a fraction, $\nu_i = 1 - \nu_q$, becomes infectious.

14.2 Classical SEIQR model. The SEIQR model in Figure 14.10 assumes that individuals only begin to quarantine when they transition from the exposed to the infectious state. How would the model in Figure 14.10 change if individuals were to quarantine when they transition from the susceptible to the exposed state? Draw the five compartments and derive the set of equations.

14.3 Quarantine modeling. Imagine all incoming travelers have to quarantine, without knowing whether they are susceptible, exposed, infectious, or recovered. How would the model in Figure 14.10 change if all travelers were to quarantine? Into which compartment would quarantined individuals transition after their quarantine is over? Draw the model with as many compartment as necessary and describe the relation of the individual compartments, either in words or through a set of equations.

14.4 Network diffusion modeling. Compare the network diffusion models of Newfoundland in Figure 14.2 of this chapter to the network diffusion model of the United States in Figure 9.3 of Chapter 10. Which elements make the models of Newfoundland simpler than the model of the United States?

    o the Newfoundland networks have less nodes
    o the Newfoundland networks have less edges
    o their connectivity can be represented through a unit matrix
    o their connectivity can be represented through a single row
    o they are only used for initial conditions
    o they neglect mobility between all regions but Newfoundland

14.5 Network SEIR modeling. For the three different network diffusion models of Newfoundland in Figure 14.2, for the Atlantic Provinces $\mathcal{G}_{AP}$, Canada $\mathcal{G}_{CA}$, and North America $\mathcal{G}_{NA}$, how many nodes, edges, and unknowns does each network SEIR model have?

14.6 Network SIR modeling. Write the discrete set of equations for the network SIR model similar to equation (14.5). For the three different network diffusion models of Newfoundland in Figure 14.2, for the Atlantic Provinces $\mathcal{G}_{AP}$, Canada $\mathcal{G}_{CA}$, and North America $\mathcal{G}_{NA}$, how many nodes, edges, and unknowns does each network SIR model have?

14.7 Interpreting outbreak dynamics. Table 14.2 summarizes the COVID-19 outbreak dynamics across North America on July 1, 2020. For some provinces and territories, the reproduction number R on this day is not defined. Explain why it is difficult to define the reproduction number for these locations. What do these locations have in common?

14.8 Estimating incoming exposed and infectious travelers. Table 14.2 summarizes the COVID-19 outbreak dynamics across North America on July 1, 2020 and the incoming air passengers from all provinces, territories, and states. Calculate the incoming exposed and infectious travelers from Florida, Texas, and Nevada. Compare your results against the Figure 14.5.

14.9 Estimating incoming exposed and infectious travelers. Table 14.2 summarizes the COVID-19 outbreak dynamics across North America on July 1, 2020 and the incoming air passengers from all provinces, territories, and states. Assume Newfoundland would open to travel within the Atlantic bubble, i.e., within all Atlantic provinces. Calculate the incoming exposed and infectious travelers to Newfoundland. Assume an extended Atlantic bubble that also includes Quebec. Comment on how do your results change.

IN THE SUPREME COURT OF NEWFOUNDLAND AND LABRADOR
GENERAL DIVISION

Citation: *Taylor v. Her Majesty the Queen*, 2020 NLSC 125
Date: September 17, 2020
Docket: 202001G2342

On 4-5 May 2020, in an effort to curtail the spread of COVID-19, the Chief Medical Officer of Health (CMOH) for Newfoundland and Labrador issued two orders pursuant to s. 28(1)(h) of the *Public Health Protection and Promotion Act* (the "*PHPPA*") restricting those permitted to enter the province. On 8 May 2020, the CMOH denied Kimberley Taylor entry from Nova Scotia to attend her mother's funeral.

Ms. Taylor challenges s. 28(1)(h) of the *PHPPA* as outside the legislative authority of the province, and the decision to refuse her entry pursuant to the travel restriction as contrary to her ss.6 and 7 *Charter* rights to mobility and liberty, respectively.

Ms. Taylor's s. 6(1) *Charter* right to mobility was infringed, albeit fleetingly, when she was denied entry into the province. Ms. Taylor's s. 7 right to liberty was not engaged.

The infringement of Ms. Taylor's right to mobility was demonstrably justified under s. 1 of the *Charter* in response to the COVID-19 pandemic.

In the result, the application is dismissed.

Fig. 14.11 Newfoundland travel ban. Summary page of the outcome of a constitutional challenge that questioned the Province of Newfoundland and Labrador's ability to restrict domestic travel in response to the COVID-19 pandemic in August 2020.

14.10 Dynamic reproduction number. In equation (14.3), we have chosen a Gaussian random walk type ansatz for the dynamic reproduction number. Plot the dynamic reproduction number $R(t)$ as the result of a Gaussian random walk with a drift $\eta = 0$, daily step width precision $\tau^* = 1$, and smoothing parameter $s = 0$ for a time window of $t = 1, ..., 100$ days. Compare your plot against the hyperbolic tangent type ansatz from equation (12.2) in Figure 12.3 and against the logarithmic Gaussian random walk type ansatz in Figure 13.10. What are the advantages and disadvantages of a Gaussian random walk versus hyperbolic tangent type ansatz for R?

14.11 The Newfoundland court case. The study of quarantine violation and partial reopening [18] in Section 14.2 was performed in support of evidence for a court case in which a private citizen, joined by the Canadian Civil Rights Association, challenged the Newfoundland travel ban. Figure 14.11 shows extracts from the summary statement of the case in which the judge ruled in favor of the travel ban. Discuss arguments for and against banning travel to Newfoundland in view of what you have learned in this chapter. Focus specifically on the special role of Newfoundland.

14.12 The Newfoundland travel ban. A year after the beginning of the COVID-19 outbreak in Newfoundland and Labrador, the total cumulative case number in the province reached 1012. Figure 14.12 illustrates the outbrak dynamics throughout the

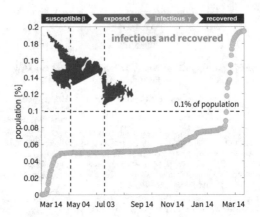

Fig. 14.12 COVID-19 outbreak dynamics in Newfoundland and Labrador. Reported infectious and recovered population for the time period of one year, from March 14, 2020 to March 14, 2021. Dashed vertical lines mark beginning of travel ban, May 4, 2020, and opening to the Atlantic bubble, July 3, 2020; dashed horizontal line marks a cumulative infectious population of 0.1%, passed on February 10, 2021.

first year of the pandemic, with the travel ban in place during the period between the dashed vertical lines. Estimate the effective reproduction number R during this period. Comment on the success of the travel ban.

14.13 Limitations of data-driven network SEIR model. Compare the predicted outbreak dynamics in Figures 14.6 and 14.7 and the predicted case numbers in Table 14.3 against the true outbreak dynamics in Figure 14.12. Identify at least four potential limitations of the model, that could have lead to these discrepancies.

14.14 Lessons of the Pandemic. In the 1919 Science publication"Lessons of the Pandemic" about the Spanish flu, George A Soper writes "There is one and only one way to absolutely prevent it and that is by establishing absolute isolation" [29]. Discuss this statement in view of what you have learned in this chapter. Identify three different strategies of preventing the spread of the COVID-19 pandemic and compare them against each other.

References

1. Balcan D, Colizzan V, Goncalves B, Hu H, Jamasco J, Vespignani A (2009) Multiscale mobility networks and the spatial spreading of infectious diseases. Proceedings of the National Academy of Sciences 106:21484-21489.
2. Berry I, Soucy JPR, Tuite A, Fisman D (2020) Open access epidemiologic data and an interactive dashboard to monitor the COVID-19 outbreak in Canada. Canadian Medical Association Journal 192:E420.
3. Brauer F, Castillo-Chavez C, Feng Z (2019) Mathematical Models in Epidemiology. Springer-Verlag New York.

4. Chinazzi, M, Davis JT, Ajelli M, Gioanni C, Litvinova M, Merler S, Piontti A, Mu K, Rossi L, Sun K, Viboud C, Xiong X, Yu H, Halloran ME, Longini IM, Vespignani A (2020) The effect of travel restrictions on the spread of the 2019 novel coronavirus (COVID-19) outbreak. Science 368:395-400.
5. Dehning J, Zierenberg J, Spitzner FP, Wibral M, Pinheiro Neto J, Wilczek M, Priesemann V (2020) Inferring COVID-19 spreading rates and potential change points for case number forecasts. Science 369:eabb9789.
6. Fornari S, Schafer A, Jucker M, Goriely A, Kuhl E (2019) Prion-like spreading of Alzheimer's disease within the brain's connectome. Journal of the Royal Society Interface 16:20190356.
7. Hellewell J, Abbott S, Gimma A, Bosse NI, Jarvis CI, Russell TW, Munday JD, Kucharski AJ, Edmunds WJ (2020) Feasibility of controlling COVID-19 outbreaks by isolation of cases and contacts. Lancet Global Health 8:e488-496.
8. Hethcote HW (2000) The mathematics of infectious diseases. SIAM Review 42:599-653.
9. Hoffman MD, Gelman A (2014) The No-U-Turn sampler: adaptively setting path lengths in Hamiltonian Monte Carlo. Journal of Machine Learning Research 15:1593-1623.
10. International Air Transport Association (2020) https://www.iata.org. accessed: July 9, 2020.
11. Jha PK, Cao L, Oden JT (2020) Bayesian-based predictions of COVID-19 evolution in Texas using multispecies mixture-theoretic continuum models. Computational Mechanics 66:1055–1068.
12. Johns Hopkins University (2021) Coronavirus COVID-19 Global Cases by the Center for Systems Science and Engineering. https://coronavirus.jhu.edu/map.html, https://github.com/CSSEGISandData/covid-19 assessed: June 1, 2021.
13. Lange KL, Little RJA, Taylor MG (1989) Robust statistical modeling using the t distribution. Journal of the American Statistical Association 84:881-896.
14. Lauer SA, Grantz KH, Bi Q, Jones FK, Zheng Q, Meredith HR, Azman AS, Reich NG, Lessler J (2020) The incubation period of coronavirus disease 2019 (COVID-19) from publicly reported confirmed cases: estimation and application. Annals of Internal Medicine 172:577-582.
15. Li Q, Guan X, Wu P, Wang X, ... Feng Z (2020) Early transmission dynamics in Wuhan, China, of novel coronavirus-infected pneumonia. New England Journal of Medicine 382:1199-1207.
16. Linka K, Peirlinck M, Sahli Costabal F, Kuhl E (2020) Outbreak dynamics of COVID-19 in Europe and the effect of travel restrictions. Computer Methods in Biomechanics and Biomedical Engineering 23: 710-717.
17. Linka K, Peirlinck M, Kuhl E (2020) The reproduction number of COVID-19 and its correlation with public heath interventions. Computational Mechanics 66:1035-1050.
18. Linka K, Rahman P, Goriely A, Kuhl E (2020) Is it safe to lift COVID-19 travel restrictions? The Newfoundland story. Computational Mechanics 66:1081–1092.
19. Linka K, Goriely A, Kuhl E (2021) Global and local mobility as a barometer for COVID-19 dynamics. Biomechanics and Modeling in Mechanobiology 20:651–669.
20. Linka K, Peirlinck M, Schafer A, Ziya Tikenogullari O, Goriely A, Kuhl E (2021) Effects of B.1.1.7 and B.1.351 on COVID-19 dynamics. A campus reopening study. Archives of Computational Methods in Engineering. doi:10.1007/s11831-021-09638-y.
21. MacGregor S (2020) What you need to know about Atlantic Canada's new coronavirus travel bubble. https://www.forbes.com/sites/sandramacgregor/2020/07/11/what-you-need-to-know-about-atlantic-canadas-new-coronavirus-travel-bubble/\#283b494559ee Forbes. published & accessed: July 11, 2020.
22. Moin P (2000) Fundamentals of Engineering Numerical Analysis. Cambridge University Press.
23. Osvaldo M (2018) Bayesian Analysis with Python: Introduction to Statistical Modeling and Probabilistic Programming Using PyMC3 and ArviZ. Packt Publishing, 2nd edition.
24. Newman M (2010) Networks: An Introduction. Oxford University Press, New York.
25. New York Times (2020) Coronavirus COVID-19 Data in the United States. https://github.com/nytimes/covid-19-data/blob/master/us-states.csv assessed: June 1, 2021.
26. Peirlinck M, Linka K, Sahli Costabal F, Bendavid E, Bhattacharya J, Ioannidis J, Kuhl E (2020) Visualizing the invisible: The effect of asymptomatic transmission on the outbreak dynamics of COVID-19. Computer Methods in Applied Mechanics and Engineering 372:113410.

27. Salvatier J, Wiecki TV, Fonnesbeck C (2016) Probabilistic programming in Python using PyMC3. PeerJ Computer Science 2:e55.
28. Sanche S, Lin Y, Xu C, Romero-Severson E, Hengartner N, Ke R (2020) High contagiousness and rapid spread of severe acute respiratory syndrome coronavirus 2. Emerging Infectious Diseases 2020;26(7):1470-1477.
29. Soper GA (1919) The lessons of the pandemic. Science XLIX 501-506.
30. Zlojutro A, Rey D, Gardner L (2019) A decision-support framework to optimize border control for global outbreak mitigation. Scientific Reports 9:2216.

Index

acceptance ratio, 239, 245, 246
adaptation time, 260
adjacency matrix, 174, 196, 200, 290
affected population, 132, 137, 159
age of infection, 52
agent, 7
agent-based model, 20, 23
AIDS, 4
aleatoric uncertainty, 222
antibody isotypes, 284
antibody test, 10, 11
antigen test, 10
assembly, 182, 186
asymptomatic, 148
asymptomatic fraction, 148, 270, 279
asymptomatic transmission, 13, 148–150, 270, 272, 277, 278, 280, 285
attack ratio, 50, 149

backward Euler method, 83, 89, 101, 119
basic reproduction number, 11, 12, 14, 34, 38, 42, 45, 54–56, 62, 64, 127, 149, 258
Bayes factor, 227, 243
Bayes' theorem, 223, 233, 253, 273, 293
Bayes, Thomas, 223
Bayesian approach, 224, 244
Bayesian statistics, 13, 222, 223, 252
Bernoulli, Daniel, 8
Black Death, 4

carrying capacity, 35
case rate, 18
characteristic equation, 63
chickenpox, 61
common cold, 33, 61
community spread, 132
compartment model, 20, 21

computational epidemiology, 6, 8, 13
conditional probability, 225
conjugate priors, 224
connectivity, 174, 200
consistency, 85
constant rate recovery, 34, 42, 62
contact period, 12, 34, 39, 42, 45, 55, 57, 62
contact rate, 64
contingency table, 16, 18, 19
continuous probability distribution, 223
continuous variables, 228
convergence, 85, 102, 121
COVID-19, 4, 12, 22, 62
COVID-19 variants, 15, 27
credible interval, 239, 241, 246, 247, 265, 277, 278, 283, 295
cumulative cases, 26

daily new cases, 25
data uncertainty, 222
data-driven epidemiology, 221
data-driven modeling, 14, 23, 222
degree matrix, 174, 196, 200, 291
Delaunay triangulation, 186, 187
diagnostic test, 10
diffusion tensor, 170, 200
discrete probability distribution, 223
discrete variables, 225
disease, 148
disease surveillance, 9
disease-free equilibrium, 37
divergence, 170
dynamic contact rate, 128, 152, 250, 280, 290
dynamic reproduction number, 13, 128, 129, 136, 138, 139, 250, 252, 253, 256, 259, 270, 277–279, 286, 290, 305
dynamic SEIIR model, 152, 270, 277

Printed in the United States
by Baker & Taylor Publisher Services